Postsocialist Shrinking Cities

This book provides a comparative analysis of shrinking cities in a broad range of postsocialist countries within the so-called Global East, a liminal space between North and South. While shrinking cities have received increased scholarly attention in the past decades, theoretical, and empirical research has remained predominantly centered on the Global North. This volume brings to the fore a range of new perspectives on urban shrinkage, identifying commonalities, differences, and policy experiences across a very diverse and vivid region with its various legacies and contemporary controversial developments. With chapters written by leading experts in the field, insider views assist in decolonizing urban theory. Specifically, the book includes chapters on shrinking cities in China, Russia, and postsocialist Europe, presenting comparative discussions within countries and cross-national cases on theoretical and policy implications.

The book will be of interest to students and scholars researching urban studies, urban geography, urban planning, urban politics and policy, urban sociology, and urban development.

Chung-Tong Wu is Honorary Professor, University of Sydney, and Emeritus Professor, University of New South Wales and Western Sydney University, and is the inaugural Chair of the Advisory Committee, Halloran Research Trust (Henry Halloran Trust), University of Sydney.

Maria Gunko is a Senior Research Fellow at the Institute of Geography Russian Academy of Sciences and Lecturer at the Faculty of Geography and Geoinformation Technologies, National Research University Higher School of Economics, Russia.

Tadeusz Stryjakiewicz is a Chair Professor of Geography and Head of the Department of Economic Geography at Adam Mickiewicz University in Poznań, Poland.

Kai Zhou is an Associate Professor of Urban Planning and Head of the Urban Planning Department in the School of Architecture and Planning, Hunan University, China.

Routledge Contemporary Perspectives on Urban Growth, Innovation and Change

Series edited by

Sharmistha Bagchi-Sen, *Professor, Department of Geography and Department of Global Gender and Sexuality Studies, State University of New York-Buffalo, Buffalo, NY, USA and* **Waldemar Cudny**, *Associate Professor. Working at The Jan Kochanowski University (JKU) in Kielce, Poland.*

Urban transformation affects various aspects of the physical, social, and economic spaces. This series contains monographs and edited collections that provide theoretically informed and interdisciplinary insights on the factors, patterns, processes and outcomes that facilitate or hinder urban development and transformation. Books within the series offer international and comparative perspectives from cities around the world, exploring how 'new life' may be brought to cities, and what the cities of future may look like.

Topics within the series may include: urban immigration and management, gender, sustainability and eco-cities, smart cities, technological developments and the impact on industry and on urban societies, cultural production and consumption in cities (including tourism, events and festivals), the marketing and branding of cities, and the role of various actors and policy makers in the planning and management of changing urban spaces.

If you are interested in submitting a proposal to the series please contact Faye Leerink, Commissioning Editor, faye.leerink@tandf.co.uk.

City Branding and Promotion
The Strategic Approach
Waldemar Cudny

Urban Events, Place Branding and Promotion
Place Event Marketing
Edited by Waldemar Cudny

Place Event Marketing in the Asia Pacific Region
Branding and Promotion in Cities
Edited by Waldemar Cudny

Postsocialist Shrinking Cities
Edited by Chung-Tong Wu, Maria Gunko, Tadeusz Stryjakiewicz, and Kai Zhou

Growth and Change in Post-socialist Cities of Central Europe
Edited by Waldemar Cudny and Josef Kunc

The Interstitial Spaces of Urban Sprawl
Geographies of Santiago de Chile's *Zwischenstadt*
Cristian A. Silva

Postsocialist Shrinking Cities

**Edited by
Chung-Tong Wu, Maria Gunko,
Tadeusz Stryjakiewicz, and Kai Zhou**

Routledge
Taylor & Francis Group

LONDON AND NEW YORK

First published 2022
by Routledge
2 Park Square, Milton Park, Abingdon, Oxon OX14 4RN

and by Routledge
605 Third Avenue, New York, NY 10158

Routledge is an imprint of the Taylor & Francis Group, an informa business

British Library Cataloguing-in-Publication Data
A catalogue record for this book is available from the British Library

Library of Congress Cataloging-in-Publication Data
A catalog record has been requested for this book

ISBN: 978-0-367-41523-5 (hbk)
ISBN: 978-1-032-21281-4 (pbk)
ISBN: 978-0-367-81501-1 (ebk)

DOI: 10.4324/9780367815011

Typeset in Times New Roman
by KnowledgeWorks Global Ltd.

Contents

Figures

Tables

Acknowledgements

Conference organizer

The genesis for this book was the two sessions on shrinking cities held at the IGU Thematic Conference in Moscow in 2018. The sessions were jointly organized by Ekaterina Mikhailova and Chung-Tong Wu. It was chiefly due to Ekaterina's network of researchers that the sessions successfully included researchers engaged in recent research on shrinking cities in Russia. Additional contributors were included post-conference to make sure this book has broad coverage amongst postsocialist countries.

Chinese language references

We want to acknowledge Routledge's commitment to incorporate the original Chinese language citations in the relevant reference entries. This is a remarkable advancement on the conventional practice of Pinyin romanization entries which do not provide the precise reference that an interested readers would need. With this change, the interested reader has the precise Chinese language reference to follow up. Hopefully, if other publishers adopt this practice, it will open up the large Chinese language literature on urban development to interested scholars.

Open Access Fee

We are grateful to the following contributors to the open access fee to ensure this book is accessible to all who are interested in the topic of shrinking cities.

Regional Development Research Center of Northeast China, Northeast Institute of Geography and Agroecology, Chinese Academy of Sciences (IGA, CAS).
School of Architecture and Urban Planning, Huazhong University of Science and Technology, China.
Department of Urban Planning, School of Architecture and Planning, Hunan University, China.

Department of Human Geography and Regional Development, University of Ostrava, Czechia.
Adam Mickiewicz University, Poznan, Poland.
Faculty of Natural Sciences, Comenius University, Bratislava, Slovakia.
Associate professorship on Qualitative Health Research, Institute for Interdisciplinary Health Research, European University at Saint-Petersburg, Russia.
Institute of Geography, Russian Academy of Sciences.
The editors group.

Contributors

Evgenii Antonov is Research Fellow at the Institute of Geography Russian Academy of Sciences, Moscow (Russia). Evgenii Antonov holds a Master's degree in Geography from the Lomonosov Moscow State University and a PhD in Human Geography. His PhD thesis was focused on socio-economic development and labor markets of Ural, Siberian, and Far Eastern Russian cities for the period 1990s–2010s. Currently, Evgenii is a Research Fellow at the Department of Social-economic Geography, Institute of Geography RAS (Moscow, Russia). His research interests include processes occurring in depopulated and structurally weak regions of Siberia, the Far East, as well as the Russian Arctic.

Ksenia Averkieva is Senior Research Fellow at the Institute of Geography Russian Academy of Sciences, Moscow (Russia). Ksenia Averkieva is a socio-economic geographer. She holds a Specialist degree in Geography from the Lomonosov Moscow State University (2009) and a PhD in Human geography from the Institute of Geography Russian Academy of Sciences (2012). Ksenia's PhD thesis focused on the transformations taking place in rural areas of the Russian Nonchernozen region and included numerous field studies. In 2013, she took part in the project *A Journey from St. Petersburg to Moscow. 222 Years Afterwards* researching small towns located between St. Petersburg and Moscow—their history, architectural appearance, as well as processes of abandonment and shrinkage. Throughout her career, Ksenia was also engaged in various research initiatives of Russian company towns, including those located in the Eastern and Northern regions of the country. Within this thread, her main research interest was on the trajectories of their demographic development, especially under conditions of structural crises.

Elena Batunova is an urban and regional planner and researcher at RWTH Aachen University, Germany (formerly with Politecnico di Milano, Italy). She holds a Master's degree in Architecture and in Professional Education from the Rostov State Academy of Architecture and Art, Russia. In 2017, she obtained a PhD in Urban Planning at Politecnico di Milano, Italy. Her PhD thesis explores urban planning and policy in small and medium-sized shrinking cities in Russia. Elena's main research interest is in examining new approaches, mechanisms, and tools for planning under conditions of shrinkage, as well as searching for development alternatives for places that do not grow. She led the international research project CRISALIDE '*City Replicable and Integrated Smart Actions Leading Innovation to Develop Urban Economies*' (ERA.Net RUS Plus). Elena's professional experience includes the development of projects at different levels of territorial planning and urban design: from regional plans to cities' master plans and urban design in different regions of Russia (from the Western city of Kaliningrad to the Far Eastern Kamchatka peninsula). She is the co-author of more than 50 projects in regional and urban planning, urban design and zoning, five of which were awarded at international competitions. She also co-authored educational programs in urban development for municipal employees in Russia.

Branislav Bleha is Professor and Head of Section of Demography and Population Geography within the Department of Economic and Social Geography, Demography and Territorial Development at the Comenius University in Bratislava. At present, he is a Vice-Dean responsible for the development and IT at the Faculty of Natural Sciences, and Vice-President of the Slovak Statistical and Demographic Society. His research is oriented on population forecasting, population geography and population policies.

Ján Buček is Professor and Head of Department of Economic and Social Geography, Demography and Territorial Development at the Comenius University in Bratislava. His research focuses on various governance and development issues at the local/urban and regional levels. He was the Chair of the International Geographical Union Commission on Geography of Governance (2008–2016). Within the last decade, he has co-edited several books on local government, governance, urban governance, and local fiscal issues published in Ashgate/Routledge, and Springer.

Donatas Burneika studied geography at Vilnius University and defended doctoral thesis in the field of economic urban geography in 2000. At present Dr. Donatas Burneika leads the Department of Regional and Urban Studies at the Lithuanian Centre for Social Sciences in Vilnius. He is also a Professor at Vilnius University, Department of Geography and Spatial Planning. He is an author of some 80 peer-reviewed publications mostly in the fields of regional and urban geography, concentrating on issues of socio-economic development in Lithuania and the Baltic Sea region, namely, the development of growing metropolitan regions and related processes of peripheralization and spatial differentiation in Lithuania are the most recent topics of his research. Regional Studies Association board elected Donatas Burneika as the new RSA Ambassador for Lithuania in 2015.

Yangui Dai is a doctoral student of Urban Planning, Urban Planning Department in the School of Architecture and Planning, Hunan University, China. Her research focuses on exploring the application and development of "smart growth" and "smart shrinkage" in China. And she has participated in practical projects such as small towns and villages planning with Chinese characteristics.

Daria Chigareva is a consultant at Transaction Real Estate department, Ernst & Young, Moscow (Russia). She holds a Master's degree in Urban Planning and Spatial Development from the National Research University Higher School of Economics (2020). Currently, Daria is an urban planner working as a Consultant at the Transaction Real Estate department of Ernst & Young (Moscow, Russia). Her professional experience lies in the realm of spatial planning at different levels. She has participated in various territorial development projects for government agencies and private companies in Russia.

Zhiwei Du is Assistant Professor in the Guangzhou Institute of Geography, Guangdong Academy of Sciences, China. His research focuses on regional development and city planning and urban growth and shrinkage in China. He is managing two research projects from the National Natural Science Foundation of China and the GDAS' Special Project of Science and Technology Development.

Vera Efremova is Bibliographer of the Moscow State University Scientific Library, Department of Geography.

Zhe Gao is Assistant Professor, College of Urban & Environmental Sciences, Central China Normal University, China. He is a young scholar of human geography, as well as a city planner. His research focuses on urban growth and shrinkage in Central China, gentrification and marginalization in Chinese communities, and the application of open data in urban analysis. He has two research projects, respectively, supported by the National and Hubei Provincial Natural Science Foundation of China. He holds degrees from Sun Yat-sen University, China (BSc and MSc) and KU Leuven, Belgium (PhD).

Maria Gunko is Senior Research Fellow Institute of Geography Russian Academy of Sciences, Moscow (Russia); Lecturer Faculty of Geography and Geoinformation Technologies, National Research University Higher School of Economics, Moscow (Russia). Since 2021, DPhil candidate at the Centre on Migration, Policy and Society (COMPAS) University of Oxford (UK) working on the project "Emptiness: living capitalism and democracy after postsocialism" (European Research Council). Maria holds a Master's degree in Human Geography (specializing in land use management) from the Lomonosov Moscow State University (2012). She was a visiting research fellow at the Leibniz Institute for Regional

Geography (Leipzig, Germany) in 2015 and at the Géographie-cités Lab CNRS / Université Paris 1 Panthéon-Sorbonne (Paris, France) in 2019.

Maria's prime research interests are related to urban planning and policymaking, placemaking, and urban shrinkage in small and medium-sized cities. She is the PI of projects *The role of local initiatives in regeneration of Russian peripheral small and medium-sized cities* (2018–2020, Russian Fund for Basic Research), *Strategies and tactics of urban planning in depopulating Russian cities* (2018–2020, Ministry for Science of the Russian Federation), *Enhancing liveability of small shrinking cities through co-creation* (2021-2022, ERA.Net RUS Plus / Russian Fund for Basic Research). Maria's research is featured in such journals as *Cities*, *European Planning Studies*, *Tijdschrift voor Economische en Sociale Geografie*, and *Geografiska Annaler, Series B: Human Geography*.

Emilia Jaroszewska is a Researcher at the Faculty of Human Geography and Planning at Adam Mickiewicz University in Poznań (Poland). Her main research interests include population geography, urban planning, and social geography, in particular the process of urban shrinkage, including regeneration strategies of post-industrial shrinking cities and revitalization of industrial heritage.

She participated in the COST Action *Cities regrowing smaller—fostering knowledge on regeneration strategies in shrinking cities across Europe*. Currently she is involved in the EU Horizon 2020 ITN project RE-CITY '*Reviving shrinking cities—innovative paths and perspectives towards livability for shrinking cities in Europe*'. She is the author (or co-author) of many research publications which focus on urban shrinkage (including "Shrinkage of old industrial cities and the mitigation of its negative effects" book in Polish—PhD dissertation, "Urban Shrinkage and the Post-Socialist Transformation: The Case of Poland"—*Built Environment*, "The process of shrinkage as a challenge to urban governance"—*Quaestiones Geographicae*, "Drivers, scale and geography of urban shrinkage in Poland and policy responses"—*Journal of Urban Planning and Development*.

He Li is an Associate Professor at the Regional Development Research Center of Northeast China, Northeast Institute of Geography and Agroecology, Chinese Academy of Sciences (IGA, CAS). As a human geographer, his research focuses on a range of issues related to the urban transition of old industrial cities, sustainable urbanization, and low-carbon economy. He was educated at Inner Mongolia Normal University for his BSc (2004) in Geography and at IGA, CAS for his PhD (2009) in Environment Science, and then joined IGA, CAS in 2009. He was a visiting scholar at the School of Geography, University of Melbourne (2014/8-2015/2) and School of Geography, Earth & Environmental Sciences, University of Birmingham (2015/9-2016/9). He serves as a member of the Urban Geography Committee of Geographical Society of

China, a member of Shrinking City Research Network of China and as an editorial board member of Journal of Asian Energy Studies.

Helin Liu is Professor of Planning and working in the School of Architecture and Urban Planning, Huazhong University of Science and Technology, China. His research interests are in computational modelling of urban and regional planning based on big data and artificial intelligence, climate change mitigation, and adaptive planning. He holds degrees from Nanjing University (BSc and MSc) and the University of Cambridge (PhD).

Ekaterina Mikhailova is a Postdoctoral Researcher at the Department of Geography and Environment, University of Geneva (Switzerland), and a Visiting Researcher in the Karelian Institute, University of Eastern Finland (Finland). Ekaterina's work lies at the crossroads of Urban Studies, Border Studies and Eurasian Studies.

As a specialist in border cities Ekaterina was awarded with the Swiss Government Excellence Scholarship in 2020-2021 and joined the Department of Geography and Environment at the University of Geneva where she later engaged in the topics of critical toponymy and city diplomacy. In the fall semester 2021/2022 Ekaterina taught a course 'City Diplomacy: the Rise of Cities as Global Actors' within the Interdisciplinary Master Programmes of the Graduate Institute of International and Development Studies (Geneva, Switzerland). Before Switzerland, Ekaterina worked as a research fellow at the University of Eastern Finland (Finland), Lomonosov Moscow State University (Russia) and the University of Tromsø - Arctic University of Norway (Norway). Ekaterina co-edited two edited volumes: Twin Cities: Urban Communities, Borders and Relationships over Time (Routledge, 2019) and Twin Cities across Five Continents: Interactions and Tensions on Urban Borders (Routledge, 2022) and is the author of over 40 English and Russian publications on border cities, city's internationalisation and city diplomacy, cross-border communities, transfrontier cooperation, governance, migration and border tourism. Most of Ekaterina's research has relied on in-depth semi-structured interviews with various stakeholders – mayors, representatives of regional and national governments, NGOs and corporate sector.

Anastasia Novkunskaya Associate Professor of Qualitative Health research, Institute for Interdisciplinary Health Research, European University at Saint-Petersburg, Saint-Petersburg (Russia). Anastasia was awarded a Master's degree in Sociology at the European University at Saint-Petersburg in 2013. In 2015, she was a Visiting Research Fellow at the University of North Carolina at Chapel Hill (Chapel Hill, NC, USA); in 2019–2020, she was a visiting research fellow at the Aleksanteri Institute, University of Helsinki. She has defended her doctoral thesis at the Faculty of Social Sciences, at the University of Helsinki in March 2020. Her PhD thesis was devoted to the arrangement of maternity care services in Russian small towns.

Anastasia's current research projects focus on the organization of healthcare, health professionals' working experience, as well as on patients' strategies to obtain the necessary healthcare services. She is coordinator and researcher of the projects *The routine of the remoteness: childbearing in the context of small-town Russia; Patient-centered care in Russian healthcare: organizational challenges and professionals' opportunities* (2019–2022, supported by the Russian Science Foundation), *Medicine and professionals in the context of COVID-19 pandemic: inconsistencies, new rules and strategies, COVID Diaries: the chronicles of everyday life, and Serological study of COVID-19 prevalence in Saint Petersburg* (2020–2021, EUSP), *Babies Born Better* (2020–2021, University of Central Lancashire).

Gintarė Pociūtė-Sereikienė studied geography at Vilnius University. At present, she is Research Fellow at the Lithuanian Centre for Social Sciences, Department of Regional and Urban Studies. Her academic research interests encompass regional and urban geography, spatial differences, urban shrinkage, depopulation, peripheralization, and the post-transition period in Lithuania and CEE countries. In 2014 she received her PhD from Vilnius University and Nature Research Centre. During 2017–2020, she was working on a postdoctoral case study project concerning the urban shrinkage topic (project named "Shrinking cities in the post-communist context: the case of Šiauliai city"). Previously, Gintarė has been working in the EU institution Committee of the Regions, in Brussels. She is active as an author and co-author of more than 20 scientific peer-reviewed publications, participant of national and international projects, attendant of conferences, seminars, training, and workshops around the globe. She had traineeships in scientific institutions in Hungary (Institute of Regional Studies Great Plain Research Department in Bekescsaba) and Germany (Leibniz Institute for Regional Geography (IfL) and Helmholtz Centre for Environmental Research (UFZ) in Leipzig). Gintarė is a Member of the Lithuanian Geographical Society (board member), Regional Studies Association (RSA), Shrinking Cities International Research Network (SCiRN), and Cities After Transition (CAT) network. She is also secretary of the Lithuanian scientific journal "Geographical Yearbook".

Marek Richter completed his PhD studies with a dissertation focusing on shrinkage in Slovak cities, primarily on its impact on social care, education, and housing. Currently, he works as a Senior Manager for the non-governmental Pontis Foundation. He is also active in local policymaking as City Councillor in the City of Žilina.

Petr Rumpel is Associate Professor in Economic Geography and Regional Development and Head of the Centre of City and Regional Management, which is a part of the Department of Human Geography and Regional Development, University of Ostrava, Czechia. His research interests are local and regional development theory and practice, especially city and regional marketing, restructuring of old industrial regions, governance of shrinking and regrowing cities, and smart sustainable cities and regions.

Ondřej Slach is Assistant Professor of Economic Geography and Regional Development and a Member of the Centre of City and Regional Management, which is a part of the Department of Human Geography and Regional Development, University of Ostrava, Czechia. His research interests include cultural / creative industries, regional and urban governance (planning), and restructuring of old industrial cities and regions. He also works as an External Consultant for the Ministry of Regional Development of the Czech Republic and the Technological Agency of the Czech Republic. He was a leader of various strategical documents preparation at the regional and city levels (e.g., cultural strategy of the city of Ostrava).

Tadeusz Stryjakiewicz is the Professor of Geography and Head of the Department of Economic Geography at Adam Mickiewicz University in Poznań (Poland). His main research interests encompass socio-economic geography, organization and dynamics of socio-economic space (including creative industries), regional and local development (including the process of urban shrinkage and its spatial consequences), urban and regional policy, and contemporary socio-economic spatial transformations.

He participated in many international research projects, including: (1) ACRE "Accommodating Creative Knowledge. Competitiveness of European Metropolitan Regions within the Enlarged Union", implemented under the 6th Framework Programme of the European Union and (2) the COST Action "Cities regrowing smaller—fostering knowledge on regeneration strategies in shrinking cities across Europe". Currently, he is involved in two EU Horizon 2020 ITN projects: (1) RURACTION "Social Entrepreneurship in Structurally Weak Rural Regions: Analyzing Innovative Troubleshooters in Action" and (2) RE-CITY "Reviving shrinking cities—innovative paths and perspectives towards livability for shrinking cities in Europe". He is the author (or co-author) of 280 research publications (including "Urban Shrinkage and the Post-Socialist Transformation: The Case of Poland"—*Built Environment*, "The process of shrinkage as a challenge to urban governance"—*Quaestiones Geographicae*), editor or co-editor of several books (including "Urban Shrinkage in East-Central Europe" in Polish), expert reports and member of many research councils and institutions.

Chung-Tong Wu is Honorary Professor, University of Sydney, Emeritus Professor, University of New South Wales, and Western Sydney University. He is the Chair of the Advisory Board, Halloran Research Trust (Henry Halloran Trust), University of Sydney. His publications and research topics include regional development, cross-border development, migration, special economic zones, and shrinking cities.

Minwei Zhang is a Postgraduate Student in the School of Architecture and Urban Planning, Huazhong University of Science and Technology, China. Her research interest is in the growth and shrinkage of resource-based cities.

Kai Zhou is Associate Professor in Urban Planning, Head of the Urban Planning Department in the School of Architecture and Planning, Hunan University, China. He is a Fulbright Scholar, a Chartered Urban Planner of China, and a Member of the Shrinking Cities International Research Network (SCiRN) and the Shrinking City Research Network of China (SCRNC). His research focuses on understanding shrinking cities in the context of China's rapid urbanization, interpreting the value-laden questions of social justice in planning decision-making, and developing ICT tools and GIS methods for urban planning. He holds degrees from Nanjing University (BSc and MSc) and the University of Manchester (PhD).

Part I

Shrinkage in the postsocialist countries: Concepts and theory

1 Introduction

Urban shrinkage in the postsocialist realm

Maria Gunko, Chung-Tong Wu, Kai Zhou, and Tadeusz Stryjakiewicz

Objectives and origin of the book

Shrinking cities, in general, have received increasing scholarly attention since the early 2000s, but the geographic focus of case studies and theory building has remained predominantly restricted to the Global North (Pallagst et al., 2014; Haase et al., 2014; Hollander, 2018). Shrinking cities in the so-called postsocialist "Global East" (Chan et al., 2018; Müller and Trubina, 2020) are in danger of "double exclusion", positioned outside both mainstream urban studies and postcolonial debates (Tuvikene, 2016). Moreover, there is a lack of in-depth comparisons between shrinking cities within this vast, contested, and diverse region.

In one of the first studies of its kind, Kubes (2013) identified the lopsided coverage of postsocialist shrinking cities. Even though the criterion he used excluded studies of Chinese cities and the Asian part of Russia, he identified over 180 articles published on Leipzig, which at the time was far more than any other postsocialist shrinking city (Kubes, 2013). More recently, Doringer and others examined 100 case studies of shrinking cities in the European Union (EU) and Japan, with about 30 percent of the case studies covering postsocialist Europe (Doringer et al., 2019). They noted a paucity of comparative studies. Bajerski (2020), investigating which countries and institutions have been contributing to the study of postsocialist cities, found a dominance of research from countries in Central and Eastern Europe (CEE), but his coverage identified an increasing number of articles on China and a notable absence of Russian institutions among the top 20 research institutions publishing on the topic (Bajerski, 2020). The above is indicative that there is a need for comparative studies of shrinking cities that include a broad range of postsocialist countries to identify commonalities, differences, and policy experiences. This book is an attempt to remedy this situation through contributions from researchers based in 15 institutions of the postsocialist "Global East" to offer a view from the inside and help to decolonize knowledge. Specifically, we have organized the book to include chapters on shrinking cities in China, Russia, and postsocialist Europe, offering a comparative discussion within countries and cross-national cases on the theoretical aspects

DOI: 10.4324/9780367815011-2

and policy implications. With this, we believe that the book partly responds to Hollander's call: "if there were more, better, and especially cross-national research on shrinkage, the on-the-ground truth might turn out to be more complex and interesting" (Hollander et al., 2009, p. 230).

The idea for this book originated from two sessions on shrinking cities organized as part of the 2018 04–06 June International Geographic Union (IGU) thematic conference dedicated to the centennial of the Institute of Geography, Russian Academy of Sciences, *Practical Geography and 21st Century Challenges* held in Moscow. The roster of authors has been expanded to ensure the book has an adequate geographic and thematic coverage. Thus, the authors are situated in diverse institutions but have relevant backgrounds.

The postsocialist label

Early discussions of urban changes within the researched region tend to use the terms "postsocialist"(Andrusz et al., 1996) or "postcommunist" (Pickles and Smith, 1998; Sykora and Bouzarovski, 2012) to group countries together either for the sake of convenience or based on a set of ideas about the specific nature of a socialist city that ultimately predefined the specific nature of the postsocialist one (this assumption is currently debated by some scholars (see, for example, Hirt et al., 2016)). Referring to cities in eastern Europe, Szelenyi (1996: 294) asserts that there are qualitative differences between socialist cities and capitalist cities. He posits socialist countries tend to have low urbanization and less spatial concentration, a lack of functional diversity and are uneconomical in the use of space. However, even in the mid-1990s, there was recognition of vast diversity among the former socialist states. In a book with much broader geographic coverage, including CEE, Russia, China, Vietnam, and others, Pickles and Smith emphasized the need to take into account the historic context and regionally uneven development when examining the impacts of the postsocialist transformations (Pickles and Smith, 1998). Pickles and Smith were the few researchers who included China and Russia in their coverage of postsocialist transformations and its impacts on urban development.

Until the early 2000s, there were few studies of postsocialist shrinking cities, while the few available tended to focus on the European cases (Stanilov, 2007). An analysis of publications on postsocialist shrinking cities noted the lopsided interest, even within Europe, with one city in former East Germany receiving the most attention, although the author of this study was careful to note that his criteria excluded China and the Eastern part of Russia (Kubes, 2013). Further research on postsocialist shrinking cities has also had a strong geographical bias focusing on Europe; though its comparative nature drawing on various cases from this region and the strive toward enhancing the concept of "urban shrinkage" with the postsocialist perspective, should be acknowledged (Haase et al., 2016).

The recent discussion on postsocialist cities, including shrinking ones produced within different disciplines, tends to focus on the stricture that the "postsocialist" label places on theory-building, at the same time recognizing that it is unhelpful to examine them with the use of models based on "western" experience (Humphrey, 2001; Tuvikene, 2016). Stenning and Horschelmann (2008) argue that there are multiple postsocialisms. Framing the application of the term postsocialism in the discourse of postcolonialism has raised issues on how and if the term may be used purely for marking a time period, a spatial area (second world in development studies), or is more divisive in terms of knowledge production (Cervinkova, 2012). Numerous scholars have voiced their objections to marginalizing the postsocialist cities as either "cases unto themselves" or "deviations" from the universalistic western "grand models" of urban development (Roy and Ong, 2011; Robinson, 2011; Gentile, 2018; Peck, 2015; Sjoberg, 2014). They argue for "multipolar, cosmopolitan, and comparative modes of urban theory making" (Peck, 2015: 160). Furthermore, Müller has declared "goodbye postsocialism" as a way to say that it is no longer relevant as a reference point since issues such as neoliberalism, globalization, and mass migrations are much more important for shaping the current urban form (Müller, 2019). Though we do see the debates on abandoning the "postsocialist" label as having a point, since the early scholars did tend to use postsocialism as a predefining condition; however, it would be remiss to altogether deny the importance of various socialist legacies and path-dependencies that still explicitly or implicitly play a role in contemporary urban development within China, Russia, the postsocialist European countries and other countries that experience state socialism.

Some early studies of postsocialist cities in Europe assert these cities have similar spatial restructuring issues in spite of variations in the national context (Stanilov, 2007; Ferencuhova, 2016, Sykora and Bouzarovski, 2012). However, neither all postsocialist cities change in the same way nor do they necessarily have a similar approach toward urban shrinkage within the country or compared to other postsocialist countries. For instance, abandoned housing and large tracks of brownfield sites are not necessarily common in shrinking cities if we draw our attention to China and Russia. Furthermore, as demonstrated by the cases from postsocialist European countries and Russia, what may appear to be the same urban transformation process has very different actors, causes, and outcomes—the pro-growth orientation in Russian cities, which appears at first glance to be led by an entrepreneurial pro-growth "coalition", may instead be a group of state agencies appointed by the government to implement development projects (Kinossian, 2012; Müller, 2011). At the same time, the diversity in postsocialist outcomes does not erase their past shared experience of state-owned means of production, residential control, and planned economies — all elements that contribute to the way their urban systems evolved and still evolve today.

So, does the postsocialist label help or obscure? It is a convenient way for a geographical grouping of countries that shared a similar politico-economic history, though it is always important to note that within these few aspects, there was a wide range of experiences which some scholars have characterized as "hard" socialism (Marcinczak et al., 2013). But most importantly, for our current discussion, we highlight the pace and extent of transformation from state socialism to varieties of capitalism that ultimately impacted on urban development, causing or reinforcing the tendencies of urban shrinkage. If the diversity is recognized and the label is not used as a simplistic way to make group comparison as per postsocialist vs capitalist, it could indeed be helpful for developing theories that are more grounded and engaged with the experiences of societies that have encountered great transformations over the last three to four decades. Thus, we acknowledge that there was neither a single socialist experience nor a single trope of transformation. Postsocialist shrinking cities possess qualities that often puzzle urban scholars since they can neither be measured against the perceived "normality" of the urban in the Global North (Gentile, 2018) nor thoroughly scrutinized from a postcolonial perspective of urban experience from the Global South.

This book is an attempt to generate knowledge based on case studies and observations on the ground with references but without unnecessary universalizations of models developed in other economic, social, and cultural settings. We take on a pragmatic approach to examine postsocialist shrinking cities with reference to their past and its "stubborn urban structures" (Drummond and Young, 2020), but most importantly to the forces that shape their present and future.

Organization of the book

We have organized the book according to geographic principle because the three main groups/countries have significant differences due to a combination of socialist experience, the starting point of their transition, their history, their institutional legacies, and their experiences since transition. China, although still calling itself "socialism with Chinese characteristics" embarked on its transition a decade or more before postsocialist European countries and Russia. The Central Eastern European (CEE) countries in our case studies became members of the EU in 2004, which enabled massive international migration with significant impacts on urban development, whereas some countries of South-Eastern Europe (SEE) still remain in a political and economic "vacuum". Russia has had the longest experience of a state socialist regime within the studied region. Here the tangible and intangible legacies of state socialism remain particularly strong—the state dominates in shaping strategic priorities, creating specific incentives for cities, and intervening directly in urban development, effectively substituting markets and market actors (Orttung and Zhemukhov, 2017; Gunko et al., 2021).

Political, social, and economic transformations in the postsocialist context are sometimes portrayed as the "Eastern branch" of the global neo-liberalization project (Golubchikov, 2016). However, despite the influence of neoliberal ideology and the proliferation of urban entrepreneurialism and competitiveness, the universal application of these trends across geographical space and scales needs to be viewed critically. Several scholars, including Myant and Drahokoupil (2011), Mykhnenko (2007), Hall and Soskice (2001), and Knell and Srholec (2007) have all elaborated on the variety of capitalisms that have emerged in the postsocialist countries. The latter characterized these variations as "state capitalism" in China, "patrimonial capitalism" in Russia, and "imported capitalism" in some CEE countries. We argue that such divergent economic transformations and associated political and social processes have specific impacts on urban changes within each group. Thus, we need to first understand shrinking cities in their own settings to make comparisons. Given the relatively sparse literature on shrinking cities in China, Russia, and parts of postsocialist Europe, keeping the regional organization of the book sections contributes to a deeper understanding of urban shrinkage in each of these geographical areas. With that knowledge, comparative themes become more meaningful.

There is also a practical issue of specific nomenclatures that are used in China and Russia that may be confusing to the reader unfamiliar with the context. Each section's introduction explains these terminologies to minimize the repetition of the same in each chapter. Furthermore, each section's introduction contains a map identifying the case studies so that readers can gain a sense of the geographic coverage.

In keeping with our aim to highlight research from each geographic region, we are including the original language of references so that local researchers' work can be given the prominence they deserve and to make it easier for readers who wish to follow up on the relevant reference have less difficulties in precisely identifying the specific reference. This is specially the case for Chinese language references.

References

Andrusz, G., Harloe, M. & Szelenyi, I. (eds.) 1996. *Cities After Socialism: Urban and Regional Change and Conflict in Post-Socialist Societies*. Oxford: Blackwell.

Bajerski, A. 2020. Geography of Research on Post-Socialist Cities: A Bibliometric Approach. *Urban Development Issues*, 65, 5–16.

Cervinkova, H. 2012. Postcolonialism, Postsocialism and the Anthropology of East-Central Europe. *Journal of Postcolonial Writing*, 48(2), 155–163.

Chan, K. W., Gentile, M., Kinossian, N., Oakes, T. & Young, C. 2018. Editorial–Theory Generation, Comparative Analysis and Bringing the "Global East" into Play. *Eurasian Geography and Economics*, 59(1), 1–6.

Doringer, S., Uchiyama, Y., Penker, M. & Kohsaka, R. 2020. A Meta-Analysis of Shrinking Cities in Europe and Japan. Towards an Integrative Research Agenda. *European Planning Studies*, 28(9), 1693–1712

Drummond, L. B. W. & Young, D. (eds.) 2020. *Socialist and Post-Socialist Urbanisms: Critical Reflections from a Global Perspective*. Toronto, Canada: University of Toronto Press.

Ferencuhova, S. 2016. Explicit Definitions and Implicit Assumptions about Post-Socialist Cities in Academic Writings. *Geography Compass*, 10(12), 514–524.

Gentile, M. 2018. Three Metals and the "Post-Socialist City": Reclaiming the Peripheries of Urban Knowledge. *International Journal of Urban and Regional Research*, 42(6), 1140–1151.

Golubchikov, O. 2016. The Urbanization of Transition: Ideology and the Urban Experience. *Eurasian Geography and Economics*, 57(4–5), 607–623.

Gunko, M., Kinossian, N., Pivovar, G., Averkieva, K. & Batunova, E. 2021. Exploring Agency of Change in Small Industrial Towns through Urban Renewal Initiatives. *Geografiska Annaler, Series B: Human Geography*, 103(3), 218–234.

Haase, A., Rink, D. & Grossmann, K. 2016. Shrinking Cities in Post-Socialist Europe: What Can We Learn from Their Analysis for Theory Building Today? *Geografiska Annaler: Series B Human Geography*, 98(4), 305–319.

Haase, A., Rink, D., Grossmann, K., Bernt, M. & Mykhnenko, V. 2014. Conceptualizing Urban Shrinkage. *Environment and Planning A*, 46(7), 1519–1534.

Hall, P. A. & Soskice, D. 2001. *Varieties of Capitalism: The Institutional Foundations of Comparative Advantage*. Oxford: Oxford University Press.

Hirt, S., Ferencuhova, S. & Tuvikene, T. 2016. Conceptual Forum: The "Post-Socialist" City. *Eurasian Geography and Economics*, 57(4–5), 497–520.

Hollander, J. B., Pallagst, K. M., Schwarz, T. & Popper, F. J. 2009. Planning Shrinking Cities, Progress in Planning, 72(4), 223–232.

Hollander, J. B. 2018. *A Research Agenda for Shrinking Cities*. New York: Edward Elgar Publishing.

Humphrey, C. 2001. Does the Category "Postsocialism" Still Make Sense? In Hann, C. M. (ed.) *Postsocialism: Ideals, Ideologies and Practices in Eurasia*. London: Routledge.

Kinossian, N. 2012. "Urban Entrepreneurialism" in the Post-Socialist City: Government-Led Urban Development Projects in Kazan, Russia. *International Planning Studies*, 17(4), 333–352.

Knell, M. & Srholec, M. 2007. Divergent Pathways in Central and Eastern Europe. In Lane, D. (ed.) *Varieties of Capitalism in Post-Communist Countries*. London: Palgrave Macmillan.

Kubes, J. 2013. European Post-Socialist Cities and Their Near Hinterland in Intra-Urban Geography Literature. *Bulletin of Geography: Socio-Economic Series*, 19, 19–43.

Marcinczak, S., Gentile, M., Rufat, S. & Chelcea, L. 2013. Urban Geographies of Hesitant Transition: Tracing Socioeconomic Segregation in Post-Ceausescu Bucharest. *International Journal of Urban and Regional Research*, 38(4), 1399–1417.

Müller, M. 2011. State Dirigisme in Megaprojects: Governing the 2014 Winter Olympics in Sochi. *Environment and Planning A*, 43(9), 2091–2108.

Müller, M. 2019. Goodbye, Postsocialism! *Europe-Asia Studies*, 71(4), 533–550.

Müller, M. & Trubina, E. 2020. The Global-Easts in Global Urbanism: Views from Beyond North and South. *Eurasian Geography and Economics*, 61(6), 627–635.

Myant, M. R. & Drahokoupil, J. 2011. *Transition Economies: Political Economy in Russia, Eastern Europe and Central Asia*. New York: John Wiley.

Mykhnenko, V. 2007. Poland and Ukraine: Institutional structures and economic performance. In Lane, D. & Myant, M. R. (eds.) *Varieties of Capitalism in Post-Communist Countries*. Basingstoke: Palgrave Macmillan.

Orttung, R. & Zhemukhov, S. 2017. *Putin's Olympics. The Sochi Games and the Evolution of Twenty-first Century Russia.* New York: Routledge.

Pallagst, K., Wiechmann, T. & Martinez-Fermandez, C. (eds.) 2014. *Shrinking Cities: International Perspectives and Policy Implications.* New York: Routledge.

Peck, J. 2015. Cities Beyond Compare? *Regional Studies*, 49(1), 160–182.

Pickles, J. & Smith, A. (eds.) 1998. *Theorising Transition: The Political Economy of Post-Communist Transformations.* New York: Routledge.

Robinson, J. 2011. Cities in a World of Cities: The Comparative Gesture. *International Journal of Urban and Regional Research*, 35(1), 1–23.

Roy, A. & Ong, A. (eds.) 2011. *Worlding Cities: Asian Experiments and the Art of Being Global.* Hoboken: Wiley-Blackwell.

Sjoberg, O. 2014. Cases Onto Themselves? Theory and Research on Ex-Socialist Urban Environments. *Geografie*, 119(4), 299–319.

Stanilov, K. (ed.) 2007. *The Post-Socialist City: Urban Form and Space Transformations in Central and Eastern Europe after Socialism.* Dordrecht, The Netherlands: Springer.

Stenning, A. & Horschelmann, K. 2008. History, Geography and Difference in the Post-socialist World: Or, Do We Still Need Post-Socialism? Antipode, 40(2), 312–335.

Sykora, L. & Bouzarovski, S. 2012. Multiple Transformations: Conceptualising the Post-Communist Urban Transition. *Urban Studies*, 49(1), 43–60.

Szelenyi, I. 1996. Cities under socialism—and after. In Andrusz, G., Harloe, M. & Szelenyi, I. (eds.) Cities After Socialism, (pp. 286–317). Oxford, Blackwell.

Tuvikene, T. 2016. Strategies for Comparative Urbanism: Post-Socialism as a De-Territorialized Concept. *International Journal of Urban and Regional Research*, 40(1), 132–146.

2 Postsocialist shrinking cities in a triple geopolitical and socioeconomic context

Chung-Tong Wu, Maria Gunko, Kai Zhou, and Tadeusz Stryjakiewicz

Urban shrinkage—Literature review

Despite the growing interest in the topics of urban shrinkage/shrinking cities, a widely agreed definition of urban shrinkage and a common theory is lacking. Attempts to develop definitions and conceptualizations of urban shrinkage have met with critique and raised new questions (Haase et al., 2014, Olsen, 2013, Bernt, 2016). The concepts of urban shrinkage/shrinking cities are widely used, remaining at the same time ambiguous (Olsen, 2013). Depending on which aspects are investigated and from which perspective, the term "shrinking city" may refer to quite different entities and problems: from the city's economic competitiveness at a global scale to troubling social issues such as marginalization, segregation, crime, and poverty. According to Wiechmann and Bontje (2015), the topic of "shrinking cities" has become an interdisciplinary field of research that considers the complex issues of urban decay, as well as regeneration from demographic challenges and structural crisis. Within this line of thought, urban shrinkage refers to a trajectory of development opposite to "normal" and "desirable" growth (Haase et al., 2014, Bernt, 2016, Hollander, 2018). Furthermore, it is widely agreed that significant and persistent population loss is a defining feature of urban shrinkage (Oswalt and Rienets, 2006, Grossmann et al., 2013), which happens against the preservation of the city's physical footprint–borders and urban pattern (Right-sizing America's shrinking cities, 2007). A logical consequence of the above is the alteration of the cityscape and infrastructure requirements, which initially were planned for a different population size (Weaver and Knight, 2018). Structural crises and deindustrialization are often mentioned in relation to shrinking cities (Hollander, 2018); however, their occurrence and intensity differ largely among various national and regional contexts.

Policymaking and planning in shrinking cities are usually structured around addressing the negative effects of this phenomenon such as underused and deteriorating infrastructure, vacancies, and abandonment. This has led to the emergence of voids and "hollowing out" of the urban fabric, a decrease in the quality of the cityscape, and the safety of the urban environment. However,

DOI: 10.4324/9780367815011-3

while the problems that arise in shrinking cities have a lot in common in different regional and national settings (Audirac and Alejandre, 2010, Pallagst et al., 2014), the responses to address them vary widely (Hollander et al., 2009, Ryan, 2012, Neill and Schlappa, 2016). Which policies, approaches, and instruments are employed largely depends on the local, regional, and national planning cultures and legislation, the knowledge of planners and policymakers, legacies, as well as the dominant discourses of desirable urban development (Dormois and Fol, 2017, Pallagst et al., 2021).

This chapter connects the arguments of the current book to the literature on shrinking cities in postsocialist countries. The key topics raised in the chapters form the basis for this discussion. As noted in Chapter 1 of this volume, despite the generalist "postsocialist" label, there is a great diversity of urban shrinkage among and within the countries included in this book. While there is sufficient literature on shrinking cities in the Central and Eastern Europe (CEE) countries, the literature on this topic based on examples from Russia and China is still nascent, though rapidly changing, especially in the case of China. Prior to discussing what aspects of the literature on urban shrinkage this book aspires to contribute to, it is important to highlight the issue of scale.

The urban administrative system (how we measure shrinkage and manage shrinkage)

To facilitate a comparative study of shrinking cities, we propose accepting a widely used definition of a shrinking city as an urban area with 10,000 or more inhabitants and with population decline over at least two consecutive years. Refinements to this definition are made by scholars based in the European Union (EU) to streamline EU funded studies of shrinking cities and associated policies (Wolff and Wiechmann, 2018). At the same time, while this definition is useful in the European setting with a few megacities and where urban administrative systems tend to be well defined, several chapters of this book point to the anomalies it can create. Even more challenging are the cases of Chinese cities with several levels of administrative definitions of the urban and where within each city there are further subdivisions. A city, such as Wuhan (Chapter 7), has a total population of 11 million and within its 7 central districts, the urban core, the population ranges from 1.5 m (Hongshan district) to 1.2 m (Wuchang district) to 485,000 (Qingshan district), with densities ranging from 17,863 km^2 to 3,228 km^2, all of them well over the minimal size of cities generally used to study urban shrinkage. Within megacities, there could be urban shrinkage in subdistricts even though the city as a whole may be growing.

The administrative definition of a city in China includes both rural and urban areas and within the urban areas, its components can be large in population size with comparative high densities relative to international counterparts. Consequently, a study of urban change in Wuhan, for example, would completely miss what may be taking place at the district level. Making international comparisons of Chinese cities at the city level could yield results that may lead to misleading conclusions like comparing apples and oranges. In the cases of a growing Paris metropolitan area and Wuhan, the researchers looked at the issue of "perforated" shrinkage (Albecker and Fol, 2014). The same holds true, for instance, for Detroit, which is often regarded as a showcase of urban shrinkage; however, its metropolitan area is sprawling and remains stable in terms of population, while shrinkage is predominantly confined to the core (Benfield, 2011).

In addition to the need for care when making international comparisons in terms of size and scale, it is also a reminder that evolving administrative systems impact on the ability of local governments to plan for and manage urban shrinkage. Studies of several cities in postsocialist Europe demonstrated that urban shrinkage is partly due to suburbanization. Here new housing estates are developed just outside existing urban administrative boundaries partly because the land is cheaper and less development controls are in force. However, these suburbs and their inhabitants are closely tied to the economic life of the city. In terms of functional space, they are closely linked to each other. This phenomenon, observed by researchers (Ferencuhova, 2016), emphasizes the importance of administrative boundaries and scales when identifying urban shrinkage, as well as the responsibilities, available financial resources, and policymaking capacity of the relevant government unit.

Contributions to the literature

Our motivation for including case studies drawn from several postsocialist countries with diverse histories, cultural contexts, and political systems is rooted in the desire to contribute toward a more comprehensive understanding of shrinking cities in the postsocialist realm and beyond. We reject the idea that cities in postsocialist countries, or as they are often referred to now, the postsocialist Global East, are a theoretical category or categories of their own utterly defined by the legacies of state socialism, as well as the idea that they are necessarily transiting toward an urban model of the Global North (Kovacs, 1999). All cities transform over time. Indeed, a close reading of the case studies in this book will confirm the rich diversity based on cross-national, national, and subnational "historic legacies and contemporary conjunctures" (Harloe, 1996, p. 10) intersecting with mega-trends at the national and global levels. In line with Robinson, we argue that postsocialist cities should be treated as "ordinary cities" (Robinson, 2002) to find what "lessons of wider import" can be learned from them (Sjoberg, 2014). With this, we further contribute to the "heuristic models" for understanding shrinking cities' development and change (Haase et al., 2016a).

It is evident that the emergence of shrinking cities in any country is a result of multiple factors that are interconnected and interact with each other over time. Responding to calls for studies that take into account specificities and temporal aspects (Eder, 2019, Doringer et al., 2019), we have identified two major themes to organize this discussion: Legacies and Temporality. Under the theme of Legacies, we will discuss the significance of prior existing policies, established spatial systems, and cultural context on the emergence and evolution of urban shrinkage and policy responses. Under Temporality, four aspects will be discussed—the period when the transformation from state socialism to capitalism began; the transformation process and the governance system that emerged impacting on natural population development and migration, the built environment and infrastructure, as well as land management.

Legacies

Policies

All socialist states implemented a variety of policies that regulated population movement or spatial development either as a strategy to promote industrialization or to fit into some preconceived idea about city size. These policies and programs set the stage for urban shrinkage as political and economic reforms set in and progressed.

Common to each of the countries included in this book were policies of residence regulations—a system of household registration (*hukou*) and employment-related ration coupons in China; residential permits (*propiska*) and policies that limited the growth of large cities in Russia and other former Soviet states (for example, Lithuania in this volume); policies to control internal migration in, for example, Poland and former Czechoslovakia. Associated with the above was that urban housing was either provided through the work unit or through state allocations which often meant shortages, overcrowding, and poor quality of housing and utilities. In countries that had not done so earlier, residence permits were abolished or relaxed after reforms were introduced, unleashing demand for better and spacious housing, as well as intensifying rural-urban population movement to regions with better economic and social prospects.

As housing became commodified in postsocialist countries, new housing developments mushroomed to cater for the demand for a higher quality of residential space. In many postsocialist European cities, suburbanization became important, defining the shrinkage of central cities even though the overall metropolitan area may have been growing. Though similar in its outcomes, emptying the city cores of postsocialist cities is not due to the same factors as are at play in the cities of the Global North, where emptying is largely a result of socioeconomic status, class, and racial segregation. Poor quality housing and a lack of renewal programs are more relevant for understanding suburbanization in postsocialist cities. Furthermore, suburbanization and associated urban shrinkage do not necessarily leave behind empty urban housing as found in cases within the Global North. Overcrowded residential space, which existed in many socialist cities, meant that a shrinking population led to an emergence of smaller households with less congestion and the potential for better housing for those left behind. A substantial literature on the socioeconomic consequences of suburbanization in postsocialist countries shows that the pattern of resulting segregation differs from that found in the Global North, especially in the United States and Canada (Marcinczak et al., 2015, Kubes, 2013, Marcinczak et al., 2013). It would be wrong, however, to totally exclude segregation as a social phenomenon in postsocialist countries. During state socialism, a mild version of it did exist, but the defining factors were occupation and social status (Vendina, 1997, Gunko et al., 2018). After the transition, social segregation strengthened with the emergence of gated communities or in other forms (Krisjane and Berzins, 2012, Shen and Xiao, 2020, Yang and Zhou, 2018, Li, 2005). Though in the Chinese case, due to the existence of very large city regions, suburbanization may not be recognized as such because of the vast administrative boundaries within which the population is being redistributed.

Legacy regional development strategies

The collapse of state socialism did not erase the existing spatial system or uneven regional development patterns already present in postsocialist countries. Heider (2019) noted that both pre-World War II and postsocialist urban growth in Eastern Germany have been "remarkably persistent over time" even though the policies of "prioritization" under centralist planning were not successful (Heider, 2019).

Most socialist countries implemented policies that favored specific locations, often due to military purposes, operational or resource availability. This led to the development of new, single-function cities that are generally referred to in the literature as "company towns". The phenomenon of company towns emerged during the rapid industrialization of the 19th century for the purpose of sustaining manufacturing or natural resource extraction and securing labor forces in remote, isolated areas (Garner 1992). Under state socialism, company towns were planned, developed, and managed by the state. Although state-run company towns (in Russia—*monogorods*) also existed in capitalist economies, primarily to protect strategic resources, e.g., in Argentina, Brazil, and the United States (Borges and Torres, 2012), the difference lies in the scale of their distribution, as in capitalist economies such towns were an exception. The socialist policies of industrialization led to the establishment of company towns as centers of heavy industry, coal mining, and forestry in many socialist states. Moreover, company towns were also logistical centers (Russia) and military outposts (Russia and China). In the Russian case, many such towns are linked to forced migration and the GULag labor camps, through which remote parts of the country with harsh natural conditions were developed (Barenberg, 2014).

Researchers have stated that even before the collapse of state socialism, urban shrinkage was underway, but the socialist transformations hastened the process (Haase et al., 2016b). In Russia and some CEE countries such as Lithuania and Czechoslovakia (now the Czech Republic), past policies of developing company towns based on state subsidies created regional development that became unsustainable in the market-oriented economy, resulting in urban shrinkage as economic transformation deepens. The same is true for China, where the Northeast heavy industrial and energy base, fervently promoted as a model during the 1960s, became a rust belt and many of China's resource-depleted cities became moribund as reforms began to take hold two decades later. These issues are discussed in detail in Chapters 4 and 5, which examine the trajectories of resource depleted cities in China. Furthermore, in China, the transition to a more market-oriented economy included a set of spatial policies designating cities and regions for foreign investment—the impacts of these policies will be examined in a subsequent section. The pattern of urban shrinkage and the policies that the Chinese government put in place to deal with the aftermath of economic and social transformation are discussed in two chapters. Chapters 17, 18, and 19 in the Postsocialist Europe section also deal with the process and aftermath of "shock therapy" on coal mining cities—some of which emerged precisely because of past industrial policies and industrial town development.

Patterns of uneven regional development are found to have deepened in all the cases studies included in this book though the outcomes are diverse (Pickles and Smith, 1998). In the case of postsocialist Europe and Russia, the drift of population toward an elevated pattern of regional concentration focussing on a few metropolitan centers and the decline of many others is now obvious. Most of the large centers have been enjoying economic growth since the socialist era; however, recently, the divide between the growing and the shrinking cities deepened. Overall, drawing on all said in this subsection, we support the idea that to understand and interpret the present urban trends, there is a need to carefully analyze the history and legacy of urban development in postsocialist countries.

Cultural, social, and demographic legacy

Economic reforms have led to urban shrinkage in many former mining cities in all postsocialist countries, but not all such cities have managed the impact of the reforms in the same way. While the ceasing of mining operations and resulting massive unemployment seem to be the common experience, there are examples where the process is managed differently. Chapter 18, which focuses on shrinking mining cities in Slovakia, gave the example of how the social and cultural support of mining allowed unions to demand a managed and phased closure of mines. This is an aspect seldom discussed in the literature on urban shrinkage and mine closures. Furthermore, Chapters 10 and 14, concerned with urban shrinkage in Russian mining cities Vorkuta and Kirovsk, highlight the role of agency in managing shrinkage caused by a complex interplay of factors (economic restructuring being one of them) and the latter effects on the cityscapes.

As noted by Wolff and Wiechmann (2018), both economic and demographic factors and their complex interplay shape urban shrinkage. Thus, despite the importance of economic restructuring in causing and shaping urban shrinkage in the postsocialist realm, demographic structures, and natural population change should not be underestimated. The postsocialist European countries and Russia are currently witnessing a decline in birth rates and a rise in aging (Batunova and Perucca, 2020, Kashnitsky et al., 2020) which began before the collapse of state socialism and are caused by a global trend—demographic transition (van de Kaa, 1987). The profound difference between the two regions lies in the high mortality rate and poor state of health in Russia, resulting in lesser life expectancy there (Eberstadt, 2010). Moreover, the negative population trends in Russia were inevitable because of the historical events of the 20th century that profoundly transformed the national population. Chapters 10 and 11 in this volume will further elaborate on the natural population development in Russia and policies employed by the state.

Temporality

Doringer and others noted the absence of consideration of temporality in most case studies of shrinking cities (Doringer et al., 2019). Others have argued that urban shrinkage is not new and the waxing and waning of a city's fortunes and its size can be observed historically (Turok and Mykhnenko, 2007). Contemporary studies of cities such as Leipzig (Kubes, 2013) point out how the fortune of this notable shrinking city has turned around. The temporal aspects of shrinking cities noted by the above scholars are chiefly in relationship to the process of how urban shrinkage has unfolded in a particular country or city. We posit that two other aspects: (a) the year a country embarks on the transition from socialism and (b) the pace of transition needs to be considered to fully comprehend how these processes and outcomes cause or reinforce urban shrinkage.

Timing of transformation

Whereas most calls for considering the temporal aspects of shrinking cities refer to the aspects of change of shrinking cities over time, we argue that for the postsocialist countries, it is equally important to first understand the era (global context) when they embarked on their postsocialist turn. This timing frames the opportunities available for policies that could have significant impacts on the emergence of urban expansion and shrinkage, the amount of funding, and the capacity for decision-making that became available to policymakers.

China embarked on its transition in 1979, slightly over a decade before postsocialist European countries and Russia. Several features are significant for China's postsocialist turn—its early adoption at the juncture of expanding globalization, the benefits of a head start to take advantage of what globalization had to offer and opportunities to readjust with few international competitors. Thus, China was provided with opportunities to experiment and adjust policies to those that were attractive to foreign investors. Being first off the block and achieving rapid industrial growth allowed China to start dominating the international consumer goods market at the turn of the millennium. It became possible, for example, for shrinking border regions in Northeast China to consider economic revival through border trade, as discussed in Chapter 8. China's national development policy since the 1990s has been city-centered, and in recent decades, there has been more focus on urban renewal (Lin 2021), which in turn has had major impacts on urban expansion and shrinkage. The combination of the state monopoly of land, in turn, drives a faster and larger scale of urban demolitions, partly driven by state-owned enterprise reforms leading to urban renewal projects with implications for urban shrinkage (Zhou et al., 2019). Due to its early start, within the course of three decades, China experienced rapid industrialization and urbanization as well as urban shrinkage. More recently, some Chinese regions are facing industrial restructuring through deliberate policies resulting in urban shrinkage (see Chapter 6 on Guangdong and Chapter 7 on Wuhan).

The postsocialist European countries embarked on their transformation in 1989, and within fifteen years, some became members of the EU, which enabled massive international migration with significant impacts on urban development including the emergence of urban shrinkage. Russia came late to the postsocialist turn: it was not able to benefit from EU membership, nor could it take advantage of the early stages of rapid globalization. Moreover, unlike in the case of most postsocialist European countries, the collapse of state socialism was less anticipated and was followed by drastic economic crises in the early 1990s and in 1998 in particular. The implicit outcomes of the latter are the strive for "stability", as well as suspicion around democratization and (neo)liberalization. From the vantage point of the 2020s, when the era of globalization is under question, the conjunction of political geography and temporal shifts make it imperative to grasp the temporal differences between geographical regions to understand the emergence and impacts of urban shrinkage.

Neoliberalism, state capitalism, and urban transformation

While most postsocialist European countries and Russia tend to regard their postsocialist transformation as "shock therapy", referring to the abrupt shift from a centrally planned economy to a market one accompanied by the collapse or hollowing out of state institutions, this was not the case for China (Roland, 2018). The Chinese party-state, through a variety of policy shifts and pilot schemes, made the transition with less ruptures and huge economic success. Although there are divergent views on the results of the neoliberal plunge among postsocialist European countries, an increasing number of scholars are questioning the application of the neoliberal model to explain the variety of urban and regional transformations in postsocialist Europe, Russia, and China.

Some researchers suggest China should be regarded as a developmental state, much as South Korea and Singapore are regarded (Gabusi, 2017, Bolesta, 2014), but others have argued that both China and Russia are contemporary "state capitalisms" (Tsai and Naughton, 2015). Most discussions, however, focus on Chinese state-capitalism, in which the one-party state has managed to use the market economy to reinforce its authority and prestige (Roland, 2018). Those who concentrate on the urban development outcomes point to the role of the entrepreneurial state, on its own or in alliance with the private sector, in transforming the land market and urban development paths (Zhou et al., 2019, Tang, 2019, Tsai and Naughton, 2015, He and Wu, 2009). The significance of the institutional outcomes of the postsocialist transformation on urban shrinkage requires further exploration.

Postsocialist transformation, governance, institutions, and urban shrinkage

The literature on postsocialist European cities refers to the "hollowing of institutions" in the aftermath of the collapse of socialist regimes when old institutions were either dismantled or stripped of the authority to manage urban development (Haase et al., 2016b). In Russia, the decade after the collapse of the Soviet Union is usually referred to as the "decade of turmoil" or the "lost decade" when the old was swept away or transformed, but the state institutions did not have a firm grip on the social and economic processes that took place. Since the 2000s, Russia has been undergoing a profound re-centralization of governance and rising authoritarianism (Gelman, 2018, Gelman and Ryzhenkov, 2011), with more state control and interference in urban development (Kinossian, 2012, Kinossian, 2017). Emerging governance practices have, to some extent, mirrored the Soviet past (Zupan and Gunko, 2019). The current urban development in Russia cannot be framed either as purely neoliberal or as state led. Thus, it hybridlike embraces both, which complicates addressing urban shrinkage.

This was not the case in China where the reforms were first introduced in 1979 in the rural areas (Wu and Ip, 1982) then evolved into bold policies on opening designated coastal cities (open cities) for foreign investment and the establishment of special economic zones (SEZs) (Yeung and Hu, 1992, Yeung et al., 2009, Wu, 1989). The first SEZs, announced in 1980, represented a policy of "zoning for growth"(Ong, 2004), which established magnets for internal migration and, in turn, urban shrinkage in many of the sending areas.

Chapters 4, 5, and 6 of this book touch on this issue in their respective topics with illustrations of how state policies have significant impacts on the advent of shrinking cities and how local governments manage or make use of urban shrinkage. These examples demonstrate the crucial role of the state and not just the market in creating the issue, as well as finding a solution or ways to adapt. It would be wrong to assume the state did not stumble, but the point is that the Chinese state, at all levels, is much more active as a participant than the neo-liberalism framework would indicate (Wu, 2010). Indeed, Xu argues that regional decentralization and competition is a key feature of Chinese reforms (Xu, 2011). The institutional arrangements that have evolved and their impacts on urban shrinkage are best examined under the topics of migration, land, and infrastructure.

The transformation process that has been described as a "collapse" in the postsocialist European countries and Russia leaving a vacuum is in contrast to implementing "an economy strategy to strengthen the monopoly power of the Chinese communist party" (Roland, 2018, p. 585). At the same time, the transformation process in all postsocialist countries included in this book led to substantial urban shrinkage. The key difference is that much of the urban shrinkage in China is the result of specific and deliberate development policies—not just a shift away from a planned economy. Whether it is shrinkage associated with reforms of state-owned enterprises or the "export" of labor from the economic backward cities and counties, these are largely the result of deliberate policies. These policies may not have been well-integrated or well-formed, however, the institutions of the state were in control. The policymakers were neither necessarily aware of the likely impacts on urban development nor were they willing to acknowledge shrinkage as an issue subsequently because the paradigm was "growth". Anything that did not fit the dominant paradigm was regarded as a failure and ignored, fuelled by the pro-growth cadre promotion system (Zhang et al., 2019). It was only in mid-2019 that the Chinese government officially acknowledged the issue of shrinking cities in a document on urbanization policy (National Development and Reform Commission State Council People's Republic of China, 2019). In comparison with Chinese or Russian shrinking cities, those located in Central and Eastern Europe (CEE) have an additional advantage: they can create or join supranational networks and participate in EU assistance programs, including access to significant financial support. Unfortunately, shrinking cities located in the majority of postsocialist countries of South-Eastern Europe due to the break-up of the former Yugoslavia must rely mainly on their own scarce national resources, taking into consideration the recent war destruction (cf. Chapter 16).

From the dynamic perspective, a few stages in the reaction of governance bodies to urban shrinkage can be distinguished: first—ignoring shrinkage; second—attempts at reversing unfavorable changes; third—gradually adapting to conditions of shrinkage; and fourth—"reinventing" the city (at this stage the role of so-called "soft" strategies such as placemaking or performative planning, is gaining importance). Such an adjustment of governance systems to urban shrinkage in a temporal dimension is known in Western Europe and well described in literature (Danielzyk et al., 2002, Hospers, 2014, Hospers and Reverda, 2012). The postsocialist cities of Central, Eastern, and part of South-Eastern Europe seem to follow the same pattern (Wiechmann, 2009, Jaroszewska and Stryjakiewicz, 2020).

Cross-border migration and urban shrinkage

There are numerous studies of cross-border interactions, particularly within the EU, and their impacts on shrinking regions and shrinking cities. Studies of cross-border exchanges between postsocialist countries are comparatively uncommon. As Chapter 8 points out there are few studies of these issues between Russia and its neighbors because the borders are much less porous due to policy. Between other postsocialist countries, the obstacles to intense cross-border interactions and their potential impacts on urban shrinkage are quite different.

Labor migration between CEE countries that are members of the EU and the rest of the EU have profound impacts on urban shrinkage in the originating countries such as Poland, Lithuania, Latvia, and others (Bulgaria and Romania). Indeed, countries such as Latvia and Lithuania are overall shrinking because so many of their small population have migrated to other parts of the EU for better employment opportunities (see Dzenovska (2018) and Chapter 20 in this volume). For these countries, the connections with the EU have become defining features of how shrinking cities and regions have emerged and the long-term implications of these connections. In addition to the lack of studies of cross-border interactions and their implications on urban shrinkage between postsocialist countries, there are few studies that compare the return migration policies, if any, of the sending countries.

Policymakers in most countries tend to have the perspective that urban shrinkage should be avoided, consequently the discussion is on how to manage urban shrinkage with policies of revival, adaptation, and regrowth (Hospers, 2014), but there are communities that consider labor outmigration and its resultant urban shrinkage to be opportunities. Chapter 6 on growth and shrinkage in Hunan and Guangdong presents the case of one county-level city in Hunan that enthusiastically managed and encouraged labor outmigration to deal with poverty issues.

The nexus between demographic change and urban shrinkage has another dimension. The predominant input of natural population decline to urban shrinkage is now the norm for many countries. Russia, Lithuania, and Poland are examples included in this book. In Russia, population decline is the result of a high mortality rate and relatively low fertility. This population dynamic is also found in Lithuania, Poland, and other European countries. Japan, South Korea, and Taiwan are notable examples from Asia. China is also expected to enter the phase of absolute population decline by the year 2030. So, broad population dynamics, variously called the "population bomb", "population burden", and other terms, should not be ignored in studies of urban shrinkage.

Social infrastructure

Comprehensive research into the infrastructure issues (including public services) associated with urban shrinkage is still lacking though much attention has been paid to the impacts of population decline and abandoned city districts on urban hardware such as water and sewerage systems due to lack of usage and the cost of maintenance or through studies of sustainability (Hoornbeek and Schwarz, 2009). While those issues are important, especially for city administrations, what is needed is a perspective on how infrastructure, including public services, facilitates or hinders the lives of the residents. Chapter 13 delves into the issue of public services restructuring, specifically health care provisions in a shrinking city/region in Russia that demonstrates the lack of institutional knowledge, comprehensive assessment of urban shrinkage, and its impacts on residents.

Land ownership and managing urban shrinkage

Irrespective of whether postsocialist transformation led to neoliberal, state capitalism, or another hybrid, some institutional issues are in sharp contrast. The local government's ability to manage land and constraints on managing urban shrinkage provides a valuable example.

A key differentiator is land ownership. Since land is state or collectively owned in China, the local government's main source of revenue, in addition to transfer payments from the province or the central government, is the "sale" of land and the designation of land for urban uses. Hence the state is an integral part of urban development (Lin, 1999) with a number of scholars pointing to the role of the state in uneven urban development and the emergence of gentrification (Yang and Zhou, 2018, Liu et al., 2012).

Russia and some postsocialist European cities face similar issues. Russian municipal governments cannot undertake demolition or redevelopment of brownfield sites that are not municipally owned, but the chaotic aftermath of the collapse of state socialism is that the ownership of many properties is uncertain. Redevelopment of derelict or abandoned properties of uncertain ownership can become an expensive and lengthy legal process, thus presenting obstacles to how Russian cities are able to manage urban shrinkage (Chapter 14). However, variations of gentrification are, for instance, shared within the postsocialist "Global East". (Müller, 2020, Badyina and Goulubchikov, 2005, Gentile, 2018). While city governments in Lithuania (Chapter 20) cannot dispose of publicly owned land without permission from national authorities, other postsocialist European countries have more complex and diverse systems.

References

Albecker, M. F. & Fol, S. 2014. The restructuring of declining suburbs in the Paris region. In Pallagast, K., Wiechmann, T. & Martinez-Fermandez, C. (eds.) *Shrinking Cities: International Perspectives and Policy Implications.* New York: Routledge.

Audirac, I. & Alejandre, J. A. 2010. *Shrinking Cities South/North.* Mexico City: Juan Pablos Editor.

Badyina, A. & Goulubchikov, O. 2005. Gentrification in Central Moscow—A Market Process or a Deliberate Policy? Money, Power and People in Housing Regeneration in Ostozhenka. *Geografiska Annaler: Series B, Human Geography,* 87(2), 113–129.

Barenberg, A. 2014. *Gulag Town, Company Town: Forced Labor and Its Legacy in Vorkuta.* New Haven, CT: Yale University Press.

Batunova, E. & Perucca, G. 2020. Population Shrinkage and Economic Growth in Russian Regions 1998–2012. *Regional Science Policy and Practice,* 12(4), 595–609.

Benfield, K. 2011. Detroit: The "Shrinking City" That Isn't Actually Shrinking. *The Atlantic.* Boston: Emerson Collective, The Atlantic Media Inc.

Bernt, M. 2016. The Limits of Shrinkage: Conceptual Pitfalls and Alternatives in the Discussion of Urban Population Loss. *International Journal of Urban and Regional Research,* 40(2), 441–450.

Bolesta, A. 2014. *China and Post-Socialist Development.* Chicago: University of Chicago Press.

Borges, M. & Torres, S. B. 2012. Company towns: Concepts, historiography, and approaches. In Borges, M. & Torres, S. B. (eds.) *Company Towns: Labor, Space, and Power Relations Across Time and Continents.* New York: Palgrave Macmillan.

Danielzyk, R., Mielke, B. & Zimmer-Hegmann, R. 2002. *ILS Beiratsbericht Demographische Entwicklung—Schrumpfende Stadt (Demographic Development—Shrinkage City).* Dortmund: Institut fur Landes- und Stadtentwicklungsforschung des Landes Nordrhein-Westfalen.

Doringer, S., Uchiyama, Y., Penker, M. & Kohsaka, R. 2019. A Meta-Analysis of Shrinking Cities in Europe and Japan: Towards an Integrative Research Agenda. *European Planning Studies,* 28(9), 1693–1712.

Dormois, R. & Fol, S. 2017. Urban Shrinkage in France: An Invisible Issue? *Metropolitics,* 13 September. https://metropolitics.org/IMG/pdf/met-dormois-fol-eng.pdf

Dzenovska, D. 2018. Emptiness and Its Futures: Staying and Leaving as Tactics of Life in Latvia. *Focaal–Journal of Global and Historical Anthropology,* 80(1), 16–29.

Eberstadt, N. 2010. *Russia's Peacetime Demographic Crisis: Dimensions, Causes, Implications.* Seattle: National Bureau of Asian Research.

Eder, J. 2019. Peripheralization and Knowledge Bases in Austria: Towards a New Regional Typology. *European Planning Studies,* 27(1), 42–67.

Ferencuhova, S. 2016. Explicit Definitions and Implicit Assumptions about Post-Socialist Cities in Academic Writings. *Geography Compass,* 10(12), 514–524.

Gabusi, G. 2017. "The Reports of My Death Have Been Greatly Exaggerated": China and the Developmental State 25 Years After Governing the Market. *The Pacific Review,* 30(2), 232–250.

Garner, J. 1992. *The Company Town: Architecture and Society in the Early Industrial Age.* Oxford: Oxford University.

Gelman, V. 2018. Bringing Actors Back In: Political Choices and Sources of Post-Soviet Regime Dynamics. *Post-Soviet Affairs,* 34, 282–296.

Gelman, V. & Ryzhenkov, S. 2011. Local Regimes, Sub-National Governance and the "Power Verticle" in Contemporary Russia. *Europe Asia Studies*, 63(5), 449–465.

Gentile, M. 2018. Gentrifications in the Planetary Elsewhere: Tele-Urbanization, Schengtrification, Colour-Splashing, and the Mirage of "More-Than-Adequate" Critical Theory. *Urban Geography*, 39(10), 1445–1464.

Grossmann, K., Brontje, M., Haase, A. & Mykhnenko, V. 2013. Shrinking Cities: Notes for the Future Research Agenda. *Cities*, 35, 221–225.

Gunko, M., Bogacheva, P., Medvedev, A. & Kashnitsky, I. 2018. Path-dependent development of mass housing in Moscow, Russia. In Hess, D. B., Tammaru, T. & van Ham, M. (eds.) *Housing Estates in Europe: Poverty, Ethnic Segregation and Policy Challenges*. Springer.

Haase, A., Bernt, M., Grobmann, K., Mykhnenko, V. & Rink, D. 2016a. Varieties of Shrinkage in European Cities. *European Urban and Regional Studies*, 23(1), 86–102.

Haase, A., Rink, D. & Grossmann, K. 2016b. Shrinking Cities in Post-Socialist Europe: What Can We Learn from Their Analysis for Theory Building Today? *Geografiska Annaler: Series B Human Geography*, 98(4), 305–319.

Haase, A., Rink, D., Grossmann, K., Bernt, M. & Mykhnenko, V. 2014. Conceptualizing Urban Shrinkage. *Environment and Planning A*, 46(7), 1519–1534.

Harloe, M. 1996. Cities in the Transition. In Andrusz, G., Harloe, M. & Szelenyi, I. (eds.) *Cities After Socialism*. Oxford: Blackwell.

He, S. & Wu, F. 2009. China's Emerging Neoliberal Urbanism: Perspectives from Urban Redevelopment. *Antipode*, 41(2), 282–304.

Heider, B. 2019. What Drives Urban Population Growth and Shrinkage in Postsocialist East Germany? *Growth and Change*, 50, 1460–1486.

Hollander, J. B. 2018. *A Research Agenda for Shrinking Cities*. New York: Edward Elgar Publishing.

Hollander, J. B., Pallagast, K., Schwarz, T. & Popper, F. J. 2009. Planning shrinking cities. *Progress in Planning*, 72(4), 223–232.

Hoornbeek, J. & Schwarz, T. 2009. *Sustainable Infrastructure in Shrinking Cities: Options for the Future*. Cleveland, OH: Center for Public Administration and Public Policy and Cleveland Urban Design Collaborative, Kent State University.

Hospers, G.-J. 2014. Policy Responses to Urban Shrinkage: From Growth Thinking to Civic Engagement. *European Planning Studies*, 22(7), 1507–1523.

Hospers, G.-J. & Reverda, N. 2012. *Krimp, het nieuwe denken: bevolkingsdalingin theorie en praktijk (Shrinkage, new thinking: population decline in theory and praxis)*. Den Haag, Boom Lemma.

Jaroszewska, E. & Stryjakiewicz, T. 2020. Drivers, Scale, and Geography of Urban Shrinkage in Poland and Policy Responses. *Journal of Urban Planning and Development*, 146(4), 1–12.

Kashnitsky, I., De Beer, J. & Van Wissen, L. 2020. Economic Convergence in Ageing Europe. *Tijdschrift voor Economische en Sociale Geografie*, 111(1), 28–44.

Kinossian, N. 2012. "Urban Entrepreneurialism" in the Post-Socialist City: Government-Led Urban Development Projects in Kazan, Russia. *International Planning Studies*, 17(4), 333–352.

Kinossian, N. 2017. State-Led Metropolisation in Russia. *Urban Research & Practice*, 10(4), 466–476.

Kovacs, Z. 1999. Cities from State-Socialism to Global Capitalism: An Introduction. *GeoJournal*, 49(1), 1–6.

Krisjane, Z. & Berzins, M. 2012. Post-Socialist Urban Trends: New Patterns and Motivations for Migration in the Suburban Areas of Riga, Latvia. *Urban Studies*, 49(2), 289–306.

Kubes, J. 2013. European Post-Socialist Cities and their Near Hinterland in Intra-Urban Geography Literature. *Bulletin of Geography: Socio-Economic Series*, 19, 19–43.

Li, S.-M. 2005. From work-unit compounds to gated communities: Housing inequality and residential segregation in transitional Beijing. In Ma, L. J. C. & Wu, F. (eds.) *Restructuring the Chinese City*. New York: Routledge.

Lin, G. L. S. 1999. State Policy and Spatial Restructuring in Post-reform China, 1978–1995. *International Journal of Urban and Regional Research*, 23(4), 670–696.

Lin, G. L. S. 2021. Drawing up the missing link: State-society relations and the remaking of urban landscapes in Chinese cities. Environment and Planning A, 53(5), 917–936.

Liu, Y., He, S. & Wu, F. 2012. Housing Differentiation Under Market Transition in Nanjing, China. *The Professional Geographer*, 64, 554–571.

Marcinczak, S., Gentile, M., Rufat, S. & Chelcea, L. 2014. Urban Geographies of Hesitant Transition: Tracing Socioeconomic Segregation in Post-Ceausescu Bucharest. *International Journal of Urban and Regional Research*, 38(4), 1399–1417.

Marcinczak, S., Tammaru, T., Novik, J., Gentile, M., Kovacs, Z., Temelova, J., Valatka, V., Kahrik, A. & Szabo, B. 2015. Patterns of Socioeconomic Segregation in the Capital Cities of Fast-Track Reforming Postsocialist Countries. *Annals of the Association of American Geographers*, 105(1), 181–202.

Müller, M. 2020. In Search of the Global East: Thinking between North and South. *Geopolitics*, 25(3), 743–755.

National Development and Reform Commission State Council People's Republic of China 2019. 2019. *New Urbanization Development Key Tasks 2019* 年新型城镇化建设重点任务. Beijing: State Council People's Republic of China.

Neill, W. J. V. & Schlappa, H. 2016. *Future Directions for the European Shrinking City*. New York and London: Routledge.

Olsen, A. 2013. Shrinking Cities: Fuzzy Concept or Useful Framework. *Berkeley Planning Journal*, 26(1), 107–132.

Ong, A. 2004. The Chinese Axis: Zoning Technologies and Variegated Sovereignty. *Journal of East Asian Studies*, 4(1), 69–96.

Oswalt, P. & Rieniets, T. (eds.) 2006. *Atlas of Shrinking Cities*. Ostfildern, Germany: Hatje Cantz Publishers.

Pallagst, K., Fleschurz, R., Nothof, S. & Uemura, T. 2021. Shrinking Cities: Implications for Planning Culture? *Urban Studies*, 58(1), 164–181.

Pallagst, K., Wiechmann, T. & Martinez-Fermandez, C. (eds.) 2014. *Shrinking Cities: International Perspectives and Policy Implications*. New York: Routledge.

Pickles, J. & Smith, A. (eds.) 1998. *Theorising Transition: The Political Economy of Post-Communist Transformations*. New York: Routledge.

Urban Design Center of Northeast Ohio. 2017. Right-sizing America's shrinking cities: Results of the policy charette and model action plan. Cleveland, Ohio.

Robinson, J. 2002. Global and World Cities: A View from off the Map. *International Journal of Urban and Regional Research*, 26(3), 531–554.

Roland, G. 2018. The Evolution of Post-Communist Systems: Eastern Europe vs. China. *Economics of Transition*, 26(4), 589–614.

Ryan, B. 2012. *Design After Decline: How America Rebuilds Shrinking Cities.* Philadelphia: University of Pennsylvania Press.

Shen, J. & Xiao, Y. 2020. Emerging Divided Cities in China: Socioeconomic Segregation in Shanghai, 2000–2010. *Urban Studies*, 57(6), 1138–1356.

Sjoberg, O. 2014. Cases Onto Themselves? Theory and Research on Ex-Socialist Urban Environments. *Geografie*, 119(4), 299–319.

Tang, W.-S. 2019. Introduction: Urban China Research Is Dead, Long Live Urban China Research. *Eurasian Geography and Economics*, 60(4), 369–375.

Tsai, K. S. & Naughton, B. 2015. Introduction: State capitalism and the Chinese economic miracle. In Naughton, B. & Tsai, K. S. (eds.) *State Capitalism, Institutional Adaptation, and the Chinese Miracle.* Cambridge: Cambridge University Press.

Turok, I. & Mykhnenko, V. 2007. The Trajectories of European Cities, 1960–2005. *Cities*, 24(3), 165–182.

van de Kaa, D. 1987. Europe's Second Demographic Transition. *Population Bulletin*, 42(1), 1–59.

Vendina, O. I. 1997. Transformation Processes in Moscow and Intra-Urban Stratification of Population. *GeoJournal*, 42(4), 349–363.

Weaver, R. & Knight, J. 2018. Can Shrinking Cities Demolish Vacancy? An Empirical Evaluation of a Demolition-First Approach to Vacancy Management in Buffalo, NY, USA. *Urban Science*, 2(3), 69.

Wiechmann, T. 2009. Conversion strategies under uncertainty in post-socialist shrinking cities—The example of Dresden in Eastern Germany. In Pallagst, K., Aber, J., Audirac, I., Cunningham-Sabot, E., Fol, S., Martinez-Fernandez, C., Moraes, S., Mulligan, H., Vargas-Hernandez, J., Wiechmann, T. & Wu, C.-T. (eds.) *The Future of Shrinking Cities: Problems, Patterns and Strategies of Urban Transformation in a Global Context.* Berkeley: IURD.

Wiechmann, T. & Bontje, M. 2015. Responding to Tough Times: Policy and Planning Strategies in Shrinking Cities. *European Planning Studies*, 23(1), 1–11.

Wolff, M. & Wiechmann, T. 2018. Urban Growth and Decline: Europe's Shrinking Cities in a Comparative Perspective 1990–2010. *European Urban and Regional Studies*, 25(2), 122–139.

Wu, C.-T. 1989. The Special Economic Zones and the Development of the Zhujiang Delta Area. *Asian Geographer*, 8(1&2), 71–88.

Wu, C.-T. & Ip, D. F. 1982. Regional Autonomy and Rural Development in China: Recent Directions in Shandong and Guangdong. *Rural Development Alternatives*. New Delhi: Concept Publishing Co.

Wu, F. 2010. How Neoliberal is China's Reform? The Origins of Change during Transition. *Eurasian Geography and Economics*, 51(5), 619–631.

Xu, C. 2011. The Fundamental Institutions of China's Reforms and Development. *Journal of Economic Literature*, 49(4), 1076–1151.

Yang, Q. & Zhou, M. 2018. Interpreting Gentrification in Chengdu in the Post-Socialist Transition of China: A Sociocultural Perspective. *Geoforum*, 93, 120–132.

Yeung, Y. M. & Hu, X. W. (eds.) 1992. *Chinese Coastal Cities: Catalysts for Modernization.* Honolulu: University of Hawaii Press.

Yeung, Y.-M., Lee, J. & Kee, G. 2009. China's Special Economic Zones at 30. *Eurasian Geography and Economics*, 50(2), 222–240.

Zhang, X., Cheung, D. M.-W., Sun, Y. & Tan, J. 2019. Political Decentralization and the Path-Dependent Characteristics of the State Authoritarianism: An Integrated Conceptual Framework to Understand China's Territorial Fragmentation. *Eurasian Geography and Economics*, 60(5), 548–581.

Zhou, Y., Lin, G. C. & Zhang, J. 2019. Urban China Through the Lens of Neoliberalism: Is a Conceptual Twist Enough? *Urban Studies*, 56(1), 33–43.

Zupan, D. & Gunko, M. 2019. The "Comfortable-City" Mode: Researching Russian Urban Planning and Design Through Policy Mobilities. *Gorodskie Issledovaniya i Paktiki*, 4(3), 7–22.

Part II
China

3   Introduction to the China section

Kai Zhou and Yangui Dai

The scientific inquiries into the shrinking cities of China have emerged quite recently and rapidly from the critical reflections on its rapid urbanization process since the 1980s. Comparing the demographic data collected from the fifth and sixth national censuses at various geographical scales, researchers (Wu et al., 2015a; Long and Wu, 2016) revealed that a large number of administrative units (about 180 cities or counties and 39,000 districts, towns, or villages) lost population between 2000 and 2010, in sharp contrast to the general image of China's prosperity and growth during the same period. To uncover this so-called *other facet of urbanization* of China (Long and Gao, 2019), a loosely organized researcher group, the Shrinking City Research Network of China (SCRNC), was established in 2015 by young scholars enthusiastic about this topic. Their investigations identified multiple factors that have caused the depopulation of cities or regions in China, including (1) a slowing economic growth at both regional and local levels (Wu, 2019; Wu and Wang, 2020); (2) fluctuations in the global market caused by either financial crises or changes of supply chains (Wu and Wang, 2020; Li, Du and Li, 2015); (3) a sharp drop in the fertility rate as a result of the one-child policy introduced in the early 1980s (Zhou, Qian and Yan, 2017); and (4) uneven regional impacts of changing national, regional, or local development policies (Zhou, Yan and Qian, 2019). These studies predicted that maintaining the double-digit economic growth and rising population may be too optimistic for most areas in China. Regardless of a prevailing pro-growth mindset, the alternative development scenario featured by a stagnating or shrinking population in the near future seems inevitable for some big cities, many medium-sized and small cities, and almost all rural areas.

The China section of this book starts with this introduction, which serves (1) to provide an overview of the urban shrinkage in China; (2) to briefly review the literature of shrinking city studies of China; (3) to explain the complex terminology used in Chinese studies, for example, the jargon in demographic statistics; and (4) to describe the structure of China's unique administrative system, which is essential for understanding the status quo of its demography. This is to avoid repetition of similar material or explanation

DOI: 10.4324/9780367815011-5

in each of the following chapters and, more importantly, to identify common themes across the various studies.

Overall demography trends

Demographic turning points of China have been predicted by both national and international research institutes. In 2007, the Chinese government published the National Population Development Strategy Research Report, in which the total population was predicted to reach its peak in 2033, and the working population was predicted to decrease from 2016 (Research Group on National Population Development Strategy, 2007). In 2019, the forecast of the Department of Economic and Social Affairs Population Division of the United Nations indicates that China's total population will reach its peak in 2031 (approximately 1.5 billion, excluding Hongkong, Macau, and Taiwan). The total population was projected to shrink by 31.4 million (−2.2%) between 2019 and 2050. Meanwhile, the working population (aged 15–64) is already declining. This number was also projected to decrease by 173.8 million (−17.2%) between 2019 and 2050 (United Nations, Department of Economic and Social Affairs, Population Division, 2019). International out-migration is another crucial factor that affects demographic change. At the global level, Chinese-born international migrants were the third-largest foreign-born population in the world in 2019 after Indians and Mexicans, with nearly 11 million Chinese migrants living outside of China, while the number of foreign immigrants in China (about 1 million in 2019) only accounts for 0.1% of the total population (International Organization for Migration, and United Nations, 2000). It is against such demographic trends that Chinese academics started to pay attention to the cities and regions that have experienced or show signs of depopulation.

Measuring shrinkage with census data

In light of the shrinking tendency, an immediate question was how to measure shrinkages using the data available in China. The national censuses are still the main sources of data used by most Chinese researchers on shrinking cities, although the latest one dated back to 2010. To better understand the use of demographic data in the following chapters, several key features of China's census data are worth noting.

First, considering the high mobility rate and frequency in China nowadays, to accurately measure the total population of cities in a given time period is a challenging task. The Hukou registration system, a legacy of the planned economy implemented to control unauthorized domestic migrations, continues to provide the number of *"registered population"*, also called the "de jure population" (Chan, 2007), "hukou population", or "household registration population". It refers to the population numbers recorded in the permanent household registration (i.e., Hukou), covering those who were locally born and have not changed their Hukou even though they are residing elsewhere. By definition, the registered population leaves out a large number of newly attracted immigrants in the developing regions where job opportunities are abundant, and it overcounts those who have just left their "hometowns", residing, studying, or working elsewhere.

To complement the Hukou registration, the census also provides the headcounts of the *"resided population"* from surveys, also called the "de-facto population", "permanent residents", "permanent population", or "resident population". It refers to the number of the population actually residing for longer than a half year, according to the records of the police departments. Contrary to the registered population, which counts the number of citizens who have officially registered in an administrative unit, the resided population counts the number of residents who are actually living and working there.

Both registered population and resided population are useful for identifying shrinking cities or regions. Because migrants commonly keep their household registrations at home until they become property owners in their place of residence, the change of registration usually lagged behind the change of residency. Therefore, the difference between the two could indicate that: (1) a city is usually gaining population if the number of resided population is larger than that of the registered population; or (2) it is a strong signal of losing population vice versa.

Second, there are limited reliable resources of demographic data when comparing population numbers between administrative units or across time. The censuses were conducted decennially in the years 2000, 2010, and 2020, covering all households at a national scale. So far, these provided the most cited headcounts of administrative units. Between the censuses, the long gap was filled by data of the 1% population sampling surveys collected in the years 1995, 2005, and 2015, which were organized by the central government, using a proportionate stratified method to draw out samples at the national level; therefore, the estimated results lack accuracy at the local scale.

Furthermore, the statistic bureaus in all levels of local government publish the statistical yearbooks every year. The numbers of the *registered population* from those sources are generally correct and consistent because these are exported directly from the national household registration information system. However, the quality of the *resided population* statistics varies hugely between administrations, where different methods are used to collect data. The data of the resided population in the statistical yearbooks published annually by the local governments are also estimated results, using the numbers of the resided population in the decennial censuses that cover all households as benchmarks for estimating numbers of all other years. The methods of estimation used by provincial statistical bureaus are also different. While some collect data through a 1‰ sampling survey, others simply combine relevant data from different sources and make a forecast based on the figures of the benchmark year. So far, the numbers of the resided population from the decennial censuses are the most accountable sources for identifying shrinking cities in China.

Complexities of administrative divisions

China's unique administrative system adds another layer of complexity in understanding the Chinese shrinking cities (Chan and Hu, 2003). On the one hand, within the hierarchy of the Chinese government, each local administration is under the supervision of an upper-level one while also overseeing its own lower-level sub-units. On the other hand, the city governments, which evolved from the feudal era, are not only in charge of the management of the core cities (i.e., the urbanized areas) but also responsible for the development of a wider region (i.e., the region within its administrative boundary).

The administrative division of China is currently based on a six-tier system, dividing the nation into provinces, prefectures, counties, towns or townships, and villages or communities (Figure 3.1). At the provincial level, the nation consists of 23 provinces, 5 autonomous regions, 4 municipalities directly under the Central Government (i.e., Beijing, Shanghai, Tianjing, and Chongqing), and 2 special administrative regions (i.e., Hongkong and Macao). The four municipalities directly under the Central Government are cities of national importance economically and politically and large metropolitan areas with urban populations of more than 10 million people. Municipalities directly under the Central Government, which have the same authorities, autonomy, and responsibilities as provincial governments, are also divided into districts and counties. At the prefectural level, a province or an autonomous region is subdivided into prefecture-level cities or autonomous prefectures and prefectures in ethnic regions. Prefecture-level cities oversee lower-level counties and county-level cities, the so-called *shi-dai-xian* system (or "city leading county" system, 市带县). Prefecture-level administrations, counties, and autonomous counties usually govern a core city with an urban population between 100,000 and 10 million. The urbanized area of large prefecture-level cities can be further divided into districts. At the county level, a county or an autonomous county is subdivided into towns and townships (with a total population usually less than 100,000), each of which governs a number of villages. Districts consist of subdistricts (also known as *Jiedao*, "街道"), which are further divided into communities. Autonomous regions, autonomous prefectures, and autonomous counties are all ethnic autonomous areas.

A province or autonomous region in China is a regional administration that does not directly govern any urbanized area. Towns and townships are mostly rural settlements, whose urbanized cores (with total residents less than 10,000) are usually called downtowns ("镇区") rather than cities. In a broad sense, all other administrations (gray-toned in Figure 3.1) are

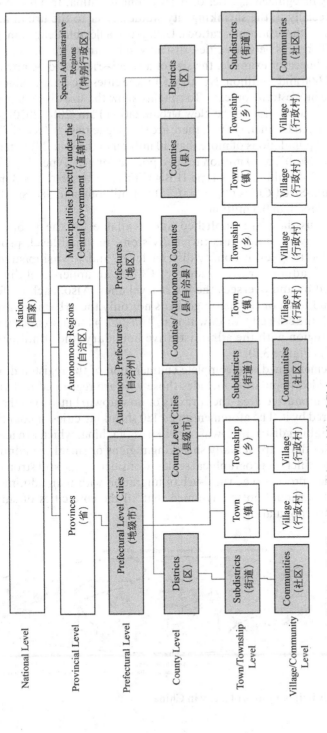

Figure 3.1 Administrative divisions of people's republic of China.

commonly recognized as cities or parts of one in China. In a narrow sense, cities, especially in the shrinking city studies, refer to the urban areas (or core cities) of those administrations. Using such a flexible definition, the size or area differences between cities, districts, subdistricts, or communities in China are huge; for example, the size of a subdistrict could be any number between 1,000 and 1 million. Notably, Chinese cities are usually much larger than those in Western contexts. To roughly show the size distribution of cities in China, according to the New Urbanization Plan (2014–2020) released by the National Council in 2014, there are six Megacities ("超大城市", which had resided populations of more than 10 million in the year 2010), ten Large Cities ("特大城市", 5–10 million resided population), 21 Type I Big Cities ("I 型大城市", 3–5 million), 103 Type II Big Cities ("II型大城市", 1–3 million), 138 Medium-Sized Cities ("中等城市", 0.5–1 million), and 386 Small Cities ("小城市", less than 0.5 million).

The census data are published and available for the public at the upper five layers of administrative divisions: national level, provincial level, prefectural level, county/district level, and township/subdistrict level. In the administrative system of China, an upper-level division is constituted by and oversees the lower-level ones. Also, each prefecture, county, and town or township governs not only the urbanized area (the municipal, the core city, or downtown, "市/镇区") but also the surrounding rural region with the administrative boundary (the administration, "市/镇域") (Figure 3.2).

As aforementioned, the complete enumeration of the registered population or resided population includes those who are registered or living not only in the urbanized area (the core city or downtown) but also in the region administered by it. The aforementioned 180 shrinking cities or counties and 39,000 shrinking districts, towns, or villages of China, which are identified using census data, are actually shrinking regions or shrinking administrative units at different geographical scales. Considering the still strong in situ urbanization process (i.e., rural-urban migration within an administrative boundary), it should not be assumed that all the core cities of shrinking regions are depopulating.

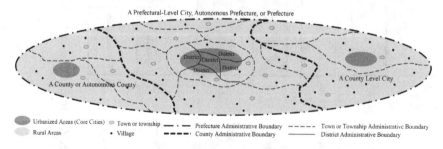

Figure 3.2 Administrations of a city in China.

Regional variation and typology

The mist of the quantitative evidence in the census data of shrinking cities was partially clarified by several empirical studies published in recent years, which illustrated the geography of shrinking cities in China.

Case studies show that the northeast region has the most typical shrinking cities in this country. The old industrial base was first established during WWII and rebuilt with Soviet aid immediately after the foundation of the People's Republic of China (P.R. China) has suffered from severe declines economically, physically, and demographically in the economic reform after the 1980s (Gao and Long, 2017; Liu et al., 2018). This is due to the combined influences of the social transformation of China from the centralized planned economy to the market economy introduced by the opening up policy of the 1980s.

Developing southeastern China is also not entirely immune to shrinkage. Studies have shown that those booming metropolitans in the Yangtze River Delta, Pearl River Delta, and Beijing-Tianjin-Hebei regions are also surrounded by a number of shrinking medium/smaller-sized cities and rural areas (Li, Du and Li, 2015; Wu, Long and Yang, 2015b). Du and Li (2017), Du et al. (2019), and Li, Li and Deng (2019) revealed the impacts of the 2008 financial crisis on the manufacturing sectors in southern China, showing that the "world factory" was vulnerable to the fluctuations in the global market. Those export-oriented cities could easily lose millions of migrant workers during recessions.

In central and western China, the shrinkages of the population are associated with the outflowing of surplus labor forces, either from rural areas to urban areas or from inland provinces to coastal provinces. Cities of the central and western regions were striking the balance of gaining population from its hinterland and losing population to the coastal regions (Zhou, Qian and Yan, 2017). As a result, (1) depopulations are often observed in some county-level cities, many towns and townships, and most rural areas; (2) sites of "perforated shrinking" were found in some districts of the growing large cities of central and western China, such as Wuhan, Chengdu, and Chongqing; and (3) the "siphon effect" has caused the loss of population of towns and cities in the outer ring to the core in metropolitan areas. Uneven distribution of job opportunities, aging population due to the labor outflow, and sometimes difficult natural conditions or bad policies shape the "regional shrinkage", "peripheral shrinkage", and "perforated shrinkage" in cities in the central and western regions (Zhang, Zhang and Xiao, 2018; Liu and Zhang, 2017).

Notably, it is misleading to suggest that the typology of shrinkage in China is strictly geographically bounded. As legacies of the socialist planned economy of China, there are a number of resource-based cities and single-industrial cities dispersed throughout the country (Li and Mykhnenko, 2018), which are now known as shrinking "resource-depleted cities" (or resource-exhausted cities) or "declining old industrial bases", awaiting policy interventions. Chapter 4 of this book, written by He Li, investigates the characteristics of this type of population shrinkage through a robust analysis of the multiple drivers of this phenomenon under China's special political and economic background. Chapter 5, written by Minwei Zhang and Helin Liu, further elaborates this topic by measuring the economic development levels of 41 resource-exhausted cities using a composite economic performance index and composite economic growth index. Resource-depleted cities are the most discussed cases in China, which are the shrinking cities that the Chinese government has developed programs and policies for, even though the policymakers were mainly concerned about the economic revival rather than the management of shrinkage. Similar mining cities and mono-industry cities of post-socialist countries elsewhere are also discussed in other sections of this book.

The hollowed-out settlements that include depopulating villages, townships, towns, small-median-sized cities, and districts within major metropolitans were not only a phenomenon of the northeastern, western, or central China but also a common outcome of promoting further concentration of population in rapid urbanization. Shrinkages caused by the changing market are also not limited to those manufactural hubs in coastal China but also experienced by some inland cities that rely on trade and exchange. Chapter 6, by Kai Zhou, Zhiwei Du, and Yangui Dai, discusses the shrinking cities and regions caused by the migration flows driven by economic motivations, showing the parallel logic of growing and shrinking in China. Chapter 7, by Zhe Gao, further elaborated this argument with a case study at the city-region level, discussing the "perforated shrinking" and "siphon effect" of Wuhan's metropolitan area.

Finally, the economic stimulating programs at the national level were mainly initiated by the top-down administrative structure in China (He et al., 2017; Yang and Dunford, 2017). Considering the uneven provision of such preferential policies (Chien and Wu, 2011), the unsuccessful central-led interventions may have created a new type of shrinkage in China: shrinking by planning (Zhou, Yan and Qian, 2019; Li and Mykhnenko, 2018; Jin and Sui, 2020). Ekaterina Mikhailova and Chung-Tong Wu contributed to Chapter 8 of this book, providing their original insights on the core periphery-dependent relationships of the trade hubs in shrinking border regions between China and Russia.

One interesting aspect of urban shrinkage in China is that it combines several phenomena one can identify elsewhere. First, the story of shrinking industrial bases in northeast China, which is similar in some respects to the Rust Belt of the United States, can also be found in other countries that have endured shifts of investment in globalization, although in China, the initial cause is the economic transition from planned economy to market economy. The resource-depleted cities nationwide are shrinking due to similar factors. Second, uneven economic development is causing urban shrinkage in the economically less developed regions of China due to the extraction of the labor force to fast developing regions as a result of rapid industrialization. Similar cases can be found in the studies on the shrinking cities of eastern Germany or Central Europe, following a core-periphery model. Third, the shrinkages caused by the industrial restructuring of China—for example, Guangdong and other rapidly industrialized regions having to restructure to move up the industrial/economic ladder—are somewhat similar to the de-industrialization process that occurred in the developed countries in the 1970s to 1990s. However, the Chinese version has different national impacts because, while some of these industries moved overseas, some have moved to the central and western regions. Shrinking cities of China appear to have similar processes but have slightly different causes in different contexts.

Fundamentally, urban shrinkages in China that were indicated by the 2000 and 2010 censuses, as discussed above, were mainly caused by the population migrating from rural areas to cities, from inland provinces to coastal regions, and from old industrial bases to new manufacturing hubs. Economic incentives as a result of new job opportunities and huge income differences between regions have played a major role in creating the prosperity of some cities and the decline of others. Although data from the 2020 census have not been released yet, it is observed that the pattern of China's domestic migration flows have become more complex between 2010 and 2020. Overcrowded megacities like Beijing, Shanghai, and Shenzhen are losing their attractiveness due to the high housing prices and competitive public services (such as education and medical care). Spill-over investments to the central and western cities have created local jobs and raised salaries in central China to a similar level as that of the coastal provinces. There are inspiring stories in the media about migrant workers who went back to their hometowns and successfully started a new business on tourism or e-commerce. Obviously, the backflowing of outmigrants became a new trend. Behind the influx and reflux of population between cities and regions nowadays, economic incentives are not the only determinative drivers of change. Other factors such as amenities, affordability, environment quality, social capital, and career prospects have also been added to the equilibrium of individual's decision for settlement or resettlement. These new features in population mobility are currently reshaping the pattern of urban growth and shrinkage in China.

Academic research and policy responses

The research projects on shrinking cities that are funded by the National Natural Science Foundation of China and the National Social Science Fund of China were first approved in 2015, and the number has been growing steadily since. Several journal papers were published between 2010 and 2020, mostly written in Chinese. Three major themes have emerged from those contributions: (1) to explore the appropriate quantitative indicators and methods using the censuses or other sources of data available (socio-economic statistics, satellite images, nighttime light, or street view photos) for the identification and measurement of shrinkage at the city or regional scale; (2) to explain the occurrence, process, typology, and spatial pattern of urban shrinkage from case studies; and (3) to recognize the unique features of the shrinking cities in China to enrich the international shrinking cities research. Researchers have been accumulating knowledge from investigating shrinking cases, collecting evidence of the consequences, and suggesting policy actions for good governance.

Academic discussions on shrinking cities have raised the awareness at the central government about the potential and existing risk of urban and regional shrinkages. Two policy guidelines recently released by the National Development and Reform Commission (i.e., the Plan for New Urbanization 2019 and the Plan for New Urbanization and Urban-Rural Integration 2020) used the term "shrinking small and median-sized cities" for the first time in officially announced policy documents, calling for overcoming the pro-growth obsession and "the slimness and fitness" (瘦身强体) of shrinking cities.

Since shrinking cities have become academic hotspots and a policy buzzword in China, current studies are gradually turning to seek potential policy implementations. On recognizing the importance of managing shrinkage, Chinese researchers are testing various theoretical models for the successful governance of Chinese shrinking cities. By reviewing the literature on a wide range of case studies and producing international comparisons among the modes of governance of shrinking cities in the United States, Germany, France, Japan, and several countries in Central and Eastern Europe, efforts are made to develop a model of good policy action for shrinking cities in China, including (1) finding a pathway for the successful economic restructuring in resource-dependent regions has been a long discussed topic by Chinese economists and geographers, who have found new arguments in recent debates on the economic revival of shrinking cities, especially in the northeast; (2) vacant houses and lands are becoming increasingly visible in shrinking cities and regions of China, and researchers have been devoted to recognizing, recovering, and reusing those abundant properties in both urban and rural areas using quantitative and qualitative methods; and (3) recognizing that migrants are now flowing in both directions (i.e., inland to coastal and vice versa) (Zhou, Yan and Zhao, 2019), and depopulating regions and cities in central and western China might have an opportunity to reclaim population due to withdrawal of global investments from southeastern China and in the formulation of new supply chains after the Covid-19 pandemic. Original research contributions included in this section of the book cover some of these aspects (Figure 3.3).

Decision makers in the public sectors, including central governments, majors, heads of the planning departments, and senior planners in design institutions, are gradually accepting the fact that maintaining all-round high growth rates is impossible and are preparing mentally and technically for a scenario without assuming growth will take place. However, "planning for growth" is still the mainstream mentality of the planning professionals in China.

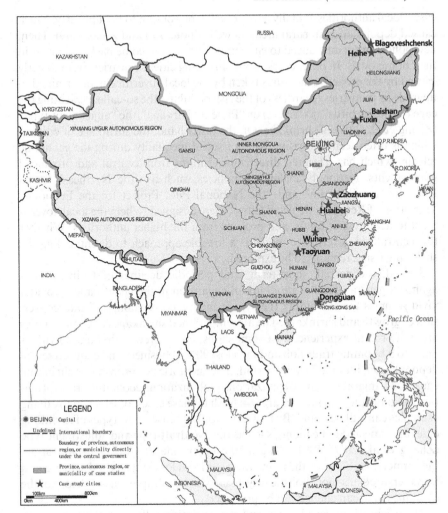

Figure 3.3 Location of case studies in China.

Potential contributions

How can academic studies and planning practices on shrinking cities of China contribute to the international shrinking city research? Existing publications indicate several potential fields.

First, China is an appropriate place to investigate the impacts of national policies and centralized planning power on the development of shrinking cities. It is claimed that public policies here were both the causes of and remedies for the shrinkage of many cities in the past decades. On the one hand, some shrinkages were caused by the local government that took bold actions on "overdrafting" its resources in terms of land, finance, or human capital, which produced shrinking landscapes, such as ghost towns or empty commercial real estate properties. Zhang, Feng and Cheng (2017) referred to this as the "overdraft-type". In some cases, local authorities actually encouraged labor output to fight poverty, which caused depopulation in rural areas as well (Zhou, Yan and Zhao, 2019). Then, new policies were introduced to encourage migrants to come back and invest in their hometowns. In other cases, some shrinking cities or districts were directly caused by the proactive measures taken by the local government, determined to upgrade its industries regardless of the risk of failure, the so-called "vacating the cage to change birds" (腾笼换鸟) or "Phoenix Nirvana", the "adjustment-type" (Zhang et al., 2017) of shrinkage. On the other hand, the shrinkage of cities in China is often seen as a rising issue of social inequality during the economic reform, which has triggered direct interventions from central and provincial governments. National rebalancing policies, such as "the Northeast China Revitalization Program" and "the National Pilot Project for the Economic Transition of Resource-depleted Cities" were designed to help the local government to acquire aid, funds, and projects from the higher authorities. Whether this centralized governance model is a feasible approach to disentangling the shrinking issues remains to be explored.

Second, the partial, short-term, and periodic declines of Chinese cities against a general trend of rapid urban expansion in the last decades provides vivid samples for examining the synchronicity, correlation, and interconversion of growth and shrinkage. The aforementioned shrinkages of Chinese cities are quite recent experiences. In some cases, a city grew from an anonymous place to a booming trade hub and then lost 90% of business and endured severe depopulation, all in a 20-year period. The memories of growing or shrinking caused by migration, industrialization, globalization, economic transformation, or good and bad policies are still fresh in these cases, of which many living examples can still be found. By looking into those cases, new types of shrinking cities are expected to be found, such as the overdraft-type and adjustment-type (Zhang et al., 2017). By looking at the other side of China's urbanization, the connections between the population gain and loss in the past decades could be understood in a more holistic way, from which a comprehensive theory of the dynamics in urban change can be produced.

Third, shrinking rural areas and depopulating villages are other significant features in the Chinese context. In a time dominated by the prosperous urban economy, rural areas and their settlements are becoming dilapidated in large parts of China. While younger generations are determined to find their future in cities, rural China has lost its vitality, becoming home to the elderly and children. The collective ownership of rural land has limited the scope of traditional agriculture to scale up. It remains unclear how to respond to the shrinking of both population and economy in rural regions of China, and unfortunately, little research has been dedicated to this question.

Acknowledgement

This research was funded by the General Program of the National Natural Science Foundation of China (project no. 52078197).

References

Chan, K. W. 2007. Misconceptions and Complexities in the Study of China's Cities: Definitions, Statistics, and Implications. *Eurasian Geography and Economics*, 48(4), 383–412.

Chan, K. W. & Hu, Y. 2003. Urbanization in China in the 1990s: New Definition, Different Series, and Revised Trends. *The China Review*, 3(2), 49–71.

Chien, S. S. & Wu, F. 2011. Transformation of China's Urban Entrepreneurialism: The Case Study of the City of Kunshan. *Cross-Currents: East Asian History and Culture Review*, 1(1), 1–28. https://cross-currents.berkeley.edu/sites/default/files/e-journal/articles/chien_and_wu_0.pdf.

Du, Z. W. & Li, X. 2017. Growth or Shrinkage: New Phenomena of Regional Development in the Rapidly-urbanizing Pearl River Delta. *Acta Geographica Sinica*, 72(10), 1800–1811. //杜志威, 李郇, 2017. 珠三角快速城镇化地区发展的增长与收缩新现象. *地理学报*, 72(10), 1800–1811.

Du, Z. W. et al. 2019. Spatiotemporal Evolution and Influences of Urban Population Shrinkage in Guangdong Since 2000. *Tropical Geography*, 39(1), 20–28. //杜志威, 等, 2019. 2000年以来广东省城市人口收缩的时空演变与影响因素. *热带地理*, 39(1), 20–28.

Gao, S. Q. & Long, Y. 2017. Distinguishing and Planning Shrinking Cities in Northeast China. *Planners*, 33(1), 26–32. //高舒琦, 龙瀛, 2017. 东北地区收缩城市的识别分析及规划应对. *规划师*, 33(1), 26–32.

He, S. Y. et al. 2017. Shrinking Cities and Resource-based Economy: The Economic Restructuring in China's Mining Cities. *Cities*, 60, 75–83.

International Organization for Migration, and United Nations. 2000. *World Migration Report*. Geneva: International Organization for Migration.

Jin, S. T. & Sui, D. Z. 2020. Do Central State Interventions Cause Urban Shrinkage in China? *Journal of Urban Affairs*, (1), 1–20.

Li, X., Du, Z. W. & Li, X. F. 2015. The Spatial Distribution and Mechanism of City Shrinkage in the Pearl River Delta. *Modern Urban Research*, (9), 36–43. //李郇, 杜志威, 李先峰, 2015. 珠江三角洲城镇收缩的空间分布与机制. *现代城市研究*, (9), 36–43.

Li, X., Li, X. F. & Deng, J. Y. 2019. The Growth and Shrinkage of China's Rapidly Urbanizing Areas from the Perspective of Property Rights: A Case Study of Dongguan. *Tropical Geography*, 39(1), 1–10. //李郇, 李先锋, 邓嘉怡, 2019. 产权视角下中国快速城镇化地区的增长与收缩——以珠江三角洲东莞市为例. *热带地理*, 39(1), 1–10.

Li, H. & Mykhnenko, V. 2018. Urban Shrinkage with Chinese Characteristics. *The Geographical Journal*, 184(4), 398–412.

Liu, F. B. et al. 2018. The Research on the Quantitative Identification and Cause Analysis of Urban Shrinkage from Different Dimensions and Scales: A Case Study of Northeast China during Transformation Period. *Modern Urban Research*, (7), 37–46. //刘风豹, 等, 2018. 城市收缩多维度、多尺度量化识别及成因研究: 以转型期中国东北地区为例. *现代城市研究*, (7), 37–46.

Liu, Y. B. & Zhang, X. L. 2017. A Study on the Shrinkage of Wuhan Metropolitan Area. *Planners*, 33(1), 18–25. //刘玉博, 张学良, 2017. 武汉城市圈城市收缩现象研究. *规划师*, 33(1), 18–25.

Long, Y. & Gao, S. Q. 2019. *Shrinking Cities in China: The Other Facet of Urbanization*. Singapore: Springer.

Long, Y. & Wu, K. 2016. Shrinking Cities in a Rapidly Urbanizing China. *Environment and Planning A, 2016*, 48(2), 220–222.

Research Group on National Population Development Strategy, 2007. Study on the National Population Development Strategy. *Population and Family Planning*, (3), 4–9. //国家人口发展战略研究课题, 2007. 国家人口发展战略研究报告. 人口与计划生育, (3), 4–9.

United Nations, Department of Economic and Social Affairs, Population Division. 2019. *World Population Prospects 2019: Highlights* (ST/ESA/SER.A/423). https://population.un.org/wpp/Publications/Files/WPP2019_Highlights.pdf

Wu, K. 2019. Urban Shrinkage in the Beijing-Tianjin-Hebei Region and Yangtze River Delta: Pattern, Trajectory and Factors. In Long, Y. and Gao, S.Q. (eds.), *Shrinking Cities in China* (pp. 43–61). Singapore: Springer.

Wu, K. et al. 2015a. Featured Graphic. Mushrooming *Jiedaos*, Growing Cities: An Alternative Perspective on Urbanizing China. *Environment and Planning A*, 47(1), 1–2.

Wu, K., Long, Y. & Yang, Y. 2015b. Urban Shrinkage in the Beijing-Tianjin-Hebei Region and Yangtze River Delta: Pattern, Trajectory and Factors. *Modern Urban Research*, (9), 26–35. //吴康, 龙瀛, 杨宇, 2015. 京津冀与长江三角洲的局部收缩: 格局、类型与影响因素识别. 现代城市研究, (9), 26–35.

Wu, K. & Wang, X. 2020. Understanding the Growth and Shrinkage Phenomena of Industrial and Trade Cities in Southeastern China: A Case Study of Yiwu. *Journal of Urban Planning and Development*, 146(4), 05020028.

Yang, Z. S. & Dunford, M. 2017. City Shrinkage in China: Scalar Processes of Urban and Hukou Population Losses. *Regional Studies*, 52(8), 1111–1121.

Zhang, J. X., Feng, C. F. & Cheng, H. 2017. International Research and China's Exploration of Urban Shrinking. *Urban Planning International*, 32(5), 1–9. //张京祥, 冯灿芳, 陈浩, 2017. 城市收缩的国际研究与中国本土化探索. 国际城市规划, 32(5), 1–9.

Zhang, X. L., Zhang, M. D. & Xiao, H. 2018. Study on Spatial Pattern and Formation Mechanism of Urban Contraction in Chengdu−Chongqing City Cluster. *Journal of Chongqing University (Social Science Edition)*, 24(6), 1–14. //张学良, 张明斗, 肖航, 2018. 成渝城市群城市收缩的空间格局与形成机制研究. 重庆大学学报(社会科学版), 24(6), 1–14.

Zhou, K., Qian, F. F. & Yan, Y. 2017. A Multi-scaled Analysis of the "Shrinking Map" of the Population in Hunan Province. *Geographical Research*, 36(2), 267–280. //周恺, 钱芳芳, 严妍, 2017. 湖南省多地理尺度下的人口"收缩地图". 地理研究, 36(2), 267–280.

Zhou, K., Yan, Y. & Qian, F. F. 2019. A Multi-scaled Analysis of the Shrinking Population in a Region with Out-Migration: A Case Study of Hunan Province. In Long, Y. and Gao, S. Q. (eds.) *Shrinking Cities in China* (pp. 25–41). Singapore: Springer.

Zhou, K., Yan, Y. & Zhao, Q. H. 2019. Planning Policy Responses in Population Contraction Scenarios: A Case Study of Hunan Province. *Beijing Planning and Construction*, (3), 12–19. //周恺, 严妍, 赵群荟, 2019. 人口收缩情景下的规划政策应对:基于湖南案例的探讨. 北京规划建设, (3), 12–19.

4 Population shrinkage in resource-dependent cities of China during transition period

He Li

Introduction

Population shrinkage, the key indicator of urban shrinkage, can be considered one of the most critical challenges of contemporary global urbanization. During the past two decades, the widespread phenomenon has been increasingly studied by academics. Numerous cases of population shrinkage have been found in cities of both North and South economies (Oswalt, 2005; Richardson and Nam, 2014). Among these cases, population shrinkage in resource-dependent cities (RCs) that developed in relation to extracting and processing natural resources is particularly acute. Different from other types of shrinking cities, the population shrinkage of RCs is, to some extent, predictable but usually unavoidable because most economic transition strategies are developed after decline has begun, and with few successful cases reported (Li et al., 2015; Martinez-Fernandez et al., 2012; Hayter and Nieweler, 2018). Prior studies have investigated the population shrinkage of RCs affected by the decline of resource-based industries, particularly in capitalist economies like North America, Europe, Australia, Japan, and some countries in Africa (Martinez-Fernandez et al., 2012; Knierzinger and Sopelle, 2019). Results indicate that long-term and significant population shrinkage is common among RCs affected by resource depletion, technological upgradation, and changing resource market conditions. In the worst cases, RCs that fail to transform their economies are abandoned and become desolate ghost towns.

RCs are an important part of China's urban system. According to the Plan of Sustainable Development for RCs in China (2013–2020) issued by the State Council (State Council of China, 2013), China has 262 RCs,[1] including 126 prefecture-level administrative units and 136 county-level administrative units—and 1/4 of them has entered a recession stage. RCs in China share many aspects with RCs in capitalist economies. They are mostly located in areas that are economically and physically peripheral (Sun and Mao, 2018), and they are vulnerable to the decline of resource-based industries (Li et al., 2009). Furthermore, they are weak in local innovation systems (Xie et al., 2017) and face environmental degradation (Zhang et al., 2011). However, there are some distinctive characteristics of China's RCs specific to China's context. First,

DOI: 10.4324/9780367815011-6

compared with those in capitalist economies, China's RCs usually have a larger population. The total population of China's 262 RCs is about 440 million, accounting for 33% of China's population (Yu et al., 2019). Second, the scattered settlement pattern is common in China's RCs, especially in prefecture-level RCs, as most of them are founded at sites of resource extraction and processing (Song and Wang, 2011). Third, most of China's RCs emerged under intensive investment from the central government in the 1950s, following an economic strategy that strongly favored heavy industrialization. These cities underwent not only the boom-and-bust economic cycles of the dominant resource-based industries but also the systemic reformation from a planned economy to a socialist market economy (Li et al., 2015).

While massive and fast urban population growth has been the leading trend of China's urbanization since the 1990s, population shrinkage is emerging in China's RCs and has been considered one of the major types of China's shrinking cities (Li and Mykhnenko, 2018; Long and Gao, 2019). A few studies have identified the population shrinkage of some of China's prefecture-level RCs since 2000, especially resource-depleted cities (He, 2014; He et al., 2017) and RCs in less developed northeast and northwest regions (Chen and Mei, 2018; Gao and Long, 2017; Woodworth, 2016). These studies have pointed out the typical drivers behind the phenomenon of population shrinkages, such as the slowdown of national economic growth, a single industrial structure, and the boom-and-bust industrial cycle (He, 2014; He et al., 2017). To date, however, the characteristics and drivers of population shrinkage in China's RCs have not been adequately examined (Li et al., 2020). First, existing studies mainly focus on prefecture-level RCs in the declining stage or in a particular region, and the overall picture of population shrinkage in China's 262 RCs and its variations across regions and stages of RC's life cycle is still unknown. Second, related literature only pays attention to the population change in China's RCs from the years 2000 onwards, without considering the potential population shrinkage of China's RCs in the 1990s. However, many RCs began to enter a transition period in the 1990s when the former top leader Jiang Zemin addressed the importance for sustainable development of RCs when he visited Daqing (known as the oil capital of China) in 1990 (Wang et al., 2014a). Third, the use of a prefecture-level city as a unit of analysis in prior studies is also problematic. Given the dispersed urban morphology of most prefecture-level RCs in China, this unit of analysis is too coarse and it cannot detect intra-city population changes. Finally, although it is argued that China's particular pathways of socialist industrialization and subsequent reform established the conditions for population change among RCs in the 2000s (Woodworth, 2016), the impact of systemic reformation on population shrinkage in China's RCs has not been widely discussed. Against this background, this chapter tries to present an overall and detailed analysis on the population shrinkage of China's RCs and discuss the distinctiveness of the phenomenon within China's specific context.

Data and methods

Measuring the precise population size of Chinese cities can be complicated, mainly for two reasons. First, cities in China are not exclusively urban entities but rather administrative units typically comprising an urban core (roughly comparable to the continuously built-up area) and the surrounding peri-urban and rural areas. Second, the criteria used to calculate the population size of Chinese cities are confusing. Estimating the population size of a city using household registration data would include many people who are registered in the locale but no longer live there and exclude those who live in the locale without a local household registration (Chan, 2003). A more accurate data source is the census, which tallies all long-term residents of a locality every ten years. Currently, there are two approaches using census data to count the long-stay population. The first approach uses the overall administrative boundary of a municipal authority (i.e., the entire "city" area), which may potentially overestimate the permanent urban population by counting the inhabitants living at the outskirts in peri-urban and rural areas of the municipality. An alternative, and ostensibly more accurate, approach involves counting the long-stay population within a primary urban area, as defined by the National Bureau of Statistics. This is the ideal way to estimate the actual population of Chinese cities. However, the main challenge to this approach is that the criteria for defining primary urban areas have not been consistent over time (Kamal-Chaoui et al., 2009). Furthermore, detailed population breakdown figures below the town/township level, which would be necessary for defining its functional boundary, are not readily or officially available.

Based on the available published population census data, the following approach was taken to estimate the population size of RCs:

1 For large RCs with urban districts (an administrative unit below municipalities), this study treated the population change in each urban district individually to discover the spatial unevenness of urban population change. Long-stay population figures, covering the administrative boundaries of 297 urban districts, were used accordingly;
2 For small RCs without urban districts (including 71 county-level cities (CLCs) and 132 counties), this study could not access all the necessary population and lower-tier boundary adjustments data to measure the long-stay population of each urban area individually. Hence, this study used the total long-stay population residing in the urban areas of specific cities and towns.

The observed period was split into two-decade-long intervals, 1990–2000 and 2000–2010, with the population data derived from the National Population Census of China (by county) in 1990, 2000, and 2010 (National Bureau of Statistics, 1992, 2002a, 2002b, 2012). In terms of statistical adjustments, this study had to deal with several significant administrative type changes occurring between 1990 and 2010, especially between counties, CLCs, and urban districts, which generated a dramatic population decline or growth accordingly. In addition, the study confronted a number of more fundamental boundary changes, which fragmented the time series data and prevented comparisons across the time period under investigation. To maintain continuity, we had to redraw the administrative boundaries of most county-level units of analysis based on official central government information concerning administrative boundary adjustments (Ministry of Civil Affairs of the PRC, 2010; XZQH, 2016) and the population data of the lower tier units (National Bureau of Statistics, 2002a). Some units had to be removed from the sample due to missing data, and the number of units analyzed in 1990–2010, 2000–2010, and 1990–2010 are 457, 499, and 457, respectively.

Results

National overview

Although China has undergone a process of rapid urbanization since 1990, the pace of urban population growth in China's RCs has slowed, down from an increase of 26.6% between 1990 and 2000 to 18.5% between 2000 and 2010. This change has been accompanied by a slowing down in absolute growth numbers overall, with the RC populations as a whole increasing by 21.4 million between 1990 and 2000—but only by 19.1 million between 2000 and 2010. The proportion of the RC population in relation to the national urban population declined from 27.4% in 1990 to 18.4% in 2010, which means that the contribution of RCs to China's urbanization is declining and RC population growth has been outpaced in other parts of China.

The analysis on the different trajectories of population change shows the diversity among China's RCs. Based on different purposes and data availability, existing literature has provided relevant methods for the classification of population change. For example, Turok and Mykhnenko (2007) identified nine of the most common trajectories of European cities by comparing the direction of population change between different points in time. Wolff and Wiechmann (2018) categorized Europe's shrinking cities into three trajectories based on the rate and persistence of population loss. The population data used in the study cannot support the detailed analysis on the persistence of population change, so the classification method proposed by Turok and Mykhnenko (2007) is adopted to give a full depiction of population change among China's RCs. The following trajectories are identified: (a) continuous growth: 323 units experienced continuous urban population growth during the period from 1990 to 2010; (b) recent resurgence: 29 units experienced negative urban population growth during the 1990s but exhibited positive growth in the 2000s; (c) recent shrinkage: 60 units experienced positive urban population growth during the 1990s, followed by negative growth in the 2000s; and (d) long-term shrinkage: 45 units experienced continuous urban population shrinkage during the period from 1990 to 2010. Overall, although most units in the RCs maintained population growth, shrinkage was identified as a rising problem in some RCs. A total of 74 units (16% of total units) had experienced urban population shrinkage in the 1990s, while in the 2000s, that figure grew by over 47% to include 109 units (22% of the total units). The prevalence of population shrinkage of RCs was much higher than the national average (6.8% in the 1990s and 10.2% in the 2000s) (Li and Mykhnenko, 2018). Moreover, population shrinkage of some RCs has become a persistent phenomenon, as 10.2% of the total units experienced long-term shrinkage during the research period, which was also higher than the national average of 2.8% (Li and Mykhnenko, 2018).

Morphologies of population shrinkage

Among the three kinds of investigated units, urban districts of prefecture-level RCs are most affected by population shrinkage, accounting for nearly 80% of the total number of shrinking units in China's RCs from 1990 to 2010. Given the dispersed spatial form of prefecture-level RCs, the study further identified the morphologies of population shrinkage in China's prefecture-level RCs (Table 4.1), including the following six different types. Type A refers to "total shrinkage". This type is quite rare but is becoming more common. Cities with total shrinkage typically have a shorter urban history, are smaller in size, and are economically relatively underdeveloped. It should be noted that none of the RCs experienced this type of population shrinkage long term, either experiencing it in the 1990s or in the 2000s. Furthermore, apart from Guangan, population loss was modest at less than 10%. Type B refers to "near-total shrinkage". This type represents cities with serious population shrinkage problems and some Type B cities, like Yichun, Shuangyashan, and Jixi in Heilongjiang, were already very close to Type A. Type B usually occurs at RCs established in the 1950s and typically involves cities that are resource exhausted. The scale of population loss is more serious than in Type A. For example, most districts of Yichun, Shuangyashan, and Jixi experienced over 10% population loss between 2000 and 2010. Type C refers to "peripheral shrinkage". This type is uncommon and typically occurs among cities that have not exhausted their resources. Type D refers to "satellite shrinkage" and involves the most common form of shrinkage. These shrinking districts are typically places of resource extraction suffering resource decline. The severity and persistence of shrinkage are also the most pronounced in Type D. Between 1990 and 2000, 12 cities experienced this type of shrinkage, and these cities continued to suffer shrinkage in the period between 2000 and 2010, with more than half of them losing more than 10% of their population each decade. Type E refers to "disjointed shrinkage". This type of RC was typically formed in the period between the 1950s and the 1970s, when urban areas were built near resource sites, contributing

Table 4.1 Urban morphology of population shrinkage in China's prefecture-level RCs

Types	Names	Number			Typical cities	Characteristics
		1990–2000	2000–2010	1990–2010		
A:	Total shrinkage	1	5	–	Chizhou, Anshun	All urban districts suffer population loss
B:	Near-total shrinkage	4	8	4	Yichun (13/15 urban districts shrinking), Jixi (4/6 urban districts shrinking)	More than half of the urban districts suffer population loss
C:	Peripheral shrinkage	3	3	1	Datong (Nanjiaoqu), Changzhi (Jiaoqu)	Urban districts located at the urban fringe suffer population loss, but the urban core continues to grow
D:	Satellite shrinkage	12	22	12	Tangshan (Guye), Fuxin (Qinghemen)	Shrinkage occurs at one or two urban districts that are spatially disconnected from the main urban core
E:	Disjointed shrinkage	7	8	2	Baishan (Jiangyuan), Ezhou (Huarong)	One district in RCs with loosely connected urban districts suffer population shrinkage
F:	Partial shrinkage	12	11	2	Liaoyuan (Xian), Jiaozuo (Zhongzhan)	Less than half of the urban districts suffer population shrinkage

Note: Grey areas stand for shrinking areas, and black areas are growing areas. In the column "Typical cities" under types C/D/E/F, the names in the brackets are urban districts suffering from population shrinkage.

to a highly fragmented urban footprint. Compared to other types, the population shrinkage in these RCs is typically not as severe, usually less than 10% population loss over ten years and typically not persistent. Type F refers to "partial shrinkage" and is the second most common type of shrinkage. Cities of this type represent many varieties of resource use and were built at varying times, but most of these cities are located in the northeast. We further noted that some cities, such as Fushun, changed from this type to Type B (near-total shrinkage) in the studied period as more districts began to experience population shrinkage in those cities.

Spatial differentiation

Population change varied significantly across the four economic macro-regions of China (Figure 4.1). RCs in northeast China, the industrial heartland during the era of planned economy, experienced significantly slower urban population growth and a higher prevalence of urban population shrinkage during the period from 1990 to 2010. Compared with other regions, the northeast had the highest number of population shrinking units in both time periods, with 40 shrinking units in the 1990s and 52 shrinking units in the 2000s. Between 1990 and 2010, there were 30 continuously shrinking units in northeast China, accounting for 66.7% of the national total. Some studies have contended that, since 2000, the economic situation in respect of RCs in northeast China has improved (Tan et al., 2016; Lu et al., 2016); however, any such improvement does not seem to be accompanied with population growth. On the contrary, the problem of population shrinkage of RCs in northeast China has significantly worsened, with the prevalence of population shrinkage rising—from 41% in the 1990s to 54% in the 2000s.

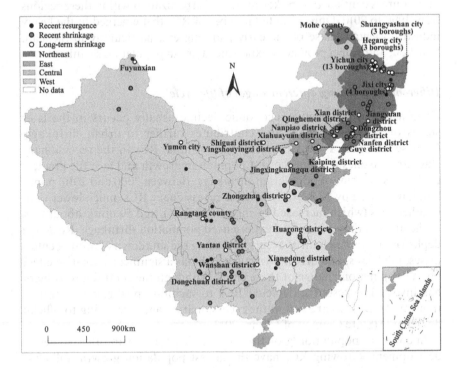

Figure 4.1 Spatial differentiation of urban population shrinkage in China's RCs.

Eastern (coastal) China comprises the country's most developed and urbanized region. With a vibrant economy, it was not expected that population shrinkage would be significant. Overall, our data show that the RCs in this region grew throughout the studied period, but the speed of growth is now declining. Shrinking units are uncommon in this region, accounting for 11% and 15% of the total units in the 1990s and 2000s, respectively, mostly located in Hebei Province. During the period from 1990 to 2010, 7 out of 26 resource-dependent units in Hebei experienced urban population loss. These shrinking units involved districts isolated from the urban core (Type D), including Zhangjiakou (Xiahuayuan district), Chengde (Yingshouyingzi district), Tangshan (Guye district), Handan (Fengfengkuangqu district), and Shijiazhuang (Jingxingkuangqu district). Among them, Xiahuayuan district, Yingshouyingzi district, and Jingxingkuangqu district have been listed as resource-depleted cities.

For less-developed central and western China, the urban population growth rate of RCs remained at a relatively high level despite some slowing down. In the period from 1990 to 2000, population growth in RCs in central and western China accounted for 63.8% of the total RC growth, and the figure rose to 71.2% in the period from 2000 to 2010. The prevalence of population shrinkage within RCs in these regions is less severe, slightly lower than that of the east in the 1990s and 2000s. Generally, the exploitation and utilization of natural resources continued to be key drivers of urbanization within these regions from 1990 to 2010. Only a limited number of RCs that emerged in the heavy industry-oriented phase of the early planning era, particularly in Sichuan, Guizhou, and Hubei provinces, experienced urban population shrinkage.

Differentiation among different stages of life cycle[2]

From a global perspective, economic decline usually occurs in the later stages of RC development and contributes further to urban shrinkage. Chinese RCs face a similar trajectory. In China, RCs in the recessionary stage are the ones with the slowest population growth and face the greatest challenges in terms of population shrinkage. Between 2000 and 2010, population increased by 10.3% only in these recessionary RCs, much lower than in other RCs (which increased by more than 20%), and 59 units, about 45% of the units in these cities, had experienced population shrinkage. Resource depletion triggers a chain of events, including the shutdown of mining enterprises, job losses, and urban poverty, which has a significant negative effect on the population change of these cities. Although the central government has invested in resource-depleted RCs to facilitate post-resource transitions, most of them are still in the recessionary stage, according to official documents (State Council of China, 2013).

In contrast, population growth rate is higher for RCs in other stages of development. Growing RCs have the fastest population growth, followed by mature RCs and regeneration RCs (i.e., RCs transiting from resource dependency to alternative sources of growth). However, 50 units in these cities also experienced population shrinkage to varying extents, despite not facing problems with resource depletion or already transiting to a new development path. For example, some growing and mature RCs in the remote Sichuan and Guizhou Provinces, such as Nanchong, Guangan, Liupanshui, and Anshun, experienced population shrinkage in some of their urban districts. Some regenerating RCs such as Tangshan and Anshan also experienced partial population shrinkage. Overall, the prevalence of population shrinkage in growing, mature, and regenerating RCs was below 14% during the years 2000–2010. Compared with shrinking recessionary RCs, most shrinking RCs in the stages of growth, maturity, and regeneration experienced recent shrinkage. Only a few of these RCs, such as Guye district in Tangshan (Type D in Table 4.1), suffered long-term population shrinkage.

Typical drivers

Population shrinkage of RCs is usually associated with outmigration triggered by economic decline and exacerbated by the unique characteristics of RCs, such as remote location and environmental degradation (Martinez-Fernandez et al., 2012). While these factors can also be used to explain the population shrinkage process of China's RCs, the specific reasons behind these factors might be different in China than in other countries. Moreover, some unique factors, such as systemic reformation and demographic policy, should also be taken into consideration within China's specific context. A recent study about urban population shrinkage in China identified four drivers with Chinese characteristics: economic, institutional, environmental, and demographic factors (Li and Mykhnenko, 2018). Their work provides direction for analyzing the drivers for population shrinkage of China's RCs. Based on their analysis, the following four drivers are identified: lack of industrial support, maladjustment to market-oriented reformation, poor urban environment, and natural population decline.

Lack of industrial support

Since the 1990s, as China's urbanization has accelerated, the drivers of urbanization have changed fundamentally, moving away from the industrialization-driven model to a more balanced approach that involves both industrialization and service-oriented growth (Gaubatz, 1999; Cao and Liu, 2010; Gu et al., 2015). However, for most shrinking RCs, the traditional drivers of urbanization have been weakening and the rise of new drivers is not strong enough.

The first case in point, the role played by resource-based industries on urban population growth is weakening, especially for resource-depleted RCs. For RCs at other stages of development, jobs created by the resource sector are also declining. For example, while the production of coal increased from 1.08 billion tons in 1990 to 32.35 billion tons in 2010, jobs in that sector decreased from 5.39 million in 1990 to 5.27 million in 2010 (National Bureau of Statistics, 2011). Fewer workers are needed due to increased productivity per worker. Another factor contributing to population shrinkage is a reduction in production due to unfavorable market conditions and/or the government's environmental protection policy. For example, the profit margins of Daqing Oil Field have been significantly eroded by falling oil prices on the international market and an increase in production costs (Li et al., 2015). Consequently, Daqing's oil production in 2016 was only 36.56 million tons, which was a significant drop from the 1997 peak of 56 million tons. In the forestry sector, Heilongjiang's timber harvest has declined from 10.2 million m³ in 1998, the year in which the natural forest protection policy was introduced, to only 1.46 million m³ in 2011.

Second, the development of post-resource industries has been slow, with only limited capacity to stimulate population growth. RC economic development is characterized by strong path-dependency, and many RCs lack sufficient motivation to transition before entering a resource-depletion phase. Although the central government has pushed for the economic transition of RCs since 2001, experience has shown that most RCs find it difficult to develop new internet-based or service-oriented industries (Li et al., 2013). Most RCs have been seeking a pathway of reindustrialization (Wang et al., 2014a). The newly developed sectors have involved mostly capital-intensive heavy industries comprising petrochemicals, metallurgy, and equipment manufacturing, none of which provide large numbers of jobs. As of 2010, there were over 600,000 jobless miners (State Council of China, 2013), which suggests that new industries had failed to absorb the surplus labor from the resource sectors.

Maladjustment to market-oriented reformation

With the acceleration of market-oriented reformation since the 1990s, China's urbanization is shifting from a government-led model to a more balanced approach where the private-sector economy, in terms of both domestic and foreign involvement, also plays a significant role (Liu, 1992; Shen and Ma, 2005). As developed by the central government, state-owned enterprises (SOEs) used to play a leading role in the urban development of China's RCs. They have not only controlled the extraction and processing of resources but have also contributed to local jobs and fiscal budgets and have provided an array of public services such as water supply, electricity, heating, and road construction (Li et al., 2015). Local governments, on the other hand, have played a minor role in urban development. Following China's market-oriented reformation, SOEs in RCs began to experience difficulties and have downsized their staff and handed over public service responsibilities to local governments in order to increase efficiency and streamline operations (Wang et al., 2014b). Meanwhile, the long-entrenched dominance of

SOEs and their monopoly of resources have made it very difficult for private companies to thrive in these cities. For example, in Daqing, the private-sector economy only contributed approximately 20% to the city's GDP in 2010 (Li et al., 2015). The weakening of SOEs, the burdened local government, and the weak private sections created problems, such as unemployment and fiscal shortfall, which contributed to the shrinkage and slowing down of population growth in RCs.

Another aspect of market-oriented reformation of RCs is the resource pricing mechanism. During the era of planned economy, RCs contributed resource-based products under central government command and sold these products at a price solely determined by the central government, which ensured the stable development of China's RCs. However, a market-based pricing mechanism of resource-based products has been gradually introduced during the process of market-oriented reformation. Under this situation, RCs have to face the fluctuation of the resources market. As a result, the economic stability of China's RCs has been greatly reduced, which could lead to a periodical or long-term population shrinkage of China's RCs.

Poor urban environment

During the period from 1990 to 2010, urban population growth in China was mainly fueled by extensive rural-to-urban migration. According to the 2010 National Population Census, the overall size of China's floating population living and working outside their registered localities had reached 261.4 million—or 20% of China's total population. Research shows that coastal regions with more developed economies and better living environments have been more attractive to migrants. The coastal regions attracted 61.2% of total migrants in 2005 (Duan and Yang, 2009). RCs, in comparison, have been unattractive to migrants due to their poor urban environment. Several reasons help to explain this.

First, economic opportunities are limited, and salaries lower in RCs. While the central government has increased its support, especially to resource-depleted RCs, the overall economic development of RCs remains unimpressive. In 2010, the per-capita salary of prefecture-level resource-depleted RCs was 28,202 RMB/year, which was 79% of the national average for cities (Zhang et al., 2014). Second, because RCs are mainly located where resources are readily found, they are typically located in remote locations, far away from large metropolitan areas (Sun and Mao, 2018). Third, influenced by the ideal of the planning era, "production first, living second", RCs typically focus on their production function, and urban services such as education, health, housing, and transportation tend to be neglected. Finally, pollution and environmental degradation problems are more serious in RCs. A long history of resource exploitation and processing typically results in a poor environment with many pollution problems. For instance, the resource-depleted RCs' average industrial wastewater, SO_2 emissions, and smoke emissions are all ten times higher than the average city levels in China. Land subsidence is another problem, with an estimated 140,000 ha of subsidence in need of treatment before the land can be used again (Zhang et al., 2014). Because of these problems, RCs are typically not the first choice of the floating population.

Natural population decline

Internationally, an aging population and a low birth rate are key reasons for the slowing down and shrinkage of population growth, especially in developed countries such as the UK, Japan, and Scandinavia (Mykhnenko and Turok, 2008; Matanle et al., 2011). While China is a developing country, these demographic factors also play a role due to the nation's strict birth control policy, in place since the 1980s. As a result of this policy, China's natural population growth declined from 14.39‰ in 1990 to only 4.79‰ in 2010. Because the population control policy is stricter in the cities than in rural areas, the impact of population control on urban populations has had even greater significance.

The natural population growth of RCs in 2000 was 6.37‰, dropping to 4.24‰ in 2010, which was below the national average. A total of 79.8% of the studied units had a natural population growth decline. Moreover, in 2000, there were 17 units with a negative natural population growth, which had increased to 68 units in 2010. This phenomenon is especially common in the northeast. Between 2000 and 2010, of the 97 units in the northeast, 85 had experienced a decrease in natural population growth. In 2000, there were 13 negative growth units in the northeast, whereas in 2010, there were 54.

Conclusions and recommendations

Although China is still in the process of rapid urbanization, population growth has not been the only development trajectory of its cities, especially RCs. By conducting an analysis featuring a longer time scale, a larger sample of RCs, and a finer unit, this chapter shows that population shrinkage has been a short-term or long-term phenomenon in nearly 30% of China's RCs since 1990, with many more experiencing a slowing down of population growth over the years. Comparatively, RCs in northeast China and the resource-depleted RCs experienced a higher prevalence of population shrinkage and a lower population growth rate. Based on the spatial characteristics of population shrinkage, the study identified six morphologic types of population shrinkage in China's RCs. Furthermore, the study presented four typical drivers for this phenomenon in China's RCs: (1) lack of industrial support; (2) maladjustment to market-oriented reformation; (3) poor urban environment; and (4) natural population decline.

Compared with RCs in capitalist economies, population shrinkage in China's RCs manifests some unique features in terms of morphology and dynamics. First, most of China's shrinking RCs experience population growth and shrinkage in different parts of a city simultaneously. Total shrinkage is not the main type of population shrinkage in China's RCs and abandoned RCs have not been found. Overall, population shrinkage of China's RCs is still in an early stage. Most of the shrinking RCs, or at least some parts of these cities, have not completely lost their attractiveness for migrants. Thus, any attempt to identify shrinking cities in China based only on population indicators should be made cautiously because the severity and persistence of population shrinkage are sensitive to the scale at which population shrinkage is calculated. Second, while resource-depleted RCs in China were severely hit by population shrinkage, about 46% of population shrinking units in 2000–2010 were in non-recessionary RCs not experiencing problems of resource depletion or already transiting to a new development path. Such a result indicates that population shrinkage of China's RCs is not solely governed by boom-and-bust industrial cycles. Instead, institutional transition in urban governance, urban economy systems, and environmental protection are nonnegligible drivers of population shrinkage of China's RCs. Thus, a more comprehensive coping strategy should be adopted to tackle the multiple factors behind the phenomenon. This strategy should not be limited to economic transitioning toward a diversified industrial structure and should also include institutional rearrangements to make them adapt to the market economy.

With the enhancement of population mobility in China, population shrinkage of RCs might continue to intensify due to their comparative disadvantage in terms of economic development and urban environment. However, so far, this phenomenon has not been widely considered in local governments' planning for urban development. Responding to China's institutional arrangements regarding land-related finance, local officials tend to want a larger population size in order to receive more construction land quota for urban expansion. In reality, many RCs have failed to achieve their population growth targets, and the commitment to urban expansion is neither sustainable in the long run nor helping RCs address their many barriers to growth.

This study shows that population shrinkage in China's RCs has significant differences in terms of morphology, severity, and persistence. Thus, the coping strategies for these RCs should also be diversified. For some shrinking RCs, population shrinkage is just a short-term phenomenon that is reversible— even without specific interventions. For RCs suffering from long-term population shrinkages, such as near-total shrinkage and satellite shrinkage (Type B and Type D in Table 4.1), interventions should be adopted to either reverse population shrinkage or adapt to smaller populations, depending on the RCs' potential for further urbanization.

Acknowledgement

This research was funded by Strategic Priority Research Program of the Chinese Academy of Sciences (Grant No. XDA28110100).

Notes

1 Three indicators were used to identify the 262 RCs: the performance index of mining industries, the scale coefficient of resource output, and the historical and expected contribution of resource supply. For more details, see Yu et al. (2019).
2 According to the Plan of Sustainable Development for RCs in China (2013–2020) issued by the State Council of China, these 262 RCs can be classified into four types: growth, mature, recession, and regeneration (State Council of China, 2013).

References

Cao, G. Z. & Liu, T. 2010. Dynamic Mechanism of Urbanization and Its Evolution in Post-reform China. *China Soft Science*, (9), 86–95. //曹广忠, 刘涛, 2010. 中国省区城镇化的核心驱动力演变与过程模型. *中国软科学*, (9), 86–95.

Chan, K.W. 2003. Chinese Census 2000: New Opportunities and Challenges. *China Review*, 3(2), 1–12.

Chen, Y. & Mei, L. 2018. Quantitative Analysis of Population Distribution and Influencing Factors of Resource-Based Cities in Northeast China. *Scientia Geographica Sinica*, 38(3), 402–409. //陈妍, 梅林, 2018. 东北地区资源型城市人口分布与影响因素的定量分析. *地理科学*, 38(3), 402–409.

Duan, C. R. & Yang, G. 2009. Trends in Destination Distribution of Floating Population in China. *Population Research*, 33(6), 1–12. //段成荣, 杨舸, 2009. 我国流动人口的流入地分布变动趋势研究. *人口研究*, 33(6), 1–12.

Gao, S. Q. & Long, Y. 2017. Distinguishing and Planning for Shrinking Cities in Northeast China. *Planners*, 33(1), 26–32. //高舒琦, 龙瀛, 2017. 东北地区收缩城市的识别分析及规划应对. *规划师*, 33(1), 26–32.

Gaubatz, P. 1999. China's Urban Transformation: Patterns and Processes of Morphological Change in Beijing, Shanghai and Guangzhou. *Urban Studies*, 36(9), 1495–1521.

Gu, C. L., Kesteloot, C. & Cook, I. G. 2015. Theorising Chinese Urbanisation: A Multi-layered Perspective. *Urban Studies*, 52(14), 2564–2580.

Hayter, R. & Nieweler, S. 2018. The Local Planning-economic Development Nexus in Transitioning Resource-industry Towns: Reflections (Mainly) from British Columbia. *Journal of Rural Studies*, 60, 82–92.

He, S. Y. 2014. When Growth Grinds to a Halt: Population and Economic Development of Resource-depleted Cities in China. In Richardson, H. W. & Nam, C. W. (eds.) *Shrinking Cities: A Global Perspective*. Abingdon: Routledge.

He, S. Y. et al. 2017. Shrinking Cities and Resource-based Economy: The Economic Restructuring in China's Mining Cities. *Cities*, 60, 75–83.

Kamal-Chaoui, L., Leman, E. & Zhang, R. 2009. Urban Trends and Policy in China. *OECD Regional Development Working Papers 2009/1*. Paris: OECD.

Knierzinger, J. & Sopelle, I. T.-I. 2019. Mine Closure from Below: Transformative Movements in Two Shrinking West African Mining Towns. *The Extractive Industries and Society*, 6(1), 145–153.

Li, H., Lo, K. & Wang, M. 2015. Economic Transformation of Mining Cities in Transition Economies: Lessons from Daqing, Northeast China. *International Development Planning Review*, 37(3), 311–328.

Li, H., Lo, K. & Zhang, P. 2020. Population Shrinkage in Resource-dependent Cities in China: Processes, Patterns and Drivers. *Chinese Geographical Science*, 30(1), 1–15.

Li, H. J., Long, R. Y. & Chen, H. 2013. Economic Transition Policies in Chinese Resource-based Cities: An Overview of Government Efforts. *Energy Policy*, 55, 251–260.

Li, H. & Mykhnenko, V. 2018. Urban Shrinkage with Chinese Characteristics. *The Geographical Journal*, 184(4), 398–412.

Li, H., Zhang, P. & Cheng, Y. 2009. Economic Vulnerability of Mining City: A Case Study of Fuxin City, Liaoning Province, China. *Chinese Geographical Science*, 19(3), 211–218.

Liu, Y. L. 1992. Reform From Below: The Private Economy and Local Politics in the Rural Industrialization of Wenzhou. *The China Quarterly*, 130, 293–316.

Long, Y. & Gao, S. Q. 2019. *Shrinking Cities in China: The Other Facet of Urbanization*. Singapore: Springer.

Lu, C. P. et al. 2016. Sustainability Investigation of Resource-Based Cities in Northeastern China. *Sustainability*, 8(10), 1058.

Wolff, M. & Wiechmann, T. 2018. Urban Growth and Decline: Europe's Shrinking Cities in a Comparative Perspective 1990–2010. *European Urban and Regional Studies*, 25(2), 122–139.

Martinez-Fernandez, C. et al. 2012. The Shrinking Mining City: Urban Dynamics and Contested Territory. *International Journal of Urban and Regional Research*, 36(2), 245–260.

Matanle, P., Rausch, A. S. & The Shrinking Regions Research Group. 2011. *Japan's Shrinking Regions in the 21st Century: Contemporary Responses to Depopulation and Socioeconomic Decline*. Amherst, NY: Cambria Press.

Ministry of Civil Affairs of the PRC. 2010. The Administrative Divisions Code of the People's Republic of China (PRC) 1980–2018. http://www.mca.gov.cn/article/sj/xzqh//1980/. Cited 23 January 2019. //中华人民共和国民政部, 2010. 2018–1980 年中华人民共和国行政区划代码. http://www.mca.gov.cn/article/sj/xzqh//1980/, 2019 年1月23日.

Mykhnenko, V. & Turok, I. 2008. East European Cities—Patterns of Growth and Decline, 1960–2005. *International Planning Studies*, 13(4), 311–342.

National Bureau of Statistics. 1992. *China Population Statistics Yearbook 1991*. Beijing: China Statistics Press. //国家统计局, 1992. *1991 年中国人口统计年鉴*. 北京: 中国统计出版社.

National Bureau of Statistics. 2002a. *China Population by Township*. Beijing: China Statistics Press. //国家统计局, 2002a. *中国乡、镇、街道人口资料*. 北京: 中国统计出版社.

National Bureau of Statistics. 2002b. *Tabulation on the 2000 Population Census of the People's Republic of China by County*. Beijing: China Statistics Press. //国家统计局, 2002b. *中国2000年人口普查分县资料*. 北京: 中国统计出版社.

National Bureau of Statistics. 2011. China Statistical Yearbook 2011. http://www.stats.gov.cn/tjsj/ndsj/2011/indexeh.htm. Cited 20 March 2019. //国家统计局, 2011. 2011 中国统计年鉴. http://www.stats.gov.cn/tjsj/ndsj/2011/indexeh.htm. 2019年3月20日.

National Bureau of Statistics. 2012. *Tabulation on the 2010 Population Census of the People's Republic of China by County*. Beijing: China Statistics Press. //国家统计局, 2012. *中国2010年人口普查分县资料*. 北京: 中国统计出版社.

Oswalt, P. 2005. *Shrinking Cities*. Ostfildern-Ruit, Germany: Hatje Cantz.

Richardson, H. W. & Nam, C. W. 2014. *Shrinking Cities: A Global Perspective*. London and New York: Routledge.

Shen, X. P. & Ma, L. J. C. 2005. Privatization of Rural Industry and de Facto Urbanization from Below in Southern Jiangsu. *China Geoforum*, 36(6), 761–777.

Song, Y. & Wang, S. J. 2011. *Urban Space of Mining Cities: Pattern, Process and Mechanism*. Beijing: Science Press. //宋飏, 王士君. 2011. *矿业城市空间: 格局、过程、机理*. 北京: 科学出版社.

State Council of China. 2013. Plan of Sustainable Development for Resource-Dependent Cities in China, 2013–2020. http://www.gov.cn/zhengce/content/2013-12/02/content_4549.htm. Cited 23 January 2019. //中国国务院, 2013. 全国资源型城市可持续发展规划 (2013-2020 年). http://www.gov.cn/zhengce/content/2013-12/02/content_4549.htm. 2019 年1月23日.

Sun, W. & Mao, L. 2018. Are Chinese Resource-exhausted Cities in Remote Locations? *Journal of Geographical Sciences*, 28(12), 1781–1792.

Tan, J.T. et al. 2016. The Urban Transition Performance of Resource-based Cities in Northeast China. *Sustainability*, 8(10), 1022.

Turok, I. & Mykhnenko, V. 2007. The Trajectories of European Cities, 1960–2005. *Cities*, 24(3), 165–182.

Wang, M., Kee, P. & Gao, J. 2014a. *Transforming Chinese Cities*. New York: Routledge.

Wang, M. et al. 2014b. *Old Industrial Cities Seeking New Road of Industrialization: Models of Revitalizing Northeast China*. Singapore: World Scientific Publishing Company.

Woodworth, M. D. 2016. Booms, Busts and Urban Variation Among "Resource-based Cities" in China's Northwest. *Inner Asia*, 18(1), 97–120.

Xie, Y. T., Li, H. & Zou, Q. 2017. A Study on the Innovation Index of Resource-based Cities in China: Taking 116 Cities at the Prefecture Level for Example. *Journal of Peking University (Philosophy and Social Sciences)*, 54(5), 148–160. //谢远涛, 李虹, 邹庆. 2017. 我国资源型城市创新指数研究—以116 个地级城市为例. *北京大学学报: 哲学社会科学版*, 05, 148–160.

XZQH. 2016. Administrative Division Website. http://www.xzqh.org/html.

Yu, J. H., Li, J. M. & Zhang, W. Z. 2019. Identification and Classification of Resource-based Cities in China. *Journal of Geographical Sciences*, 29(8), 1300–1314.

Zhang, P. Y. et al. 2011. *Vulnerability of Coupled Human-environment System in Mining Cities—Theories, Methods and Cases*. Beijing: Science Press. //张平宇, 等. 2011. *矿业城市人地系统脆弱性—理论•方法•实证*. 北京: 科学出版社.

Zhang, W. Z. et al. 2014. *Study on Sustainable Development of Resource-Based Cities in China*. Beijing: Science Press. //张文忠, 等. 2014. *中国资源型城市可持续发展研究*. 北京: 科学出版社.

5 Examining the economic growth trajectories of China's resource-exhausted cities from 1995 to 2018 within the framework of urban shrinkage

Minwei Zhang and Helin Liu

Introduction

Resource-exhausted cities (RECs) are resource-based cities that have entered the late or end-stage of their natural resource extraction cycle. In practice, the term may also refer to cities where resources are not exhausted but where the development thereof is highly constrained by mining costs or environmental protection policies that might lead to economic or population shrinkage.

Shrinkage has not only occurred in old industrial areas in market economies such as the United States (Kahn, 1999), the United Kingdom, France (Oswalt, 2005; Wolff et al., 2013), West Germany (Häußermann and Siebel, 1988; Knapp, 1998), Australia, and Japan (Martinez-Fernandez et al., 2012a, 2012b), but also in postsocialist mining cities in Central and Eastern Europe (Stryjakiewicz, Ciesiółka and Jaroszewska, 2012; Constantinescu, 2012). In addition to the vulnerability of the industries in these cities or areas, regional conditions such as geographic and social isolation from more developed economies (Haggerty et al., 2018) and long-term dependence on welfare policies (Leadbeater, 2009) also have negative impacts on resource-exhausted economies. Moreover, in postsocialist regions, the collapse of the economic system poses even greater serious challenges to mining cities (Wirth, Mali and Fischer, 2012). Some scholars have argued that in a market economy system, the solutions proposed for resource-dependent cities are more locally oriented at a micro level, while in a socialist system, the strategies tend to be more diverse and macroscopic (Li, Lo and Wang, 2015). Shrinking cities in China share some of these socialist-market features, but at the same time, have distinctive features. Compared with cities in other countries with socialist backgrounds (Crowley, 2016; Buček, 2016), China's cities generally have the characteristics of larger population size, longer marketization reform cycles, and stronger government macro-regulation capabilities. In addition, rather than more straightforward embodiments of resource-exhaustion, such as unemployment, vacant houses, and land abandonment, in China, the effects of resource-exhaustion are more invisible and commonly include a decline in economic scale and overall competitiveness.

DOI: 10.4324/9780367815011-7

In China, RECs exemplify a concept that implies both academic definition and policy orientation. According to the *China Resource-based City Sustainable Development Plan (2013–2020)* released in 2013, China has 262 resource-based cities. Most of these cities emerged in the era of the planned economy and then experienced the reform and opening-up policy. Since the 1980s, some resource-based cities, represented by Baiyin and Fuxin, have begun to experience problems such as resource reserve depletion, closure of factories or firms, and rising unemployment, which are signs of urban economic decline. At the beginning of the new millennium, the "resource curse" dilemma for these cities became more serious. To alleviate the resultant problems and risks, in 2007, the State Council issued a document titled *Suggestions for Promoting the Sustainable Development of Resource-based Cities* urging resource-based cities to respond actively to the great socioeconomic transformation and promised to provide fiscal support to the officially approved RECs. Subsequently, after a comprehensive assessment, out of hundreds of cities that submitted applications, 69 (including 25 prefecture-level cities, 22 county-level cities, 14 municipal districts, 5 counties, and 3 special areas) were officially designated as RECs eligible to receive special subsidies and other programs.

These policy initiatives stimulated a variety of studies on the development trends and Chinese-style initiatives for China's resource-based cities. In terms of research emphasis, these studies either focus on a single resource-based city (Li, Zhang and Xu, 2018) or region (Wang, Qu and Wu, 2019) to analyze the characteristics and causes of shrinkage, or, through quantitative methods, measure, compare, or classify the shrinkage of resource-based cities in China from the perspective of the population (Li, Lo and Zhang, 2020), economy (Hu and Mo, 2016), or space (Zheng, 2014). However, none of these studies has treated RECs as a specific group to identify their common features and internal heterogeneity. A key trend in the available literature is to identify whether a city or a group of cities has shrunk in a certain period (Liu et al., 2019), while the long-term evolution trajectories, particularly the economic growth trajectories, have seldom been discussed in depth.

Although there have been a considerable number of studies on the economic issues of RECs since the implementation of the opening-up policies of the 1980s, most of these tend to be descriptive (He et al., 2017; Yang, Xiao and Zhao, 2019; Wang and Feng, 2013). Few studied the city's economic situation from a long-term perspective. None has offered A systematic analysis of the causes and mechanisms of economic changes in RECs or a reasonable explanation for the differences in the development trajectories among these cities. In this context, we raise three questions: what are the characteristics of the economic growth trajectories of China's RECs? What differences among them can be identified? What are the key factors that cause the divergence of the economic growth trajectories of these RECs?

To answer these questions, 41 cities among the RECs designated by the State Council were used as a research sample (excluding the 22 non-city areas and 6 cities that had experienced administrative boundary change or had no credible data). To capture the economic growth trajectories of the RECs from 1995 to 2018, we first constructed a composite economic performance index and composite economic growth index. The economic growth trajectories of each of these 41 cities across the 24 years were then plotted and analyzed. Based on the development trends of these trajectories, these RECs were classified into four categories: steady growth, growth slowdown, growth-shrink, and growth-shrink-rebound, and the characteristics of each category were analyzed. Finally, through a comparative study of four representative cities, we continued to investigate the causal factors that have shaped the economic growth or shrinkage trajectory of each of these four types of REC.

Methods and data

The decline in economic performance is not only an important dimension of urban shrinkage (Fol and Cunningham-Sabot, 2010; Martinez-Fernandez et al., 2012a; Zhang et al., 2019) but also an important reciprocal factor in population shrinkage (Manville and Kuhlmann, 2018; Zhang and Qu, 2020; Wang, Qu, and Wu, 2019). Unlike the methods associated with measuring population change, the metrics of economic performance tend to be more abstract and intricate. To describe urban economic recession, Western scholars generally appeal to indicators such as unemployment rates and unemployment subsidy growth, the number of factories closed down, and tax revenue declines (Friedrichs, 1993; Pallagst, Wiechmann, and Martinez-Fernandez, 2013), while Chinese scholars tend to use urban total gross domestic product (GDP) and GDP growth rates to gauge economic change (Wang, Qu, and Wu, 2019; Zhang et al., 2019). In a general sense, however, a single indicator cannot fully reflect the economic changes and problems at play within a city. For instance, there could be only a slight change in the rate of employment and population size in the process of, say, an urban industrial transition from traditional to high-tech industries in a given city, yet the city's tax revenue could drop significantly, and the labor force structure might also undergo major changes (Pallagst, Wiechmann and Martinez-Fernandez, 2013). Therefore, it is necessary to use comprehensive indicators, as explained below, to capture the features of the economic performances of RECs.

Data sources

The research sample of 41 RECs includes 24 prefecture-level cities and 17 county-level cities located in 17 provinces. Of these 41 cities, 23 were coal-based, 4 were petroleum-based, 5 were forestry-based cities, and 9 were multi-mineral-based cities. There was one city (Aershan) where the population was below 50,000, 18 where the population was between 0.1 and 1 million, 18 where the population was between 1 million and 3 million, and 4 with a population of over 3 million.

The requisite data were acquired from the *China City Statistical Yearbook (1995–2018), China County Statistical Yearbook (2000–2018),* and provincial or municipal statistical yearbooks. Local chronicles were also used as complementary materials. Missing economic growth data for a couple of years were obtained by using growth rate interpolation. To eliminate the impact of inflation, all economic data were uniformly processed using the GDP deflator and calculated based on the 1995 price.

Composite economic performance index

In this study, we chose to use a comprehensive index and an indicator system to measure the comprehensive economic performance of the RECs from four perspectives. The first of these was GDP, including total GDP and GDP per capita, which is the core indicator in China's official comprehensive economic accounting and can directly reflect a city's strength in production or consumption. The second was citizens' living standards, represented by income and consumption, using average employee wages and total retail sales of consumer goods as indicators (most of the data on residents' disposable income and consumption levels were missing). The third perspective from which we measured comprehensive economic performance was investment, and total fixed-asset investment was adopted as an indicator to reflect the level of urban infrastructure construction, the capacity of local enterprises to expand production, and the potential for attracting investment. The fourth perspective used was financial capacity, which we took to reflect the city's fiscal revenue (such as corporate taxation and land transaction income), as well as the capacity to provide public services and support new projects (through fiscal incentives and subsidies). Fiscal revenue in the local general budget is closely related to local taxation, central fiscal transfer payments, and land revenue.

Although some other indicators such as urban unemployment rate, the total number of employees, resource output, and high-tech industry output also tend to have good or even better representativeness, their data were not generally available. In the section on causal factors, where data were available, some of these indicators have been included.

The data were standardized using the range method. To circumvent the problem of potential negative values that are not appropriate for the entropy method, all standardized results were shifted by +1. The weight of each indicator was obtained using the entropy method, and then the composite economic performance index was calculated by the following formula (5.1):

$$E_t = \sum_{i=1}^{n} w_i x_{it}, \tag{5.1}$$

where E_t is the composite economic performance index of a city at time t, x_{it} is the evaluation value of indicator i at time t, n is the number of indicators ($n = 6$), and w_i is the weight value of x_i. The final value range of E_t is $[1, 2]$;

the closer it is to 2, the higher the comprehensive economic level of the city at time *t*.

Composite economic growth index

With the results of the index of formula (5.1), it was possible to use formula (5.2) to measure the composite economic growth index at each time period in each city:

$$G_t = \frac{E_t - E_{t-1}}{E_{t-1}} \tag{5.2}$$

In formula (5.2), G_t is the composite economic growth index; E_t and E_{t-1} were the composite economic performance index values at time *t* and time *t*–1, respectively. We assumed that when $G_t < 0$, it meant that the city had experienced economic shrinkage at time *t*, and it was obvious that the larger the absolute value, the more severe the shrinkage. Otherwise, it meant that the city did not shrink but grew. This formula is also more common in studies of urban population shrinkage. Normal economic development experiences a certain amount of volatility, and a single small drop may not reflect economic decline but rather, say, an abnormal increase in the previous year; accordingly, we define economically shrinking cities as cities with $G_t < 0$ for two consecutive years or $G_t < -0.1$.

Results

The overall characteristics of economic growth

Using the proposed composite economic performance index and a composite economic growth index, the overall economic growth trajectory of the RECs from 1995 to 2018 is depicted in Figure 5.1 (on the left). As shown by the bars, the RECs as a whole display an increasing trend despite a flattened tail. More precisely, if we examine the G_t represented by the line, it appears as a "∩" shape, which roughly indicates three development periods experienced by the RECs. The first is 1995–2001, a low-speed period in which most RECs maintained slow growth, during which time the average G_t was approximately 0.01. The second is 2002–2012, the high-speed period, during which the average G_t gradually increased from 0.016 (2002) to 0.061 (2012), and there were no RECs with significant shrinkage. The third is 2013–2018, a period of stagnation and decline when the average G_t decreases from 0.036 (2013) to –0.001 (2017) and 0.003 (2018) in this period. Except for 5 cities that continued to demonstrate a rapid growth trend, 36 cities experienced significant slowdowns, fluctuations, or shrinkages during this period. There were 16 cities that experienced significant shrinkage, and 7 of them did not show signs of recovery until 2018.

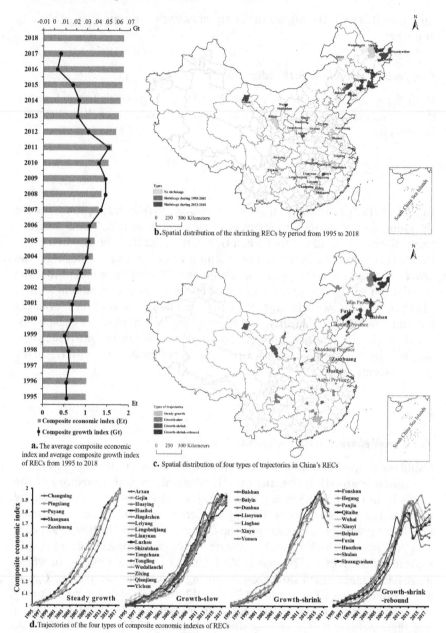

Figure 5.1 The four types of economic growth trajectories and their spatial distributions.

In the period 1995–2018, most shrinking RECs were located in the regions north of the Yangtze River, especially in the northeast region of China (section b in Figure 5.1). In the period from 1995–2001, economic shrinkages only occurred in three smaller county-level cities and did not show obvious north-south differentiation. During 2013–2018 (Beipiao shrank in both time periods), the number of shrinking RECs increased significantly, and the north-south divergence substantially expanded. Among them, the economic shrinkage of RECs was particularly prominent in the northeast, and it spread from county-level cities to prefecture-level cities. Thus, the economic shrinkage of China's RECs in recent years has been characterized by an expansion in spatial scope, an increase in intensity, and a trend of moving up from lower to higher administrative tiers.

The four types of economic growth trajectory

A more detailed examination of the economic growth trajectories of the 41 RECs reveals that the trajectory differences among these cities mainly occurred after 2013. Taking the trajectories from 2013 onward as the base, these RECs can be grouped into four types (section d in Figure 5.1 and Table 5.1):

1 Steady growth: this type of REC, including Changning, Pingxiang, Puyang, Shaoguan, and Zaozhuang, has maintained a steady growth rate without shrinkage from 1995 to 2018. All of them are located in the central and eastern regions of China (section c in Figure 5.1). This type accounts for the smallest percentage, only around 12.2% of the total cities studied;
2 Growth-slowdown: this refers to RECs that maintained growth during 1995–2013 but showed stagnant growth or a slight downwards trend after 2013. This type accounts for the largest percentage, including 18 cities making up 43.9% of the total cities studied. Most of them are in central and southwest China, with only two (forestry industry) cities, Yichun and Wudalianchi, located in the northeast of China (section c in Figure 5.1);
3 Growth-shrink: this type of REC maintained growth between 1995 and 2016 but then shrank and did not rebound. There are seven cities in total that are subject to this trend, accounting for 17.1% of the total cities studied. These include Baishan, Dunhua, Liaoyuan, Baiyin, Yumen, Lingbao, and Xinyu, of which three are in Jilin province and two are in Gansu Province (section c in Figure 5.1);
4 Growth-shrink-rebound: these RECs share a similar trend with the growth-shrink cities but rebounded before 2018 after an obvious shrinkage. This category includes 11 cities, accounting for 26.8% of the total cities studied, all concentrated in the northeast. This type of RC in Heilongjiang Province, including Qitaihe, Hegang, and Shuangyashan, shrank as early as 2013 but rebounded rapidly. The shrinkage seen in Liaoning Province is the worst (section c in Figure 5.1), and the average G_t is 0.27.

Table 5.1 Types of economic growth trajectories of China's resource-exhausted cities

Number	Types	Characteristics	Typical cities
1	Steady growth	The city maintained a stable growth trend without shrinking.	Changning, Pingxiang, Puyang, Shaoguan, Zaozhuang
2	Growth-slowdown	The city maintained stable growth from 1995 to 2013 and, after 2013, showed a trend of slowing growth, fluctuations, or slight shrinkage.	Arxan, Gejiu, Huaying, Huaibei, Huangshi, Jiaozuo, Jingdezhen, Leiyang, Lengshuijiang, Lianyuan, Luzhou, Shizuishan, Tongchuan, Tongling, Wudalianchi, Capital Xingshi, Yichun, Qianjiang
3	Growth-shrink	The city maintained growth between 1995 and 2016 and underwent a significant shrinkage since 2016, with no recovery till 2018.	Baishan, Baiyin, Dunhua, Liaoyuan, Lingbao, Xinyu, Yumen
4	Growth-shrink-rebound	It maintained growth between 1995 and 2013, experienced a significant shrinkage after 2013, and rebounded significantly before 2018.	Fushun, Hegang, Panjin, Qitaihe, Wuhai, Xiaoyi, Beipiao, Fuxin, Huozhou, Shulan, Shuangyashan

It is worth noting that for the fourth type of cities, shrinkage occurred earlier and more severely than the third type. The average G_t of the 11 cities in the fourth type was −0.21, much lower than the value of −0.15 featured by the former. Moreover, the "rebound" of some trajectories may not indicate the actual recovery of the economy but rather a flat or mild rebound after bottoming. Therefore, the occurrence of a "rebound" in a city's trajectory cannot be used as the only criterion for judging whether a city's economy has improved. The situation must be understood with more data and evidence to ensure a comprehensive evaluation. In addition, we found that, despite similar locations, the intensity of economic shrinkage of the forestry industry cities was weaker than that of cities generating mineral resources and energy resources. Of the five forestry industry cities (Arxan, Wudalianchi, Dunhua, Yichun, and Shulan) in the northeast, only Shulan experienced a decline (−0.11 in 2017) and then rebounded in 2018.

Understanding the trajectories by examining the underlying causal factors

The economic trajectories of RECs have been shaped by global, national, and regional forces as well as forces at the local level. In this sense, we define macro global economic development and labor division, China's economic and political context, and the evolution of resource-based industries as external factors. Correspondingly, local causal factors are regarded as internal factors, such as the normalization of resource exploitation, industrial structures, investment attractiveness, local government's capacity to respond, and regional economic conditions. This section starts by presenting temporal trends to provide detail on the macro background and explain the overall trend of China's RECs' economic growth trajectories. Following this, we analyze the influencing factors that lead to differences in RECs from an internal perspective.

External factors

1995–2001: the state-owned enterprise reform and the global economic crisis

In the 1990s, while postsocialist countries and regions were undergoing economic transformation, China actively explored the market economy system. First, in the context of the reform of state-owned enterprises to optimize China's economic structure, state-owned enterprises, which accounted for much of the economy of the RECs, were pressured by bankruptcy to laid-off workers. Since 1995, the central government has put forward an agenda to establish modern managerial systems for state-owned enterprises and has advocated for the merging of state-owned assets. In 1996, 675 state-owned enterprises went bankrupt, 1,022 merged, and 1.69 million became unemployed. As the localities concentrated with state-owned enterprises, the RECs were the main areas where the reform was implemented. Of the 111 pilot cities for "optimized capital structure" (including municipalities, provincial capitals, and prefecture-level cities) announced by the State Council in 1997, 12 prefecture-level cities were later identified as RECs. Furthermore, in the era of the planned economy, many RECs in China were based on resource mining quarries, as their economic backbone and urban foundation, giant state-owned mining companies were not only the main sources of local tax and employment but also key providers of public services and infrastructure (such as many affiliated communities, schools, and hospitals).

At the same time, the financial crisis that broke out in Southeast Asia in 1997 had gradually spread throughout the Asia-Pacific region, seriously impacting on China's export trade. In 1997, about 7,000 state-owned enterprises suffered losses, accounting for 39.1% of the total, and the losses suffered in 1998 were even more significant. As the main industrial production areas dominated by state-owned enterprises, RECs were severely affected, and the economy became sluggish. The average number of urban employees across the 34 RECs in 1998 decreased by 19.38% compared to 1995 (Figure 5.2, Section a). The average annual GDP growth rate of the 41 RECs in 1996–2002 was 7.83%, which is lower than the average national growth rate by 0.82%. The situation in the northeast was even more bleak, with economic growth shifting from leading to lagging behind the country. In the period 1997–2002, the average annual growth rate of GDP in the 13 RECs in the northeast region of China was lower than the national average.

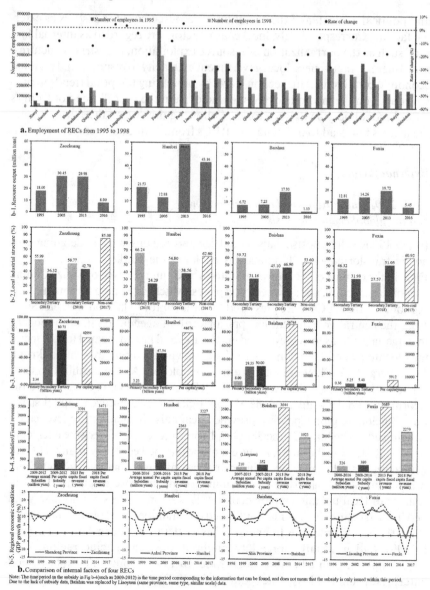

a. Employment of RECs from 1995 to 1998

b. Comparison of internal factors of four RECs

Note: The time period in the subsidy in Fig b-4(such as 2009-2012) is the time period corresponding to the information that can be found, and does not mean that the subsidy is only issued within this period.
Due to the lack of subsidy data, Baishan was replaced by Liaoyuan (same province, same type, similar scale) data.

Figure 5.2 Employment changes from 1995 to 1998 in the RECs and the mutual comparison among four representative cases.

From a long-term perspective, although the economic growth of China's RECs was relatively sluggish at this stage, the impact of domestic institutional restructuring and the international financial crisis was temporary. After the organizational reform of state-owned enterprises, most of the leading industries and enterprises in the RECs regained vitality and the ability to operate in global markets.

*2002–2012: the development of the market economy
and the prosperity of the resource industry*

Under the dual influence of the rapid development of China's market economy and the prosperity of the domestic resources market, the economy of
most RECs grew rapidly from 2002 to 2012. The average annual growth rate
across the 41 RECs was 12.8%, which was 2.34 percentage points higher
than the national growth rate. In addition, China's entry into the WTO in
December 2001 further released the potential of the market for both China
and the globe. With increasing foreign investment and demand, China's
industrial output value grew rapidly. Benefiting from a demographic dividend and well-established industrial foundation, China's position as a
"world factory" gradually emerged. The huge demand also led to the prosperity of the resource industry. For example, the period from 2002 to 2012
was hailed as the "Golden Decade of Coal", for domestic coal production,
sales, and prices had risen continually. In addition, according to data from
1995 to 2012, most of the RECs steadily maintained high resource output.
Therefore, owing to the advantages of industrial raw materials, energy supply, and heavy manufacturing, most RECs maintained a high development
rate from 2002 to 2012.

After 2007, particularly after the financial crisis of 2008, the slowing of
the international and domestic markets and the resource exhaustion of
the first RECs exposed social and economic development dilemmas, the
government became aware of the risks caused by cities' over-reliance on
resource-based industries. The State Council designated a list of RECs
and proposed supportive measures including financial transfer payments.
In addition, to deepen the reform of state-owned enterprises, the central
government mandated that "enterprises should divest their government and
social functions" so that schools, hospitals, and communities which once
belonged to enterprises were gradually transferred to local governments.
These policies fundamentally altered the financial structure and even the
social structure of the RECs. However, as most RECs had not yet entered
a period of complete resource depletion, many cities continued to aggres-
sively exploit resources while looking for transformation strategies. During
this period, the role of national transformation policies and the capacity of
local REC governments to boost the economy have yet to be fully realized.

*2013–2018: the "New Normal" economy
and the environmental protection policy*

From 2013 to 2018, the average annual growth rate of the 41 RECs was
2.16%, which was lower than China's average GDP growth rate (6.56%).
This implied that economic decline was underway. Against the backdrop
of a shrinking domestic market and the low-priced imported coal and
iron ore competition, the production, sales, and prices of major domestic
resource products had all declined since 2013. Coupled with the fact that
the cost of resources in RECs tends to be higher, the competitiveness of

these industrial firms as well as the RECs themselves were weakened. Furthermore, since 2013, with the supply-side structural reform and the goal of "ecological civilisation" (a policy focus on sustainable environment), the central government initiated a series of actions to implement higher air pollution prevention standards, to reduce excess capacity, and to completely end natural forest logging. In 2013, the State Council proposed reducing the concentration of inhalable particulate matter in cities by more than 10% by 2017 and to resolve the severe overcapacity contradiction, especially in high-consumption and high-emission industries such as steel, cement, and electrolytic aluminum. As a response, many severely polluted RECs began to shut down open-pit mines and disestablish companies. In addition, for the forestry industry RECs in the northeast, it was also required that the natural forest farms in the Xing'an Mountains must reach a no-logging goal by 2014, which also limited their industrial development.

In summary, after 2013, changes in China's political and economic contexts posed new challenges to the RECs development model and their capabilities to respond. Coping with these problems became the key tasks shaping the economic growth trajectories of these RECs.

Internal factors

Current literature shows that resource reduction, single industrial structure, and poor urban environmental quality all have a certain effect on the shrinkage of RECs. However, whether individualized or regionalized conditions of RECs play a role in the economic shrinkage process is still a question worthy of discussion. This is particularly the case in China, where top-down administrative power is a key force in the development of cities/towns at the lower tier. To minimize the impact of differences in city-level, resource-based industry, and population size, this section uses available data on four prefecture-level coal-based cities (Zaozhuang, Huaibei, Baishan, and Fuxin) with population sizes within the range of 1 million and 4 million,[1] to further explain the four types of economic development trend that occurred after 2013 (Table 5.2). Following this is a comparative study examining how the differences in the internal causal factors have produced such diverse results.

Table 5.2 Composite economic growth index of case cities in 2013–2018

City/year	2013	2014	2015	2016	2017	2018	Types
Zaozhuang	0.0400	0.0375	0.0358	0.0311	0.0085	0.0228	Growth
Huaibei	0.0213	0.0338	0.0359	0.0052	0.0046	0.0053	Slowdown
Baishan	0.0108	0.0259	0.0260	0.0239	**−0.0541**	**−0.0116**	Shrink
Fuxin	0.0451	**−0.0107**	**−0.0860**	**−0.0604**	**−0.0002**	0.0127	Shrink-rebound

Note:
Bold font indicates the year of shrinkage.

Resource exploitation status quo

Although many cities were designated as RECs in 2007, most of them did not enter their "exhausted" phase until 2013 (Figure 5.2, Section b-1). After 2013, the output value of other resource-related industries differed, owing to different local contexts, which generated fundamentally different economic growth trajectories. In Zaozhuang, the output of raw coal decreased from nearly 30 to 8 million tons from 2011 to 2016, but the output of related industries such as the coking, thermal power, and coal chemical industries had not only risen rapidly but had also become the main sectors for investment in recent years. In Huaibei,[2] the production of coal remained at 40 million tons in 2016, and its reserves could have been exploited for more than 15 years or more, indicating that Huaibei did not fully enter the period of depletion. However, production in Baishan and Fuxin had dropped significantly, from 17.93 and 16.94 million tons in 2013 to 1.14 and 5.45 million tons in 2016, respectively, leading to a serious decline of industrial output value, urban fiscal revenue, and GDP. It is evident that the predicament in the first two cities was not particularly pressing, while in the latter two, it has been looming since 2013. The resource exploitation status quo, which is closely related to the exact resource extraction cycle and the position of the relevant industries in the value chain, is still an important factor that affects a city's economic trajectory.

Local industrial structure

In order to explain how a more balanced "industrial structure" led to the diversification of the economic trajectory in each of the four RECs, we focused on the percentage make-up of secondary and the tertiary industry from 2013 to 2018 and specifically adopted 2017 as an example to analyze the proportion of non-coal industries (Figure 5.2, Section b-2). In Zaozhuang, the proportion of tertiary industry changed from 36.32% to 42.70%, the output value of the non-coal industry accounted for 85.0%, and the high-tech industry increased to 24.92%. In Huaibei, the tertiary industry changed from 24.29% to 38.56%, and the non-coal industry accounted for 62.0%, with the high-tech industries already accounting for 21.10%. In Baishan, the tertiary industry grew from making up 31.16% of the city's industry to 46.90%, and its non-coal industry accounted for 53.60%. However, in addition to the rapid growth of the tertiary industry (21.40 to 28.9 billion), the shrinkage of the secondary industry (40.05 to 32.09 billion) was also an important cause of this structural change. Yet, relying on the tourism resources of the Changbai Mountain Natural Scenic Area, Baishan's domestic tourism revenue increased from 1.25 billion (in 2013) to 18.90 billion (in 2018) yuan, which to some extent helped to fill the revenue gap caused by the shrinkage of the coal-based industries. In Fuxin, the tertiary industry changed from 31.98% to 51.05%. However, we found that the increase in the real absolute output of the tertiary industry was relatively low (from 19.67 to 22.76 billion), while the annual value added of the secondary industry shrank by more than 50% (from 28.49 to 12.30 billion). Therefore, the apparent increase in the proportion of industry made up by the tertiary sector was not actually a sign of the development of the modern service industry in Fuxin. In reality, it was a manifestation of its development dilemma. As a result of the continuous decline in the output value of Fuxin's coal industry, the proportion of non-coal industries rose to 60.92%.

Investment scale and structure

From 2013 to 2018, Zaozhuang, Huaibei, Baishan, and Fuxin's average annual growth in fixed-asset investment was 7.1%, 10.6%, 7.0%, and −18.3%, respectively. Taking 2017 as a typical year (Figure 5.2, Section b-3), Zaozhuang had attracted an investment of 179.75 billion yuan, 42,997 yuan per capita. In the primary, secondary, and tertiary industries, the investment amount was 2.34, 96.70, and 80.71 billion yuan, respectively, and investment in the modern service industry increased by 69.4%. In Huaibei, the investment was 105.58 billion yuan, 48,676 yuan per capita. The investment in the three industries was 3.23, 54.81, and 47.54 billion yuan. Baishan boasted an investment of 67.81 billion yuan, and the per capita investment was 56,744 yuan; only 29.35 and around 30 billion yuan were spared for the secondary and tertiary industries, respectively. In Fuxin, the investment reached 11.01 billion yuan total, increased by 23.7%, and resulting in 5,912 yuan per capita. For the three industries, the amount was 0.36, 5.25, and 5.40 billion yuan, and changed by 80.6%, −28.6%, and 1.7%, respectively, indicating the shrinkage of the secondary industry and the overexpansion of the primary and tertiary industries. It was also evident that the total investment and per capita investment in Fuxin were significantly less than the other three cities.

Local government's capacity to acquire fiscal revenue

In theory, the larger fiscal budget the local government has, the greater capacity it has to support socioeconomic transformation, such as improving infrastructure, offering preferential policies and subsidies for new industries, and initiating training programs for the unemployed. Since the annual fiscal transfer, payment comes from both the central and provincial governments, and the specific schemes are mediated through the provincial governments and are therefore subject to provincial and local political bargaining, the subsidy received by each REC fluctuates every year. This complex system of allocation and negotiation is a topic beyond the scope of this paper. Based on available data (Figure 5.2, Section b-4), the average annual subsidies for the four cities were 676, 482, 210, and 326 million yuan, or 590, 610, 352, and 380 yuan per capita. Among them, Zaozhuang and Huaibei obtained additional matching subsidies from the provincial government. The changes in per capita financial income of the four cities between 2013 and 2018 were

3301–3471, 2363–3227, 3644–1923, and 3689–2270, respectively; highlighting that the situation in Zaozhuang and Huaibei were better than those in Baishan and Fuxin.

Furthermore, according to the "2018 Central Government's Transfer Payment Mode for Local Resource-Exhausted Cities" issued by the Ministry of Finance, Shandong Province (where Zaozhuang is located) received a good transition assessment, but Liaoning Province (where Fuxin is located) had a poor transition assessment. Therefore, in general, Zaozhuang would have the most favorable support, while Baishan and Fuxin were disadvantaged, and Huaibei sat between the two extremes.

Regional economic conditions

Theoretically, the better the economic situation of a region, the less likely a city within this region would experience shrinkage, particularly if this region refers to the province where this city is located. The RECs located in a more prosperous economic hinterland, compared with others in worse regions, not only have more potential markets but also more possibilities to obtain infrastructural investment, new industrial transfers, and even more fiscal support. After 2013, the economic growth of Jilin Province and Liaoning Province was somewhat weaker than that of Shandong Province and Anhui Province (Figure 5.2, Section b-5). As a result, Zaozhuang (Shandong) and Huaibei (Anhui) demonstrated better economic performance than Baishan (Jilin) and Fuxin (Liaoning). Additionally, the more concentrated the RECs within a region, the higher the risk of shrinkage. RECs that are closely interconnected with neighboring areas are likely to rely on the same veins, oil fields, or forests to develop and are therefore more likely to shrink at the same time and experience fierce homogeneous competition. Due to the popularity of the self-centered development model adopted by the RECs, the higher the number of RECs in the same province, the greater the likelihood of a fragmented regional market, investment, emerging industries, and fiscal subsidies. Consequently, for each city, the narrowed path to successful transformation is exemplified by the concentration of RECs with worsened economic performance after 2013 in the northeast region of China, where Baishan and Fuxin are located.

The overall look overview of the causal factors

The above analysis demonstrates that the economic trajectories of the RECs are the result of the combination of a range of external and internal factors (Table 5.3). Three main external factors dominated the period: (1) Domestic state-enterprise reform and resource market fluctuations from 1995 to 2001; (2) the flourishing of the market economy and the revitalization of resource-based industries from 2001 to 2013; and (3) the "New Normal" context and the environmental protection policy to tackle environmental problems as well as a new wave of global labor and capital division.

Table 5.3 Dominant factors in the four case cities

Typical cities	Dominant factors in 1995–2012	Dominant factors in 2013–2018
Steady growth: Zaozhuang	China's actions to push forward the state-owned enterprise reform; China's macroeconomic and policy changes; Global economic crisis and global industrial restructuring.	Declining resources but success in resources-related industries; Developed tertiary industry and non-coal industry; Acceptable investment; Good financial support and transformation evaluation; Good regional economic conditions.
Growth-slowdown: Huaibei		Rich resources and resources-related industries; Strong secondary industry and rising tertiary industry; Acceptable investment; Good financial support; General regional economic conditions.
Growth-shrink: Baishan		Exhausted resources; Good tertiary industry; Potential investment; Less financial support; Weaker regional economic conditions.
Growth-shrink-rebound: Fuxin		Exhausted resources; Declining secondary industry; Poor investment; Less financial support transformation evaluation; Weaker regional economic conditions.

The internal factors include the normalization of resource exploitation, local industrial structures, investment scale and structure, the local government's capacity to acquire fiscal revenue, and regional economic conditions. Providing that the abovementioned five factors continue to contribute to the local economy positively, the RECs can maintain growth. However, when conditions with respect to one or two of these factors worsen, the economy of these RECs may fluctuate and shrink, and when more than two dimensions deteriorate, the RECs will suffer heavy shrinkage.

Conclusions

Based on the data from 1995 to 2018 of the 41 RECs in China, this paper conducts a study of the economic growth trajectories of these RECs within the framework of urban shrinkage. More specifically, we draw the following three main conclusions:

1 In the past 24 years, the economic growth trajectory of the 41 Chinese RECs followed a trend that begins with fair growth and ends with a slowdown. The frequency and magnitude of shrinkage after 2013 were significantly greater than those in other periods. From the perspective of spatial distribution, most of the RECs designated as shrinking cities by the State Council of China were in the northern region of China, and RECs in the Southeast Coastal area seldom experienced serious economic shrinkage. The economic growth trajectories of China's RECs can be classified into four types: cities that experienced steady growth, maintaining a relatively high growth rate; cities that experienced slowed down growth since 2013; and cities that experienced growth-shrink and those that experienced growth-shrink-rebound, both of which suffered shrinkage after 2013;

2 The economic trajectories of RECs are not only affected by the city's own process of exploiting resources and developing industries but also by macroeconomic policies and market conditions. The reform of traditional state-owned enterprises had indeed caused unemployment and even temporary economic decline in the cities where the enterprises were located, yet the more market-oriented managerial systems in these cities served to restore enterprises and provide opportunities for the city to take advantage of its own location and resources. However, whether their actions can generate the expected outcome is still dependent on the dynamic interactions between the external and internal factors;

3 Although all 41 cities are designated as RECs, the dilemmas they face are very different, and this reality is embodied in these RECs' various development status quos, such as their industrial foundations, their available support from fiscal transfers, and their regional economic conditions. Together, these factors could lead to differences in the adaptive capacity of these RECs. In this sense, supportive policies as well as local government actions need to be customized to local conditions pertaining to the five factors discussed in this chapter.

Based on the above analysis of the impact of the external macro political economy and urban internal conditions on the economic development of RECs within the framework of China's socialist market economy system, we posit that policies to deal with urban shrinkage risks should be multi-layered and forward-looking, requiring a comprehensive combination of efforts in industries, investment, and subsidies. As more resource-based cities gradually enter the final stage of resource exploitation, it is imperative and urgent to transform in advance and adapt to shrinkage in a smart way.

Acknowledgment

The authors of this work acknowledge all parties involved in the data acquisition and those scholars and colleagues who have given valuable comments. Our special thanks for funding support from the following institutions:

(1) The Smart Planning Support System Research Program (Grant No. D1218006); and (2) Technology Innovation Foundation of Hubei Province (Grant No. 2017ADC073).

Notes

1 All are at the "big city" level in the official city standard of China.
2 It is worth noting that some articles claim that Huaibei's coal reserve is still relatively high. We suppose that this may result from a new estimation based on the deep exploitation that occurred in recent years after it was rated as a resource-exhausted city.

References

Buček, J. 2016. Urban Development Policy Challenges in East-central Europe: Governance, City Regions and Financialisation. *Quaestiones Geographicae*, 35(2), 7–26.

Constantinescu, I. P. 2012. Shrinking Cities in Romania: Former Mining Cities in Valea Jiului. *Built Environment*, 38(2), 214–228.

Crowley, S. 2016. Monotowns and the Political Economy of Industrial Restructuring in Russia. *Post-Soviet Affairs*, 32(5), 397–422.

Fol, S. & Cunningham-Sabot, E. 2010. Urban Decline and Shrinking Cities: A Critical Assessment of Approaches to Urban Shrinkage. Annales de géographie, 359–383.

Friedrichs, J. 1993. A Theory of Urban Decline: Economy, Demography and Political Elites. *Urban Studies*, 30(6), 907–917.

Haggerty, J. H. et al. 2018. Planning for the Local Impacts of Coal Facility Closure: Emerging Strategies in the US West. *Resources Policy*, 57, 69–80.

Häußermann, H. & Siebel, W. 1988. *Die schrumpfende Stadt und die Stadtsoziologie, Soziologische stadtforschung* (pp. 78–94). Springer.

He, S. Y. et al. 2017. Shrinking Cities and Resource-based Economy: The Economic Restructuring in China's Mining Cities. *Cities*, 60, 75–83.

Hu, C. S. & Mo, X. R. 2016. Structural Decomposition of Economic Convergence in China's Resources Based Cities. *Resources Science*, 38(12), 2338–2347. //胡春生, 莫秀蓉, 2016. 中国资源型城市经济收敛的结构分解. 资源科学, 38(12), 2338–2347.

Kahn, M. E. 1999. The Silver Lining of Rust Belt Manufacturing Decline. *Journal of Urban Economics*, 46(3), 360–376.

Knapp, W. 1998. The Rhine-Ruhr Area in Transformation: Towards a European Metropolitan Region? *European Planning Studies*, 6(4), 379–393.

Leadbeater, D. 2009. Single-industry Resource Communities, "Shrinking" and the New Crisis of Hinterland Economic Development. In Pallagst, K. et al. (eds.), *The Future of Shrinking Cities: Problems, Patterns and Strategies of Urban Transformation in a Global Context* (pp. 89–100). Institute of Urban and Regional Development, and the Shrinking Cities International Research Network (SCiRN), IURD, Berkeley: University of California.

Li, H., Lo, K. & Wang, M. 2015. Economic Transformation of Mining Cities in Transition Economies: Lessons from Daqing, Northeast China. *International Development Planning Review*, 37(3), 311–328.

Li, H., Lo, K. & Zhang, P. 2020. Population Shrinkage in Resource-dependent Cities in China: Processes, Patterns and Drivers. *Chinese Geographical Science*, 30(1), 1–15.

Li, G., Zhang, S. Q. & Xu, B. 2018. Measurement and Mechanism Analysis of Urban Shrink Level in Huaibei City. *Journal of Huaibei Normal University (Philosophy and Social Sciences)*, 39(6), 47–53. //李刚, 张淑清, 徐波, 2018. 淮北市城市收缩水平测度及机制分析. 淮北师范大学学报·哲学社会科学版, 39(6), 47–53.

Liu, G. W. et al. 2019. Urban Shrinkage in China Based on the Data of Population and Economy. *Economic Geography*, 39(7), 52–59. //刘贵文, 等, 2019. 基于人口经济数据分析我国城市收缩现状. 经济地理, 39(7), 52–59.

Manville, M. & Kuhlmann, D. 2018. The Social and Fiscal Consequences of Urban Decline: Evidence from Large American Cities, 1980–2010. *Urban Affairs Review*, 54(3), 451–489.

Martinez-Fernandez, C. et al. 2012a. Shrinking Cities: Urban Challenges of Globalization. *International Journal of Urban and Regional Research*, 36(2), 213–225.

Martinez-Fernandez, C. et al. 2012b. The Shrinking Mining City: Urban Dynamics and Contested Territory. *International Journal of Urban and Regional Research*, 36(2), 245–260.

Oswalt, P. 2005. *Shrinking Cities, Volume 1: International Research*. Ostfildern-Ruit: Hatje Cantz.

Pallagst, K., Wiechmann, T. & Martinez-Fernandez, C. 2013. *Shrinking Cities: International Perspectives and Policy Implications*. Routledge.

Stryjakiewicz, T., Ciesiółka, P. & Jaroszewska, E. 2012. Urban Shrinkage and the Post-socialist Transformation: The Case of Poland. *Built Environment*, 38(2), 196–213.

Wang, C. J., Qu, Y. Y. & Wu, X.L. 2019. A Study on the Economic-population Shrinking Governance of Resource-exhausted Cities: A Realistic Analysis Based on Resource-exhausted Cities in Heilongjiang Province. *Macroeconomics*, 8, 156–169. //王常君, 曲阳阳, 吴相利, 2019. 资源枯竭型城市的经济—人口收缩治理研究—基于黑龙江省资源枯竭型城市的现实分析. 宏观经济研究, 8, 156–169.

Wang, L. W. & Feng, C. C. 2013. Evolutionary Mechanism and Development Strategy of Resource-based City: Taking Gansu Baiyin as an Example. *Areal Research and Development*, 32(5), 65–70. //王利伟, 冯长春, 2013. 资源型城市发展演化路径及转型调控机制—以甘肃省白银市为例. 地域研究与开发, 32(5), 65–70.

Wirth, P., Mali, B. Č. & Fischer, W. 2012. *Post-Mining Regions in Central Europe*. Digital Print Group: Nurnberg: Germany.

Wolff, M. et al. 2013. Shrinking Cities, Villes en Décroissance: Une Mesure du Phénomène en France. *Cybergeo: European Journal of Geography*, 1–33.

Yang, X. J., Xiao, N. & Zhao, B. Y. 2019. County Level Spatial Planning Strategies and Practice of Resource Based Cities in the Context of Contraction: Lueyang County, Shaanxi Province. *Planners*, 35(16), 82–88. //杨晓娟, 肖宁, 赵柏伊, 2019. 收缩语境下资源型城市县域空间规划策略与实践—以陕西省略阳县为例. 规划师, 35(16), 82–88.

Zhang, W. et al. 2019. Multi-dimensional Identification and Driving Mechanism Analysis of Shrinking Cities in China. *Urban Development Studies*, 26(3), 32–40. //张伟, 等, 2019. 我国城市收缩的多维度识别及其驱动机制分析. 城市发展研究, 26(3), 32–40.

Zhang, M. D. & Qu, J. X. 2020. A Study on the Spatial Pattern and Generating Logic of Generalized Urban Shrinkage in China: Based on the Perspectives of Total Population and Economic Scale. *Economist*, 1(1), 77–85. //张明斗, 曲峻熙, 2020. 中国广义城市收缩的空间格局与生成逻辑研究—基于人口总量和经济规模的视角. 经济学家, 1(1), 77–85.

Zheng, T. 2014. Measuring Urban Form of Resource-based Cities in China: Do They Sprawl or Not. *China Population, Resources and Environment*, 24(S3), 403–406. //郑童, 2014. 中国资源型城市的城市形态测度: 是否存在城市蔓延. 中国人口·资源与环境, 24(S3), 403–406.

6 Two sides of the same coin

City growth and shrinkage in rapidly urbanizing southern China

Kai Zhou, Zhiwei Du, and Yangui Dai

Introduction

The growth and shrinkage (or decline) of the urban population are "two sides of the same coin" (Sabot and Roth, 2013; Dubeaux and Sabot, 2018) when assessing global urbanization through a wider geographic scope or over a longer historical period. Audirac and Alejandre (2010) declared that urban growth and decline are the "yin and yang" of the cities, which are intrinsically related to the intra/inter-city and international migratory flows between regions. Through numerous case studies, the dynamic relationships between growing and shrinking cities were explained by (1) the "parasitic" nature of population gain and loss in cities among which households, businesses, and capital switch places in response to economic and political incentives (Beauregard, 2009); or (2) the "cyclical" process that follows the "boom and bust" of urban economics, which distinguishes winners and losers in global competition (Hall and Hay, 1980; Hartt, 2018). Therefore, Laursen (2016) claimed that it is impossible to examine either topic without also considering the other because a decline might be considered an aspect of growth, and vice versa. However, investigating the causal links between the "parallel patterns of shrinking city and urban growth" (Ganser and Piro, 2016) is difficult because of urban changes that usually last decades and because data for long-term or cross-regional (sometimes cross-continental) comparisons are either unavailable or difficult to collect.

China's urbanization in the last 40 years is unique because of its size, scale, speed, and pace. Since the opening up in 1979, China has had upsurges in its coastal regions and major metropolitan areas and continuous degradations of some inland urban settlements. In a condensed spatial, temporal context, this chapter aims to examine two case cities in south-central and southern coastal China to assess how migration flows, resource allocations, market changes, and policy orientations formulate the population growth and decline in cities or towns of rapidly urbanizing China. This chapter begins with (in section 2) a literature review of three theoretical models for conceptualizing urban growth and decline. Section 3 presents two distinctive case studies: each shows a unique trajectory in demographic change.

DOI: 10.4324/9780367815011-8

We used data released by the Bureau of Statistics of the national and local governments to analyze the demographic changes, and we collected master plans, government work reports,[1] and other documents to study the relevant policies of local authorities. Next, in the discussion and conclusion, we interpret the findings to conceptualize urban growth and decline in China.

Theoretical models of urban growing and shrinking

City growth and shrinkage have a long theoretical, intellectual lineage embedded in the economic, social, political, and demographic evolution of urban studies, in which there are three conceptual models in the relevant literature.

Life-cycle model

The life-cycle model, an idea that originated from Vernon's (1966) model of a product life cycle, explains the consequence of urban growth and shrinkage with the gains and losses of the population within the areas of its city core, hinterland, and city region (Wolff and Wiechmann, 2018). The four-staged cycle model produced by Van den Berg et al. (1982) has been widely used to describe the status of urban systems, in which the stages of "urbanization" and "reurbanization" usually cause population growth, and the processes of "suburbanization" and "disurbanization" produce various types of shrinkage. This model was further associated with "the long economic cycle" (45–60 years), arising from technological revolutions (Kondratieff and Stolper, 1935), and the "the medium economic wave" (15–20 years) defined by infrastructural investments (Kuznets, 1930). In either theory, growth and shrinkage are cyclical phases within the fluctuation of the population. Policy responses during the shrinking period are proposed to maintain the status quo, economically, socially, or physically (Bartholomae et al., 2016; Hattori et al., 2017; Kim, 2019); prepare for the next round of growth if tangible (Rérat, 2019); or find temporary or interim uses of idle spaces and properties (Dubeaux and Sabot, 2018).

Heuristic model

Haase et al. (2014) first conceptualized urban shrinkage by using an integrative model to map the causes and impacts from different contexts and agencies. The urban shrinkage of local development in the heuristic model is a result of global or regional forces, which trigger depopulation that could be further deteriorated by the feedback loops of its direct or indirect consequences. This model also includes the role of "governance" in response to the problems of depopulation, hinting that external assistance is necessary for regaining growth. Bernt et al. (2014) applied the classic "policy window" theory by Kingdon (1984) to explain how responses in urban governance could be formulated. He argued that the acceptance and recognition (the stream of a problem), the formulation of policy alternatives (the stream of policy), and the partnership of actors and interests (the stream of politics) determine how the issue is successfully added the agenda of authorities; based on that, Haase et al. (2017) further articulated that the circular relationship between "condition", "discourse", and "policy & action" is framing the implementation of policies that counteract challenges posed by urban shrinkage. The heuristic model treats urban shrinkage and (re)growth as competing processes and is designated to provide "diagnoses" of population decline and "prescriptions" for economic or demographic recovery.

Politico-economic model

Through the lens of politico-economic analysis, uneven development is an outcome of the structural conflict resulting from the crisis of capital (Smith, 2008; Harvey, 1973). The dual developing processes of capitalism and urbanization reinforce each other by (1) using the investment of "footloose capital" for producing profitable urban spaces; and (2) advertising consumerism as an urban lifestyle to create thriving markets for material goods or services. Harvey (2001) argued that geographical expansion (the process of "spatial fix") is a promising means to resolve the overaccumulation problem of capital. Additionally, technological innovations and globalization further extended the speed and scale of the circuit of capital. In this model, the investors' decisions to find new spaces that fulfill the requirements of a new mode of production affect the boom and bust of cities. The new investments or reinvestments of capital create urban growth, and disinvestment produces shrinking cities.

Each aforementioned model provides a unique perspective on understanding, interpreting, and conceptualizing the mechanism of urban growth and shrinkage. Whether those Western experiences are applicable to other contexts, such as developing southern China, remains unknown.

Case studies

In this section, we investigate two case cities from the south-central and southern coastal regions of China (Figure 6.1). Some researchers call the area "the Southern China Interprovincial Migration Field" because of its strong immigrating and emigrating ties between a rapidly developing coastal province (Guangdong) and less-developed adjacent inland provinces (Hunan, Jiangxi, Guizhou, Guangxi, and Fujian) (Liu et al., 2012). According to censuses, approximately 1.2 million "floating population" (mostly migrant workers) migrated into Guangdong from nearby provinces between 2000 and 2005.

The two case studies in this chapter depict city growth and shrinkage in southern China. The first case is Taoyuan, located in the outlying regions of

Figure 6.1 Location of the case studies.

the migration field, was one of the six less-urbanized counties that were losing its population and had a low economic growth rate between 2000 and 2010 in Hunan Province. Similar to other marginal counties in the migration field, it has been contributing labor and resources to the core region (Guangdong) in exchange for modest economic returns. The second case is Dongguan in Guangdong Province, which has an advantageous position, as aforementioned. Using the abundant labor supplies from the inland provinces and the investments of global manufacturing sectors, Dongguan became one of the most rapidly industrializing and urbanizing cities in China.

Taoyuan

Taoyuan is an agriculture-based county in northwest Hunan Province. Despite having the largest portion of arable lands among all counties in Hunan, its economic strength and urbanization rate have long been below average (Table 6.1). At the county level (4,400 km² including the city and adjacent rural areas), the census data show that a registered population of approximately 134,000 were working and living outside of the county in 2010. Thus, in the inland provinces, this region is a typical labor-outputting region.

In this case study, we record and explore the process and policy responses of urban/rural shrinkage and the regrowth of Taoyuan, which first experienced the depopulation caused by labor output and then experienced in situ urbanization led by the backflow of migrant workers.

Table 6.1 Economic status of Taoyuan County in Hunan Province

Year	Total population* thousands	Status**	Urbanization rate %	Status	GDP RMB billion	Status	GDP growth rate Preceding year=100	Status	GDP per capita Yuan	Status
2015	861.3	18/124	35.45	106/124	27.51	36/124	109.6	47/124	31968	66/124
2010	853.7	18/124	27.53	101/124	11.95	32/124	114.3	82/124	17401	69/124
2005	989.5	14/123	25.95	76/123	7.13	23/115	108	113/115	8407	50/115

Note

* refers to the number of the population that has resided for longer than six months; ** status shows the rank of Taoyuan County (high to low)/the number of counties and district-level administrations in Hunan Province.

Shrinkage caused by the outflow of the labor force

According to the fifth and sixth censuses (Table 6.2), the registered population of Taoyuan increased 2.5% between 2000 and 2010 because of its high birth rates. However, the registered population residing in Taoyuan decreased by 10.2% because of the outflow of labor. Balanced by minor immigration, the population that had resided for longer than six months

Table 6.2 Taoyuan's population changes between 2000 and 2010

	Year 2000	Year 2010	Change (%)
Registered population	963,400	987,344	2.5%
Population residing for longer than six months	932,771	853,662	−8.5%
Population registered and residing locally	891,164	800,391	−10.2%
Population registered locally but residing elsewhere	72,255	187,003	158.8%
Labor population (aged 15–64)	679,824	624,214	−8.2%

Regions	Population residing for longer than six months			Population registered and in locally		
	Year 2000	Year 2010	Change (%)	Year 2000	Year 2010	Change (%)
Taoyuan County/ Region	932,771	853,662	−8.5%	891,164	800,391	−10.2%
Zhangjiang/ Core City	112,497	121,377	7.9%	99,425	96,920	−2.5%
Rural Areas/ Hinterland	53,587	51,155	−4.5%	50,680	47,839	−5.6%

Source: From the fifth and sixth national censuses.

shrank 8.5% at the county level. The working-age population (aged 15–64) shrank 8.2%. During the same time, the census data shows a 158.8% increase in the population registered locally but living elsewhere. Taoyuan's number of net outmigrants, mostly in the labor force, increased to 133,732 in 2010, compared with 30,648 in 2000.

Regarding the population by region, the population of Taoyuan's core city (Zhangjiang) increased 7.9% between 2000 and 2010. In Zhangjiang, the registered population decreased during its growth period (by 2.5%); thus, the core city grew only because its depopulation (in term of the registered population) was successfully compensated by the inflow of population from other places, and in Taoyuan's case, mainly from its rural areas. All rural areas of Taoyuan shrank from 2000 to 2010 (by 4.5% on average; Table 6.2). In the worst case, the population of Guniushan township decreased by 27.22%.

Flowing out to where and why

Taoyuan was lagging compared with the rest of Hunan Province in urbanization and industrialization, which became the main driver that pushed individuals to seek better job opportunities elsewhere. A villager from Niuchehe township said, "One person gets out, one family leaves poverty; two persons get out, an affluent household is expected".

To ameliorate poverty, from 2000 to 2010, Taoyuan's local government implemented extensive policy measures to support, manage, and organize the outflow of surplus labor to the prosperous coastal regions. Those policies were designed to reshape the existing outflow (1) from a "spontaneous" flow-out to an "organized", "managed" one; (2) from "blind flows" to "government-led" and "more safe and secured" labor output; (3) from "small groups" to "large-scaled programs"; (4) from "seasonal" to "perennial" ones; (5) from "outputting labor" to "outputting technicians"; and (5) from "simply job-seeking" to "skills learning" and "wealth accumulating" (TDRC, 2005).

Before 2009, the local policies that supported the outflow of surplus labor were in an employment-promoting program named "Trinity: Tao Yuan's Urban and Rural Employment" (桃源 "三位一体" 统筹城乡就业) (TDRC, 2010). The policies were implemented to prepare the unemployed younger generation with necessary skills, deliver job information to residents of remote areas, and provide support for legal issues such as contract signing, payment delays, or compensation and benefits. The program covers the rural population and laid-off employees in urban areas.

Based on the policies, the plan of the county government was to encourage residents to migrate out of Taoyuan. The public authorities also provided subsidized job training to raise the average income of households in poverty. Four professional schools (all private owned) and 15 training centers for migrant workers were established with the support of the government. These facilities accommodated up to 1,000 workers per year and provided training for 13 job types, for example, "electrical & electronics", "CNC machine operators", "motor mechanics", "cooks", and "domestic workers". According to the local government's report, seven types of labor output emerged: (1) cobblers in southern China (Pearl River Delta, PRD); (2) domestic-service workers in the big cities of southern China (e.g., Shenzhen, Guangzhou, Zhuhai); (3) hotel attendants for nationwide hotel chains; (4) construction workers for some major, national, mega-infrastructural projects; (5) security personnel working for property management companies in southern China; (6) welders in Fujian, Shandong, and Guangdong provinces; and (7) textile workers in Guangdong and Zhejiang provinces (TDRC, 2010). Dongguan, the second case study city in this chapter, has the largest manufactural agglomeration in Guangdong Province, is geographically close to Guangzhou and Shenzhen, and benefited from the inflow of properly skilled migrant workers.

The government, at that time, was aiming to develop a nationwide reputation for its high-quality workforces, with the expectation that those workers would eventually return with their savings and technical expertise to develop their hometowns. On the one hand, the policies were successful in terms of raising household income. As reported by an official survey, Taoyuan output over 200,000 person-times of labor between April 2010 and April 2011, raising a total income of RMB 1.5 billion, which accounted for 60% of the total income of the rural population in Taoyuan County. On the other hand, the outflow of the working-age population caused a shrinkage of the local labor supply. The five-year cumulative amount of labor outflow reached 1.1 million person-times, which accounted for 38% of the total rural working-age population. In many small towns and villages, the young generation (mainly male) left for work, and elderly individuals, children, and women remained at home. The change of demographical structure in rural areas significantly reduced the productivity of local agriculture sectors and the vitality of rural society. Visiting the countryside, we observed large, barren tracks of arable land and many rural houses that remained empty most of the year.

Flowing-back and in situ urbanization

The term "flowing-back of migrant workers" (返乡农民工) first appeared in Taoyuan County's Government Work Report in 2009. According to an analysis published by the Bureau of Statistics of Taoyuan, approximately 20,000 migrant workers (9.6% of the total), who mostly worked in manufacturing sectors, returned home between September and November of 2009, after the 2008 Financial Crisis caused many factories in coastal regions to close. Since then, how to provide assistance for the re-employment of those migrant workers has become part of the policy agenda of local authorities. The government has also realized that the outflow of labor will not support the long-term development of the local economy. First, what is now clear is that the second generation of outmigrants usually maintains fewer social ties with their distant relatives and weaker emotional connections with their hometowns. Second, the endless outmigration drains the local labor pool, harming the development of the core city (Zhangjiang) of Taoyuan County.

In this context, the slogan of "promoting in situ urbanization"[2] replaced "encouraging labor-output", as the guiding principle of policy formulation in Taoyuan. The local government of Taoyuan has thus changed its course to mostly focus on two aspects: (1) improve the attractiveness of the core city; and (2) increase the capacity of local employment.

Few migrant workers remained in rural areas and resumed their lives as farmers after returning home. Compared to where they had been living and working, at home in the countryside, the job and entertainment opportunities were insufficient. Many of them leased their lands to the local cooperation or other farmers and moved to the core city. Thus, the core city Zhangjiang had to provide adequate housing, public services, and infrastructure for the newcomers, which eventually rebooted the in situ urbanization of Taoyuan. The real challenge when attempting to seize this opportunity for the local economy is to increase the capacity of employment.

One way to absorb the backflow of migrant workers is to develop local manufacturing sectors. After 2009, the aforementioned employment-promoting policies of Taoyuan County were repackaged into the project called "Aspiring Entrepreneurs for Better Employment" (创业带动就业工程). The switch shows that the local government's focus has been redirected from "finding employments elsewhere" to "creating jobs locally", leading to the following actions: (1) offering assistants to unemployed families; (2) building an incubator for small businesses; and (3) providing personal micro-finance loads. The county reported that more than 200 small businesses and 3,000 jobs were created in 2016 (Taoyuan County's Government Work Report 2016). In 2017, a program named "Return Geese Project" (回雁工程) was introduced to encourage Taoyuan-born entrepreneurs and the 230,000 migrant workers to bring investments back home, by "attracting them with nostalgia" (以浓浓的乡情引老乡、回故乡、建家乡) (MCPRO, 2017).

The other way to create jobs is to extend the value chain of local agriculture. Food processing companies, textile businesses, agricultural machinery factories, and logistics are expected to provide the necessary local employment. Tourism is the other sector expected to contribute to local employment. To attract tourists, Taoyuan's untouched rural landscapes and its name (Taoyuan "桃源" is a well-known cultural metaphor for the escapism in Chinese literature) were used to promote tourism nationwide.

Dongguan

Dongguan is a world-famous processing and manufacturing hub in south-central Guangdong Province. It is located in the heartland of the PRD's eastern estuary between the advanced Guangzhou and Shenzhen economic zones, with a total area of 2,465 km^2 and 33 towns. Before the opening up and reform of China in 1978, Dongguan had a long history as a rural county where agriculture was the main source of income. Since the 1980s, benefitting from the comparative advantages of abundant labor supply and low-cost land, Dongguan has attracted considerable investments from outside China and from foreign-funded enterprises in Hong Kong and Taiwan. Most of them were manufacturing enterprises that engaged in processing, trade, and OEM (Original Equipment Manufacturer) business, which are located at the bottom of the production "smiling curve", with the lowest value added. In 2008, the share of manufacturing in Dongguan's GDP reached 54.98% (DMBS, 2009), and the leading sectors were electronic information, electrical machinery and equipment, paper-making and printing products, plastic products, food and beverage processing, textiles, garments, headwear, and footwear.

Over the last 40 years, Dongguan has transformed from an area with a low level of urbanization (16.47%, lower than the provincial average of China) into one of the most urbanized areas, with an urbanization level of 51.82% in 2016. Its urbanization is characterized by rapid economic growth driven by export-oriented foreign investments. At the beginning of the 2000s, Dongguan became an export-oriented economy with a foreign trade dependence (calculated by the total value of exports/GDP) of 328.4% in 2000, much higher than the national level (39.6%) and provincial level (70.84%).

Economic and demographic shrinkage in Dongguan after 2008

The industrial decline that accompanied labor loss was the challenge of urban shrinkage in Dongguan after the financial crisis in 2008. In Figure 6.2, the trajectories of both the GDP from manufacturing and the number of the residential population (i.e., the population residing for longer than six months) represented sustained growth in 2000–2007. However, economic and demographic shrinkages emerged unexpectedly in 2009. The GDP from manufacturing decreased from RMB 191 billion in 2008 to RMB 186 billion in 2009; the number of migrant workers decreased from 5.52 million to 4.29 million (a decrease of 1.3 million in one year). This finding indicates that the shrinkage of Dongguan may have mainly been induced by the financial crisis in 2008.

In the gradual recovery of Dongguan's economy after 2008, its manufacturing sector rebounded quickly. The GDP of manufacturing in Dongguan continued to increase: from RMB 222 billion in 2010 to RMB 366 billion in 2017. However, the number of migrant workers was unable to fully recover and maintained a relatively low level after the crisis. The number of enterprises also did not recover. According to the statistics from the Administration for Industry and Commerce of Dongguan, approximately 12,063 manufacturing enterprises either closed or moved from 2008 to 2014, which was 14.97% of the total in 2014.

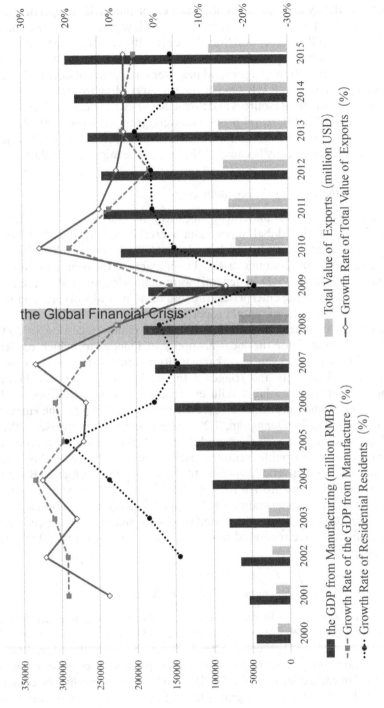

Figure 6.2 The manufacture GDP and the total value of exports in Dongguan (2000–2015).

Moreover, the increasing number of collapsed factories and unemployed workers negatively affected the rental market for industrial properties and residences, which resulted in high vacancy and low prices. Statistical data from the Human Resources and Social Security Bureau of Dongguan show that the vacancy rate of residential properties increased from 19% in 2007 to 43% in 2015, and the average rental price for an apartment declined from RMB 406 per month to RMB 296 per month. According to one of the chief officials of the Human Resources and Social Security Bureau (interviewed in Dongguan, April 2015), rental properties were generally oversupplied, and the price of a two-bedroom apartment decreased approximately 30% after the financial crisis, mainly because of the loss of the migrant population.

Global scale: the shock of the economic crisis

The export-oriented manufacturing industry in Dongguan is sitting at the terminals of the global supply chain established by the "footloose capitals" (Wang et al., 2003), which made it extremely vulnerable to fluctuations and shocks in economic crises. The impacts of the economic shocks on Dongguan (or the PRD in general) were mainly manifested in the collapse of manufacturing enterprises, the loss of industrial workers, and the increasing vacancy rate of rental properties.

The global financial crisis and worldwide economic recession in 2008 marked a watershed for the local economy in Dongguan, which caused a sharp downturn or even shrinkage in its manufacturing sectors and export-oriented markets and triggered the outflow of capital and labor, leading to increased business failures, reduced job opportunities, high unemployment, and lower salaries. According to statistics from the Human Resources Bureau in Dongguan, 1,285 manufacturing enterprises either closed or collapsed between 2008 and 2009, and the number of registered migrant workers and public rental houses decreased to 0.45 million and 18,583 houses, respectively. According to the Dongguan Municipal Bureau of Statistics (2018), the growth rate of GDP from manufacturing and the total value of exports was maintained at approximately 20%, between 2000 and 2007; in 2008, the figures decreased to −3.44% and −1.58%, respectively; and both figures have recovered but remained at less than 10% after 2012 (Figure 6.2).

National scale: diminishment of the demographic dividend

Comparative advantages formed by low-cost labor were the most attractive resources to labor-intensive export operations in Dongguan, which initiated the "exo-urbanization"[3] processes after 1979 (Sit and Yang, 1997; Lin, 2006). The concentration of labor and foreign capital provides a sustaining impetus for the "miracle of growth" in the PRD, and some believe that the contribution of the labor force (immigrants in particular) to the economy is even greater than that of the capital (Dai and Liang, 2000). However, Dongguan, after 2008, experienced a diminishment of the so-called demographic dividend[4] (Cai, 2010; Peng, 2011), which caused the reversal of labor inflow.

On the one hand, according to the social security database in Dongguan, the proportion of the working-age population (aged 15–64) peaked at 94.54% in 2010 and subsequently decreased to 93.18% by 2014. This indicates that the momentum of the large-scale rural-to-urban migration to Dongguan was gradually disappearing and that the labor supply would no longer be plentiful. As early as the spring of 2004, a shortage of migrant workers was reported by some news media, who predicted that the employment gap in Dongguan would exceed 100,000 persons, mainly in the labor-intensive sectors.

On the other hand, the minimum wage per month in these sectors increased from RMB 770 in 2008 to RMB 1,510 RMB in 2015 because of government regulations. Adding up all the other expenses, such as social security, catering, and utilities, the average cost of hiring a worker exceeded RMB 5,000 per month, indicating that the demographic dividend has diminished, which exerts great pressure on the production and operation of enterprises in Dongguan.

Local scale: changing policy orientations

Differences in the institutional environment suggest differences in the ability to absorb and create technological advances, which determine the differences in economic performance between regions (Martin, 1999; Pike et al., 2010). Policy orientations manifested by the agenda of local government set the basic conditions for promoting or suppressing the local economy and played a significant role in the shrinkage of Dongguan.

Because Dongguan's economy has long been at the lower end of the chain of production, a series of policies and regulations were introduced, including restricted commodities in processing trade, additional taxes, tightening labor laws, and environmental regulations, to force an upgrade of the existing processing trade enterprises and labor-intensive manufacturers in Dongguan. In a survey conducted by the Hong Kong Trade Development Council (2007), approximately 26.9% of the enterprises in Hong Kong claimed that the ever-tightening policy on the processing trade was the most influential issue for them.

Initiated by Guangdong Province's "emptying the cage for new birds" (腾笼换鸟) strategy in 2008, the municipality of Dongguan formulated plans for relocating traditional labor-intensive industries to less-developed areas in Guangdong, leading to massive industrial and labor transfers. Nine new industrial parks in the peripheral regions of Guangdong had been established, and more than 1,500 enterprises were relocated. Notably, some of Dongguan's towns are now confronting the restrictions of environmental regulations.

For example, a project was initiated by the Dongguan government in 2014 to relocate all the labor-intensive medium and small enterprises labeled high-pollutive, highly energy-wasting, and low-efficient. This project has tremendous impacts on the towns in northwest Dongguan, where the economy was dominated by paper-making and package-printing industries, resulting in decreasing fiscal revenue, population loss, and vacant lands. According to statistics from the local Administrative Committee, approximately 35 enterprises were relocated by this project, resulting in a loss of RMB 20.4 million in local tax revenue, the loss of 8,744 jobs, and 124.33 hectares of unused land.

Discussion

The three aforementioned models are limited in that their explanations of city growth and shrinkage in southern China are not comprehensive. The life-cycle model can be used to depict the gain and loss of the labor force in either depopulating cities such as Taoyuan or prospering cities such as Dongguan; however, it does not capture the intra-regional connections made by the outflow, inflow, and backflow of migrant workers between them. This model also focuses solely on the demographic change of urban growth and shrinkage, leaving the economic impetus of the migrations unexplained. Additionally, the heuristic model and the political-economy model have attempted to conceptualize the economic forces behind the demographic change: the former shows how the local urban development is influencing and influenced by the degradation of its population in quantity and quality; the latter examines the uneven development on a global scale, unveiling that some regions are made to be more advantageous than others in the capitalist system. However, none of the two models applies its spatial application in the city region.

We illustrate urban growth and shrinkage in rapidly urbanizing southern China by using the dynamics of "migration flows" and "economic potential" (Figure 6.3).

Our investigation indicates that the growth and decline of cities in southern China were shaped mainly by the interprovincial migration flows in the last few decades. In the first 20 years of China's economic reform, the booming manufacturing sectors attracted a large amount of outflowing labor to the cities in the coastal provinces (City A in Figure 6.3), draining the labor pool of inland cities (Cities B and C in Figure 6.3) and rural areas (other dots in Figure 6.3). At that time, Dongguan became the "world's factory" by using the comparative advantages of an abundant labor supply and foreign investments. In the same period, Taoyuan used labor output as an effective policy measure to reduce poverty, even though the population in small settlements and rural areas started to shrink. In the second 20 years, the slowing of the growth rate and the economic shocks in the global market created fluctuations in the development of coastal cities. When the economic crisis

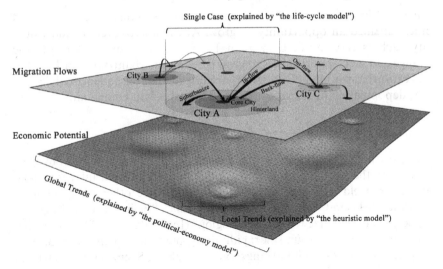

Figure 6.3 Urban growth and shrinkage in southern China.

affected China in 2008, many manufacturing enterprises closed factories and laid-off workers. Additionally, the local governments of the inland cities were implementing proactive policies to boost the local economy. Because of these push-pull factors, the economic attractiveness of the coastal cities was partially reduced, and some migrant workers flowed back to inland cities. Those world-famous processing and manufacturing hubs in the province of Guangdong are now enduring severe shrinkage, at least for a short time.

The movement of individuals in the so-called "Southern China Interprovincial Migration Field" was caused by the potential economic difference between the coastal and inland regions. Notably, the potential of individuals to create wealth varies in the developed, developing, under-developed regions, and this is similar to the following: the difference in gravitational potential moves objects from one location to another, and the difference in economic potential between cities (lower layer in Figure 6.3) is the source of the force that creates migration flows (upper layer in Figure 6.3). Based on what occurred in southern China, explanations of the changes in economic potential are possible at the global and local scales. At the macro-level, evolving trends in the global market shaped the overall spatial pattern of economic advantage and disadvantage. In the research area, the "miracle of growth" in the PRD was jointly produced by the preference of investors in globalization and the policy incentives intro-duced by the Chinese government after the reform. The global economy in favor of a low-cost labor supply created a migration field that caused thriving manufacturing coastal hubs and depopulated inland regions. The financial crisis changed this setting in 2008, our case study shows that the direction of migration flows was partially reversed immediately after the impact. When those industrial towns became shrinking cities, the inland cities regained an opportunity to grow. At the micro-level, an individual city, such as Taoyuan or Dongguan, has to adjust its urban policy to the change and seek opportunities to grow (i.e., in situ urbanization in Taoyuan and exo-urbanization in Dongguan) or mitigate the consequences of decline (i.e., depopulation in Taoyuan and economic shock in Dongguan).

Conclusion

The growth and shrinkage of cities caused by migration flows are observed beyond southern China. In addition to the "Southern China Interprovincial Migration Field", geographers have identified three other interprovincial migration fields in China: "Northeastern China", "Beijing-Tianjin-Tangshan Region", and "Yangtze River Delta Region" (Liu et al., 2012). These are adjacent provinces that surround the rapidly developing hubs of China (i.e., Beijing-Tianjin, Shanghai-Nanjing-Hangzhou, Hongkong-Shenzhen-Guangzhou), where the surplus labor migrates to seek employment. This migratory pattern is determined by the different economic potentials of regions and is also observed within provinces. The situation is similar for provincial-level administrations in China, that is, some nodes, usually the provincial capital and major cities, are gaining population while small-sized cities or towns and most of the rural areas are depopulating (Zhou et al., 2015).

"Urbanization" has been a national strategy for China's modernization in the past decades. Politically and academically, the definition of urbanization is largely oriented toward the growth of cities. Notably, the issues raised in the shrinking cities or regions, where the population and resources are from, have attracted insufficient attention. Ganser and Piro (2016) acknowledged that the spatial contexts of growth and shrinkage are challenging both planning and sustainability in urban/rural developments: the former introduces developmental pressures to the local environment, and the latter leaves degrading natural/cultural landscapes. In this perspective, city shrinkage in China, compared with urban growth, is an equally important urban issue, policy domain, and planning scenario, which requires further research.

In this chapter, we argue that the two processes, growing and shrinking, are interconnected, mutually reinforcing, and sometimes interconvertible. Using the two case studies in rapidly urbanizing southern China, we analyzed the details of such connection, interdependency, and conversion in the dynamics of demographic and economic changes of China. The results show that the growth and shrinkage of cities are two sides of the same coin. China is a fast-developing country with large-scale domestic migrations but limited international immigration. Additionally, its export-oriented economy is highly integrated into global trades, which makes it vulnerable to global market fluctuations. The condensed spatial, temporal context of China's recent development created a valuable example for studying the pattern of labor flows caused by rapid economic growth. This chapter shows that the up-surging and degrading of urban settlements in China were closely linked with migration flows, resource allocation, market changes, and policies. The labor flows discussed here could be the first step of shrinkage that is usually followed by a change of demographic structure where mortality becomes higher than the birth rate, which causes depopulation in the longer term. China's shrinking model illustrates the dynamics of "migration flows" and "economic potential"; however, whether this is also the shrinking story, or part of it, in other developing countries of the world awaits further comparative studies.

Acknowledgment

This research was funded by the General Program of the National Natural Science Foundation of China (project no. 52078197).

Notes

1 The Government Work Report is an official document released annually on behalf of the mayor of local government and reviews what the government has achieved in the previous year and maps out its major goals, policies, and agenda for the coming year. Source: http://english.www.gov.cn/.
2 In-situ Urbanization regards local cities and towns as basic nodes for urbanization, in which the rural population moves to adjacent urban settlements for better infrastructures and public services.
3 Exo-urbanization refers to the rapid urban development induced mainly by foreign investment.
4 China's population dividend refers to the economic growth potential due to the abundant supply of workers of working age since the 1980s.

References

Audirac, I. & Alejandre, J. A. (eds.) 2010. *Shrinking Cities South/North*. Florida State University and University of Guadalajara, UCLA Program on Mexico, Guadalajara, Mexico: Juan Pablo Editor.

Bartholomae, F., Nam, C.W. & Schoenberg, A. 2016. Urban Shrinkage and Resurgence in Germany. *Urban Studies*, 54(12), 2701–2718.

Beauregard, R. A. 2009. Urban Population Loss in Historical Perspective: United States, 1820–2000. *Environment and Planning A*, 41(3), 514–528.

Bernt, M. et al. 2014. How Does(n't) Urban Shrinkage Get onto the Agenda? Experiences from Leipzig, Liverpool, Genoa and Bytom. *International Journal of Urban and Regional Research*, 38(5), 1749–1766.

Dongguan Municipal Bureau of Statistics (DMBS). 2009. *Dongguan Statistical Yearbook*. Beijing: China State Statistical Press. //东莞统计局, 2009. *东莞统计年鉴*. 北京: 中国统计出版社.

Dongguan Municipal Bureau of Statistics (DMBS). 2018. *Dongguan Statistical Yearbook*. Beijing: China State Statistical Press. //东莞统计局, 2018. *东莞统计年鉴*. 北京: 中国统计出版社.

Cai, F. 2010. Demographic Transition, Demographic Dividend, and Lewis Turning Point in China. *China Economic Journal*, 3(2), 107–119.

Dai, J. L. & Liang, G. J. 2000. Reason Analysis of Economic Growth in Pearl River Delta. *Economy in the Pearl River Delta*, 2, 7–13. //代吉林, 梁国坚, 2000. 珠江三角洲经济区经济快速增长原因探析. 珠江三角洲经济, 2: 7–13.

Dubeaux, S. & Sabot, E. C. 2018. Maximizing the Potential of Vacant Spaces within Shrinking Cities, a German Approach. *Cities*, 75, 6–11.

Ganser, R. & Piro, R. 2016. *Parallel Pattern of Shrinking Cities and Urban Growth: Spatial Planning for Sustainable Development of City Regions and Rural Areas.* London and New York: Routledge.

Haase, A. et al. 2014. Conceptualizing Urban Shrinkage. *Environment and Planning A*, 46(7), 1519–1534.

Haase, A., Nelle, A. & Mallach, A. 2017. Representing Urban Shrinkage— The Importance of Discourse as a Frame for Understanding Conditions and Policy. *Cities*, 69, 95–101.

Hall, P. & Hay, D. 1980. *Growth Centres in the European Urban System.* London: Heinemann Educational Books.

Hartt, M. 2018. The Diversity of North American Shrinking Cities. *Urban Studies*, 55(13), 2946–2959.

Harvey, D. 1973. *Social Justice and the City.* Baltimore: Johns Hopkins University Press.

Harvey, D. 2001. *Spaces of Capital: Towards A Critical Geography.* Edinburgh: Edinburgh University Press.

Hattori, K., Kaido, K. & Matsuyuki, M. 2017. The Development of Urban Shrinkage Discourse and Policy Response in Japan. *Cities*, 69, 124–132.

Hong Kong Trade Development Council (HKTDC). 2007. Implications of Mainland Processing Trade Policy on Hong Kong. Hong Kong: The Greater Pearl River Delta Business Council.

Kim, S. 2019. Design Strategies to Respond to the Challenges of Shrinking City. *Journal of Urban Design*, 24(1), 49–64.

Kingdon, J. W. 1984. *Agendas, Alternatives, and Public Policies.* Boston: Little, Brown.

Kondratieff, N. D. & Stolper, W. F. 1935. The Long Waves in Economic Life. *The Review of Economics and Statistics*, 17(6), 105–115.

Kuznets. 1930. *Secular Movement in Production and Prices: Their Nature and Their Bearing upon Cyclical Fluctuations.* Boston: Houghton Mifflin.

Laursen, L. H. 2016. Urban Transformations—The Dynamic Relation of Urban Growth and Decline. In Ganser, R. & Piro, R. (eds.) *Parallel Pattern of Shrinking Cities and Urban Growth: Spatial Planning for Sustainable Development of City Regions and Rural Areas* (pp. 73–82). London and New York: Routledge.

Lin, G. C. S. 2006. Peri-urbanism in Globalizing China: A Study of New Urbanism in Dongguan. *Eurasian Geography and Economics*, 47(1), 28–53.

Liu, W. B., Wang, L. N., and Chen, Z. N. 2012. Flow Field and Its Regional Differentiation of Inter-provincial Migration in China. *Economic Geography*, 32(2), 8–13. //刘望保, 汪丽娜, 陈忠暖, 2012. 中国省际人口迁移流场及其空间差异. 经济地理, 32(2), 8–13.

Martin, P. 1999. Public Policies, Regional Inequalities and Growth. *Journal of Public Economics*, 73(1), 85–105.

Municipal Committee Political Research Office (MCPRO). 2017. *Return Geese Project: Building Home.* Taoyuan: Municipal Committee Political Research Office. http://zys.changde.gov.cn/art/2017/8/30/art_14421_1127409.html.

Peng, X. Z. 2011. China's Demographic History and Future Challenges. *Science*, 333, 581–587.

Pike, A., Rodríguez-Pose, A. & Tomaney, J. 2010. *Handbook of Local and Regional Development*. London and New York: Routledge.

Rérat, P. 2019. The Return of Cities: The Trajectory of Swiss Cities from Demographic Loss to Reurbanization. *European Planning Studies*, 27(2), 355–376.

Sabot, E. C. & Roth, H. 2013. Growth Paradigm Against Urban Shrinkage: A Standardized Fight? The Cases of Glasgow (UK) and Saint-Etienne (France). In Pallagst, K., Wiechmann, T. & Fernandez, C. M. (eds.) *Shrinking Cities, International Perspectives and Policy Implications* (pp. 115–140). New York and London: Routledge.

Sit, V. F. & Yang, C. 1997. Foreign-investment-induced Exo-urbanisation in the Pearl River Delta, China. *Urban Studies*, 34(4), 647–677.

Smith, N. 2008. *Uneven Development: Nature, Capital, and the Production of Space*. Athens: University of Georgia Press.

Taoyuan's Development and Reform Commission (TDRC). 2005. The 11th Five-Year National Economic and Social Development Plan of Taoyuan County. https://www.taoyuan.gov.cn/ in 2019.1 //桃园发展和改革委员会(TDRC), 2005. 桃源县"十一五"国民经济和社会发展规划.

Taoyuan's Development and Reform Commission (TDRC). 2010. The 12th Five-Year National Economic and Social Development Plan of Taoyuan County. https://www.taoyuan.gov.cn/ in 2019.1. //桃园发展和改革委员会(TDRC), 2010. 桃源县"十二五"国民经济和社会发展规划.

Van den Berg, L. et al. 1982. *Urban Europe: A Study of Growth and Decline*. Oxford: Pergamon Press.

Vernon, R. 1966. International Trade and International Investment in the Product Cycle. *Quarterly Journal of Economics*, 80(2), 190–207.

Wang, J. C., Luo, J. D. & Tong, X. 2003. A Comparison Between Taiwanese PC-related Industrial Clusters in Suzhou and Dongguan, *Journal of China University of Geosciences (Zhongguo Dizhi Daixue Xuebao)*, 3(2), 6–10. //王缉慈, 罗家德, 童昕, 2003. 东莞和苏州台商 PC 产业群的比较分析. 中国地质大学学报(社会科学版), 3(2), 6–10.

Wolff, M. & Wiechmann, T. 2018. Urban Growth and Decline: Europe's Shrinking Cities in a Comparative Perspective 1990–2010. *European Urban and Regional Studies*, 25(2), 122–139.

Zhou, Y.C. et al. 2015. Comparative Research on Spatial Structure of Human Urbanization of Developed Region and Undeveloped Region: A Case of Jiangsu and Hunan. *Economic Geography*, 35(2), 77–83. //周玉翠, 等, 2015. 经济发达地区和欠发达地区人口城市化空间结构比较研究—以江苏和湖南为例. 经济地理, 35(2), 77–83.

7 Shrinkage under growth

A case study of regional shrinkage in Wuhan, China

Zhe Gao

Research background

Urban shrinkage is an urban phenomenon caused by economic crises, wars, revolutions, systemic transformations, and epidemics or catastrophes with subsequent population loss (Oswalt, 2006; Rieniets, 2009). The phenomenon has been defined as occurring in densely populated urban areas with more than 10,000 residents that have faced population loss and economic problems for more than two continuous years (Wiechmann, 2008; Hollander et al., 2009; Martinez-Fernandez et al., 2012). This definition refers to population loss and economic decline that signal eventual urban decline (Beauregard, 2009; Hollander and Nemeth, 2011; Bernt, 2016). Such a decline is usually sensitive to spatial scales and geographic contexts (Martinez-Fernandez et al., 2012; Elzerman and Bontje, 2015). For example, unlike in the European context, a shrinking city in the United States has been defined as an old industrial area that has lost more than 25% of its population over a period of 40 years (Schilling and Logan, 2008). Therefore, definitions of a shrinking city may vary according to spatial scales and geographic contexts. Consequently, it would be arbitrary to apply theory and empirical evidence of urban shrinkage originating from the Western context to other geographical contexts (Zhang and Li, 2017).

China has maintained rapid urbanization since the 1990s and has been integrated into the accelerating progress of globalization (Liu and Yang, 2017). As a consequence, Chinese cities, particularly regional central cities, have witnessed exponential growth in their populations, economies, and land supplies for urban construction (Yang et al., 2015). However, changes in the economic and social environment at home and abroad, particularly as a result of the 2008 global financial crisis, have slowed the growth rate of China's gross domestic product (GDP), which fell from more than 10% to 7% or even lower in major Chinese cities in the period between 2008 and 2014 (Statistical Yearbook of China, 2008-2014). In 2014, this downturn in the GDP growth rate was officially recognized by the Chinese government as the "new normal" (Zhang and Li, 2017). In addition, the decrease in GDP was accompanied by a scalar decrease in the annual average land supply

DOI: 10.4324/9780367815011-9

for urban construction, the deceleration of urban expansion and population growth, and even population loss in some regions and cities (Zhou and Qian, 2015; Yang and Dunford, 2017).

Since urban shrinkage occurs with different magnitudes at various spatial scales (Li et al., 2015; Gao, 2017), case studies are effective ways to explore this phenomenon in Chinese cities (Zhang and Li, 2017). Most studies have concentrated on Northeast China, the Beijing-Tianjin-Hebei region, the Pearl River Delta, and the Yangtze River Delta. Wu et al. (2015) investigated the shrinkage of the Beijing-Tianjin-Hebei region and the Yangtze River Delta based on the fifth and sixth national population censuses to identify the types of shrinkage and their spatial patterns in those areas. Li et al. (2015) identified the mismatch between the demand for manufacturing and the recent labor supply as the main reason for the regional urban shrinkage in the core of the Pearl River Delta. In a study using Yichun in Heilongjiang Province, Northeast China as an example, Gao and Long (2017) identified the common problems in resource-based cities as additional causes of urban shrinkage. Overall, however, research on urban shrinkage in the cities and towns of Central China is limited. Zhou et al. (2017) quantitatively analyzed the shrinkage in Hunan Province on multiple geographic scales using population census data. Liu and Zhang (2017) used Huanggang, Hubei Province as an example to discuss the shrinkage phenomenon in the Wuhan megalopolis.

The cities and towns in Central China, a region that has experienced an economic downturn, are striving for an economic revival at present and may have a unique logic of growth and shrinkage. Among these cities and towns, Wuhan is the only subprovincial city and one of the nine national central cities. Therefore, it represents a good case study for investigation. Since 2000, the city has experienced rapid economic growth with more than 15% annual GDP growth while its urbanization rate increased from 58.88% in 2000 to 80.29% in 2018 (Wuhan Statistical Yearbook, 2019). However, contrary to this overall growth trend, there has been a population reduction in some parts of the city. Unlike the usual urban shrinkage that occurs in an economic downturn, this type of shrinkage coexists with the overall GDP growth. The coexistence of regional population loss, a decrease in the land supply, and economic fluctuation at the micro level, coupled with stable growth of the economy and urbanization at the macro level, is a new urban phenomenon that has increasingly attracted the attention of city planners, economists, urban governors, policymakers, and academics (Du and Li, 2017; Gao et al., 2019). However, such "shrinkage within growth", referred to as "regional urban shrinkage" (Li et al., 2017), has been underresearched in the Chinese context. What are the features and spatial characteristics of this type of urban shrinkage? Is it agglomerated or decentralized? What is the mechanism? Is it the same as the shrinkage that occurs in the Pearl River Delta, which has experienced similar rapid economic growth?

Study context, data, and methodology

In response to these questions, this study first identified the scope, intensity, and spatial characteristics of the growth and shrinkage in Wuhan. Based on population census data, economic data, and land supply data, this investigation was conducted at both the municipal district scale and the subdistrict scale to provide comparable insights in order to understand the regional shrinkage phenomenon within a large administrative division. With a focus on one of the key shrinking areas of Wuhan, Qingshan District, empirical research was conducted from a demographic and capital perspective to investigate this type of shrinkage.

Two points require special attention. First, in addition to the population and economic perspectives that have been adopted to measure urban shrinkage, it is critical to understand the land supply in the Chinese urban context in general and urban shrinkage in particular. Since urban land is state owned, the Chinese government has control over the urban land supply in terms of macro policies and micro practices. The scale of the land supply thereby reflects the expectations of both the government and the market for future urban development. An increase in the land supply means that a city is working well, whereas a decrease in the land supply implies that the city will be, or has been, in some kind of crisis.

Second, it is necessary to be mindful of the scale at which urban shrinkage is examined in the Chinese context. In terms of the administrative boundaries of Chinese cities at the prefecture level or above, there are three administrative zones, namely, the municipal district (qu), the subdistrict (jiedao), and the community (shequ). Each city is composed of several municipal districts, a municipal district is composed of several subdistricts, and a subdistrict is composed of several communities. Wuhan, for instance, contains 13 municipal districts, 185 subdistricts, and more than 3,000 communities. Since population census data are always released at the municipal district/county-level city[1] level in the majority of cases, most Chinese urban shrinkage studies are limited to this scale. Population in most municipal districts of Chinese cities can be huge. Hongshan District of Wuhan has a population of 1.26 million, and Qingshan District has a population of 0.54 million, for instance. These figures far exceed the size of many Western cities, making it difficult to conduct comparative studies. However, the population of most subdistricts in Chinese cities is between 10,000 and 100,000, which is closer to the size of many Western cases. Therefore, the subdistrict scale is an important analytical scale in this study.

Study area

In 2018, Wuhan, the capital city of Hubei Province, located in the central plain of China, had an urban residential population of 11.06 million, including a *hukou* (registered) population of 8.54 million (Wuhan Statistical

Yearbook, 2019). Its total area within the municipal boundary was 8,494.41 km² with 13 municipal districts, including 7 districts within the major urban area and 6 districts located in the peri-urban area. It had a GDP of Chinese Yuan (CNY) 1,484.73 billion, making the city the largest economy in Central China (Wuhan Statistical Yearbook, 2019). As the most important political, economic, transport, educational, and cultural center in Central China, Wuhan has experienced rapid urbanization and significant economic growth since the early 2000s. However, since the early 2010s, some parts of the city have also been experiencing slowing population growth and even population loss (Liu and Zhang, 2017).

We used Wuhan for our empirical study for four reasons. First, as the most important and largest capital city in Central China, Wuhan shares some common characteristics with hinterland cities in China with regard to its geographic location, industrial structures, and development trajectories. Second, some places in Wuhan have experienced slowing population growth and even population loss alongside overall economic and population growth (Liu and Zhang, 2017). This coexistence of rapid growth and regional shrinkage in Wuhan is a common urban phenomenon that has been observed in other hinterland cities in China. Third, Wuhan has been defined as a central city at the national level[2] and is the only subprovincial city in Central China. Consequently, Wuhan has profound effects on its surrounding provinces. Fourth, Wuhan has rich multidimensional data that are reliable in quality, quantity, and ease of access. Thus, Wuhan serves as a good example for the examination of regional urban shrinkage in the "new normal" era.

Data and methodology

Traditionally, demographic change, particularly changes in population size, is the main criterion for judging whether a city is shrinking (Martinez-Fernandez et al., 2012; Haase et al., 2014). For instance, Shrinking Cities International Research Network (SCIRN) defines shrinking cities as "an urban area that has a population not less than 10,000 people, experiencing a population decline for more than two years, and accompanied by structural economic transformation and crisis" (Wiechmann, 2008; Hollander et al., 2009). However, as our understanding of shrinking cities has deepened, using demographic change as the only indicator of urban shrinkage has been questioned. An improvement in the measurement method based on multidimensional criteria has been proposed (Long et al., 2015; Liu, 2016). Furthermore, if urban expansion is regarded as the flow and agglomeration of production factors such as capital and labor in urban space (Bertinelli and Black, 2004), then urban shrinkage could be understood as the outflow of these production factors. Therefore, regional urban shrinkage should be interpreted from a broader socioeconomic perspective.

Based on population, economy, and land supply dimensions, Du and Li (2017) proposed four types of urban growth and shrinkage models, namely, sustained growth, transformational growth, potential shrinkage, and all-round shrinkage. We have focused particular attention on the third type, "potential shrinkage", which has a certain similarity with the situation observed in Wuhan. The potential shrinkage occurs when "the overall population and urban land supply remain stable, but the economic growth slows". Although there is no overall decline, at a finer, subdistrict scale, a population reduction may have already taken place in the case of potential shrinkage, implying that the population stabilization can be problematic.

Briefly, data from the fifth and sixth national censuses were used to measure urban shrinkage status. Since the data are only available at a large geographic scale—for example, at the municipal district or county-level city scale—there is a research gap between large-scale quantitative studies and small-scale qualitative analyses (Lin et al., 2017; Li and Long, 2018). To address this research gap, our study identified regional urban shrinkage and explored its mechanism in Wuhan and its Qingshan District as a case study. The goals of this study were twofold: (i) to identify regional urban shrinkage from a demographic, economic, and land supply perspective; and (ii) to explore the mechanism and possible causality of regional urban shrinkage. Through this empirical case study of urban shrinkage with Chinese characteristics, the study furthers the understanding of urban shrinkage in a non-Western setting.

Our research updated the data by using fifth national census data (2000) coupled with local population sampling data (2016) to depict basic growth and shrinkage patterns in Wuhan at both municipal district and subdistrict scales. Further investigation was conducted from economic (GDP) and land supply perspectives to identify the scope and degree of regional shrinkage in Wuhan. The spatial pattern of the growth and shrinkage of Wuhan was identified through ArcGIS. A case study of a typical shrinking area of Wuhan, Qingshan District was selected for detailed analysis. The mechanism that triggers regional shrinkage is discussed comprehensively from demographic and capital perspectives. The demographic perspective focuses on the changes caused by aging and declining births. The capital perspective focuses on the investment, divestment, and reinvestment processes driven by capital appreciation behind population migration derived from the concepts of capital circulation and uneven geographical development (Harvey, 1982; Smith, 2010).

Growth and shrinkage patterns of Wuhan

The population in Wuhan maintained an overall annual growth rate of 1.6% from 2000 to 2016. Correspondingly, the urbanization rate increased from 58.88% to 79.77%. In terms of the age structure, the proportion of people older than 65 years rose sharply while the proportion of those younger than 15 years declined significantly (Table 7.1). In short, the size of the population continued to grow, providing a stable supply of working-age people, but the

Table 7.1 Changes of the age structure in Wuhan

Ages	0–15	16–64	≥65
2000	18.81	73.34	7.85
2010	10.94	78.99	10.07
2016	13.42	72.92	13.66

Source: Wuhan Statistical Yearbook (2001, 2011, 2017).

age structure changed greatly, reflecting the characteristics of aging and a lower birth rate compared with the past. At the municipal district scale, 11 out of the 13 districts in Wuhan experienced population growth from 2000 to 2016, while two districts recorded population decline. These shrinking districts are Huangpi District and Xinzhou District, whose populations fell by 0.84% and 8.25%, respectively, in the same period. Among the growing districts, Qingshan showed the slowest growth rate and rose from 443 thousands to 448 thousands by barely 1% between 2000 and 2016'.

Among the four types of urban growth and shrinkage in the model proposed by Du and Li (2017), "sustained growth" identifies close relationships between population growth, increased GDP, and the land supply. Therefore, examining the relationship between the population, GDP, and land supply can provide a deeper understanding of how "shrinkage under growth" occurs. Correlation analysis of population and GDP changes in Wuhan shows that nine out of the 11 municipal districts with an increasing population are distributed near the trend line, and only Qingshan District and Hongshan District are below the trend line. This demonstrates that the population growth in Qingshan and Hongshan did not result in matching economic growth. In particular, the 1% population growth in Qingshan drove only 0.03% of GDP growth, which is far below the level of more than 3% in other districts (Figure 7.1a).

The results from calculating the correlation between population and land supply changes indicate that 8 out of 11 municipal districts with increasing populations experienced similar trends. Only Hannan District, Dongxihu District, and Hongshan District are far from the trend line (Figure 7.1b). Hannan and Dongxihu Districts are above the trend line while Hongshan District is below the trend line, indicating that land supply in Hannan and Dongxihu grew too fast, resulting in a risk of land urbanization faster than population urbanization. In contrast, in Hongshan District, the growth in the land supply was slower than the population increase.

At the subdistrict level, complete statistical data are available for 156 out of 185 subdistricts from 2000 to 2016. Complete data are lacking for 29 subdistricts because of changes in municipal boundaries. Of the 156 subdistricts, 116 experienced population growth. Of these, 64 subdistricts had a growth rate of over 30%. The population of 40 subdistricts decreased. Of these, the populations of 29 subdistricts decreased by more than 10% and populations of 18 subdistricts decreased by more than 30% (Figure 7.2).

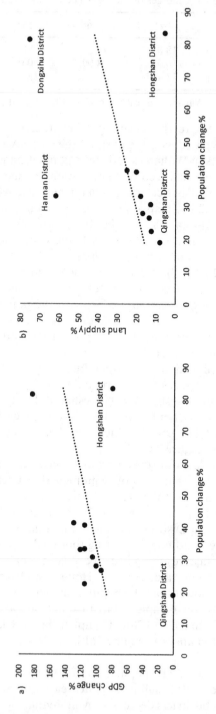

Figure 7.1 Relationship between population change (growth only) and GDP change (a) and the proportion of the land supply (b).

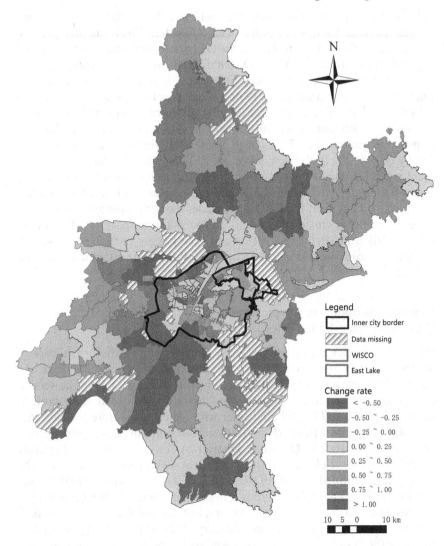

Figure 7.2 Growth and shrinkage patterns of Wuhan.

In terms of location, the rapid population growth is concentrated in the border areas between the central urban area and the new urban area, that is, along the Third Ring Road of the city, which is consistent with the main supply of new residential land. In addition, some of the new urban areas, such as Jinkou Subdistrict in Jiangxia District, Changfeng Subdistrict in Qiaokou District, and Shamao Subdistrict in Hannan District, experienced rapid population growth resulting from the launch of major industrial projects or infrastructure construction.

In contrast to this growth pattern, rapid population reduction is mainly concentrated in Changqian Subdistrict, Qingshan Town Subdistrict, and Gongrencun Subdistrict in Qingshan District; Hanzheng Subdistrict and Yijiadun Subdistrict in Qiaokou District; Yuehu Subdistrict and Qingchuan Subdistrict in Hanyang District; and Changdian Subdistrict in Caidian District. It is worth noting that 19 out of the 29 subdistricts with more than 10% population reduction are located in the central urban area, while only 10 out of 29 subdistricts are in the new urban area. In contrast, the total population increased by 37.5% from 2000 to 2016 in the central urban area but by only 18.5% in the new urban area. This indicates that the development within Wuhan's central urban area has been extremely uneven over the past 20 years. Moran's I is 0.13, and the Z value is 3.17 via spatial autocorrelation. That is, at the subdistrict scale, the shrinkage areas in Wuhan have a significant clustering trend, and the shape is "perforated". In contrast to full-scale shrinkage, perforated shrinkage is partial, dispersed on a large (urban) scale and clustered on a small (subdistrict) scale in the case of Wuhan.

Regional shrinkage in a typical area: the case of Qingshan district

Qingshan District is one of the seven districts within the urban core of Wuhan. China's first superlarge iron and steel enterprise, Wuhan Iron and Steel Corporation (WISCO), is located in this district. During the second half of the 20th century, responding to Chairman Mao's call, the establishment of WISCO and its affiliated enterprises, such as the state-owned "461" and "471" factories, attracted hundreds of thousands of people from all over the country to work in Qingshan District. A "city of steel" was therefore established in which both industrial production and people's lives had to be managed in a unified way. With the arrival of the 21st century, WISCO's net profits continued to grow, reaching a maximum value of CNY 6.526 billion in 2007, thus making Qingshan rank first in GDP among the 13 districts of Wuhan. For WISCO and the steel industry, 2008 was an important turning point. WISCO suffered heavy losses, and its net profits fell sharply because of the global financial crisis. In addition, overcapacity began to appear in the steel industry, the recession affected the broader steel industry, and multiple failed investments by WISCO eventually led to its reorganization and integration with Shanghai Baosteel in 2016. Correspondingly, the GDP ranking of Qingshan District fell to the bottom of the ranking of all districts, and the region became one of the typical shrinking areas of Wuhan.

Among the subdistricts with complete statistics in Qingshan District, more than half shrank from 2000 to 2016. Among them, the populations of Changqian Subdistrict and Qingshan Town Subdistrict dropped by 81.82% and 42.80%, respectively, and the populations of Hongwei Road Subdistrict and Gongrencun Subdistrict dropped by more than 20%. Only four

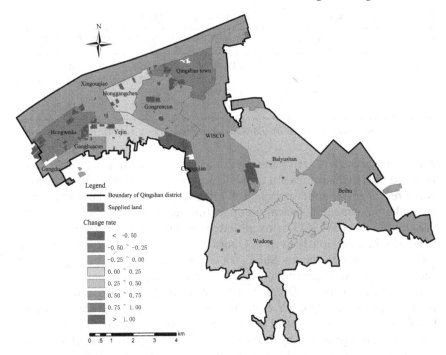

Figure 7.3 Growth and shrinkage pattern in the Qingshan District.

subdistricts continued to grow, namely, Baiyushan Subdistrict, Wudong Subdistrict, Yejin Subdistrict, and Honggangcheng Subdistrict. Of these, the growth rate of the last two is still lower than the average level of Wuhan (Figure 7.3). In spatial terms, a "perforated" spatial pattern can be observed: (i) in the eastern part, Baiyushan Subdistrict and Wudong Subdistrict continued to grow; (ii) in the central part, WISCO and its surrounding areas shrank; and (iii) in the western part, both growth and shrinkage occurred (Figure 7.3).

Growth in the East: an inflow of elderly people

Census data from 2000 to 2016 show that the population growth of Baiyushan Subdistrict and Wudong Subdistrict remained slow. However, the population structure underwent significant changes in that the proportion of people over 65 years of age rapidly increased while the proportion of those under 15 years of age rapidly decreased (Table 7.2). This implies an obvious aging and declining birth rate problem. If this trend continues, there is a high probability of population shrinkage in the area. Empirical observations confirm that on weekends, those subdistricts are frequented by elderly people. There is a serious shortage of 24-hr convenience stores and fast-food restaurants, and there are few real estate agents; however,

Table 7.2 Age structures on Baiyushan and Wudong streets

Streets	Years	0–15	16–64	65 and above
Baiyushan	2000	14.46%	80.60%	4.94%
	2016	6.17%	76.70%	17.13%
Wudong	2000	12.93%	80.39%	6.68%
	2016	5%	78.73%	16.26%

Source: Fifth National Census (2000) and Wuhan Population Sampling (2016).

many funeral parlors are operating in the area. In addition, our interviews revealed that the main reason for new local residents to move into the area is the recent urban redevelopment in Qingshan District rather than newly created employment or education opportunities. Most of the new residents are employees of WISCO or its affiliated enterprises or their family members, who had previously lived in Qingshan Town Subdistrict, Gongrencun Subdistrict, or Changqian Subdistrict. Following the redevelopment of these areas at the beginning of the 21st century, these residents were faced with a relocation choice. Because of relatively low house prices (the average price is approximately CNY 8,000/m², far lower than the average price of CNY 18,000/m² in the central and western areas of Qingshan District) and public facilities provided by WISCO (including kindergartens, elementary and middle schools, hospitals, food markets, canteens, and post offices), together with a continuing traditional state-owned enterprise (SOE) culture, Baiyushan and Wudong Subdistricts have become the first choice for the elderly low-income employees of WISCO and its affiliated enterprises. It should be noted that the SOE culture exerts a subtle influence on the choice of relocation. Our empirical interviews show that a large proportion of the respondents, especially the elderly, identify with the SOE culture. Close community relationships create a strong sense of security and belonging, and residents' economic, social, and psychological needs are met to a certain extent through relocation. Therefore, the population shift to the eastern part of Qingshan District is due to an inflow of the elderly, leading to an aging population and, ultimately, potential large-scale urban shrinkage.

Shrinkage in the Center: flight of capital

At the beginning of the 21st century, WISCO's profits and taxes continued to increase, peaking in 2007. In the same year, WISCO's cash investment reached a new high of CNY 14.82 billion. The turning point occurred in 2008 when the global financial crisis caused a significant drop in the manufacturing industry in general and the steel industry in particular. To revitalize the economy, the Chinese government launched an infrastructure investment program of approximately CNY 4 trillion, with most of the investment directed at infrastructure and real estate. This led to a substantial increase in domestic steel consumption. In response, a large number of steel projects were quickly approved, resulting in a rapid increase in China's steel production capacity in a short time.

When the short-term stimulus of CNY 4 trillion of investment ended, the domestic demand for steel quickly fell, and the overcapacity became obvious. An imbalance of steel supply and demand, together with the continued decline in international steel prices[3] (Figure 7.4a), ultimately led to a sharp decline in WISCO's profits, which fell from CNY 6.526 billion in 2007 to negative CNY 7.511 billion in 2015. This problem was further exacerbated by the loss of WISCO's own funds due to the failure of its overseas expansion. Subsequently, capital that could not achieve appreciation began to flow out of WISCO and its affiliates. The statistics show a rapid decline in investment, dropping from CNY 14.82 billion in 2007 to CNY 1.397 billion in 2015 (Figure 7.4b).

Although WISCO tried to restructure its management and implemented massive layoffs (WISCO's number of employees decreased by 35,504 from 2010 to 2015), the downturn in the steel industry and repeated investment failures eventually led to its reorganization. The crisis of overproduction faced by its steel enterprise, resulting in a decline in the profits in the primary business, led to the flight of capital from the original industry and its material space. In our case, the investment capital moved from central and eastern Qingshan District, where WISCO is located, to the secondary business of a newly built environment of higher profits and strong consumer demand. Consequently, a dilapidated material space emerged, leading to urban shrinkage in the region.

Concurrent growth and shrinkage in the West: relocation of capital

The western part of Qingshan District is close to the Inner Ring Road of the city. Metro Line 4 (in operation) and Line 5 (under construction) pass through this region and connect it with the urban core. Thanks to these transport and location advantages, this part of Qingshan District has become a prime area for urban renewal. Thus, urban construction is at its peak in the region, resulting in out-migration and short-term population decrease.

In the years following the global financial crisis, Qingshan District saw large-scale urban renewal and a boom in the construction industry. The data show that from 2010 to 2016, the added value of the construction industry increased by 51% (Wuhan Statistical Yearbook, 2017). Could this be regarded as a relocation of capital withdrawn from WISCO's rundown region? Tellingly, the empirical findings of this study reveal that most new property owners in the area are not local and do not work there. The cheaper housing prices (relative to other central urban areas) and the property appreciation potential are the main reasons for the heightened demand. Thus, private real estate investment and the return of capital thanks to new public services and commercial facilities may become a new opportunity for regional economic regeneration.

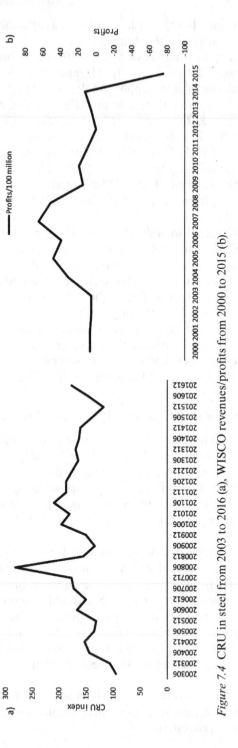

Figure 7.4 CRU in steel from 2003 to 2016 (a), WISCO revenues/profits from 2000 to 2015 (b).

The capital perspective is just one aspect of the regional shrinkage in Qingshan District. Other socioeconomic and cultural factors also play parts in this phenomenon, such as the restructuring of SOEs, the end of the welfare system, the marketization of community management, and job selection changes for WISCO's second generation of employees. Among all these factors, the capital factor can quickly respond to external changes because of its liquidity and profit-seeking nature. Thus, the capital perspective could shed light on regional shrinkage mechanisms.

Although the Qingshan case has its own distinctive features (regional economic growth depends heavily on a few enterprises and industries), what we observed is not uncommon in many Chinese cities. Because of these features, the challenges faced by Qingshan District were magnified after the global financial crisis, revealing the regional shrinkage that lies hidden in complex social changes.

Conclusion and discussion

Overall, Wuhan is a megacity that has rapidly developed over the past 20 years with a growing economy, population, and land supply. However, this high-speed growth was uneven—from 2000 to 2016, 40 of the 156 subdistricts witnessed a population decrease, resulting in regional shrinkage. In spatial terms, the regional shrinkage pattern is clustered and "perforated" in shape. As a typical shrinking area of Wuhan, Qingshan District reflects a perforated pattern of "growth in the east, shrinkage in the center and growth and shrinkage coexisting in the west". Had only aggregated data been used, this perforated pattern could have been easily missed or confused with a temporary population movement.

It should be noted that as far as we can observe in the Qingshan case, there have been major age-related changes. The changes can be seen in terms of the age structure (Table 7.2) and a new balance in regional demographics, namely, aging communities and declining birth rates. In other words, what has occurred in Qingshan is not a temporary population movement but rather a new urban phenomenon—regional shrinkage.

Another interesting topic in the Qingshan case is to where its residents relocated. Possible locations include the other parts of Qingshan District, especially those areas where urban renewal has been completed. This observation is confirmed by our interviews. Many interviewees told us that their children bought properties in the other subdistricts of Qingshan District close to the city center. This is consistent with the principles of geographic proximity and minimum cost. Other possible relocation destinations are the newly built residential areas around the city's Third Ring Road, which offer good access to transport and advanced infrastructure. The housing prices in those areas are higher than those in shrinking areas but significantly lower than those in the core areas of the city. This point of view is also confirmed by the interviews. Therefore, it can be concluded that the population outflow in the Qingshan case is mainly restricted within the administrative boundary of the city. This differs from "full-scale urban shrinkage". It is also apparent that housing prices (rent), transport, and infrastructure facilities play key roles in the choice of residence relocation. It should also be noted that regional shrinkage should not be regarded as simply the result of suburbanization. First, suburbanization causes a decrease in the population around the city center because of the emigration of the middle class; however, this population decline is temporary and will be quickly filled by new low-income families. In this regard, it differs from the long-term population decline caused by urban shrinkage. Second, although China's suburbanization has begun, almost none of the urban core areas have declined and remain growth poles. People still want to live in city centers because of their superior infrastructure, better access to public services, and greater employment opportunities. Therefore, the current suburbanization should not be considered the main reason for the large-scale population changes within Chinese megacities.

In addition, we found that Harvey's three stages of capital circulation used to explain the uneven development of space are also suitable for understanding the regional shrinkage that is occurring in megacities in China. When Chinese cities entered the late stage of industrialization, the profits produced by traditional industry began to fall, dropping further than that of emerging industries. Consequently, because of the profit-seeking nature of capital, capital moved from the original industrial space with low margins into higher-margin areas. Once such a process occurs, traditional industrial enterprises go bankrupt because of the lack of investment, and the built environment they create quickly declines. This is an explanation for the shrinkage of most old industrial areas from a capital perspective. However, the situation is different in the case of Wuhan, where state capital maintains investments in WISCO, as a key SOE, and the merger of WISCO keeps the enterprise afloat, preventing it from going bankrupt. However, the built environment in which the enterprise is located has been unable to secure sufficient capital investment, especially from the market. Therefore, the space can only be kept in a shrinking state. This makes a huge difference from the surrounding area, which can obtain continuous market investment. This is the type of regional shrinkage we observed in Qingshan District, Wuhan.

In terms of the "regional shrinkage" observed in Wuhan, we have redefined this shrinkage more accurately as "shrinkage under growth". This type of shrinkage is not the same as the shrinkage in an economic downturn. The overall economy is still growing, and there has been no sharp decline. Furthermore, it is not completely equivalent to the regional shrinkage in the Pearl River Delta region. There, the global shift of manufacturing industries from China to Southeast Asia, coupled with a similar shift of industries from the Pearl River Delta to inland cities, has led to a decrease in business and population migration across regions and cities (Li et al., 2015). Since most industrial enterprises in the Pearl River Delta are private or foreign owned, those in the transferred industrial sector in the Pearl River Delta are rapidly shrinking. However, unlike the Pearl River Delta, the economic structure of Qingshan District is dependent on SOEs. Therefore, it is more resilient than the private sector, having managed to avoid the large number of bankruptcies experienced by the Pearl River Delta region. Consequently, Qingshan District has resulted in a different type of urban shrinkage landscape. This "shrinkage under growth" is not unique to Wuhan and can also be observed in other Chinese cities with a state-owned economy.

Over the past 40 years, Chinese cities and towns have experienced rapid urbanization. Unlike European and American cities, shrinkage in Chinese cities and towns coexists with growth within the same metropolis and in the national context. How can this new phenomenon be explained? Is the regional shrinkage a staged process or a long-term process? This case study is just the beginning of research on the regional shrinkage phenomenon. Follow-up studies can focus on at least three aspects. First, there is a need for a more sophisticated identification of regional shrinkage in an economic growth context. Because of problems such as the shrinkage criteria in the Chinese context, changes to municipal boundaries, and the inconsistency of traditional data, current studies have not yet fully grasped the overall picture of urban growth and shrinkage. Using new data sources and innovative theoretical tools to identify geographic boundaries is one of the directions for future research. Second, the mechanism of regional shrinkage or "shrinkage under growth" is worthy of closer examination. The capital perspective offers a valuable direction for understanding regional shrinkage. If urban renewal (gentrification) is regarded as the return of capital to the inner city (Smith, 1979), could regional shrinkage be interpreted as the flight of capital from built-up areas? This hypothesis needs more empirical research and debate. Third, a comparative study of the growth and shrinkage in one city is necessary and valuable. Growth and shrinkage are two sides of regional development. Therefore, it is not comprehensive or systematic to discuss the regional shrinkage issues without also considering growth.

Funding

This research was funded by the National Natural Science Foundation of China (42001188), the Hubei Provincial Natural Science Foundation of China (2020CFB350), and the Fundamental Research Funds for the Central Universities (CCNU20QN031).

Notes

1 In the Chinese administrative hierarchy, municipal districts under prefecture-level cities are at the same administrative level as county-level cities.

2 National central cities are the highest level of cities in China's urban planning system. They were first proposed in 2005 by the former Ministry of Construction of the People's Republic of China in accordance with the "Urban Planning Law" when formulating a national urban system plan.
3 The CRU steel index is an international steel price index that provides comprehensive, accurate, and up-to-date price assessments.

References

Beauregard, R. A. 2009. Urban Population Loss in Historical Perspective: United States, 1820–2000. *Environment and Planning A*, 41(3), 514–528.

Bernt, M. 2016. The Limits of Shrinkage: Conceptual Pitfalls and Alternatives in the Discussion of Urban Population Loss. *International Journal of Urban and Regional Research*, 40(3), 441–450.

Bertinelli, L. & Black, D. 2004. Urbanization and Growth. *Core Discussion Papers Rp*, 56(1), 1–96.

Bureau of Statistics of Wuhan, 2001, 2011, 2017, 2019. *Wuhan Statistical Yearbook*. Beijing: China State Statistical Press. //武汉统计局, 2001, 2011, 2017, 2019. 武汉统计年鉴. 北京: 中国统计出版社.

Bureau of Statistics of Wuhan, 2016. *Wuhan Population Sampling*. //武汉统计局, 2016. 武汉人口抽样调查.

Du, Z. W. & Li, X. 2017. Growth or Shrinkage: New Phenomena of Regional Development in the Rapidly-urbanizing Pearl River Delta. *Acta Geographica Sinica*, 72(10), 1800–1811. //杜志威, 李郇, 2017. 珠三角快速城镇化地区发展的增长与收缩新现象. 地理学报, 72(10), 1800–1811.

Elzerman, K. & Bontje, M. 2015. Urban Shrinkage in Parkstad Limburg. *European Planning Studies*, 23(1), 87–103.

Gao, S. Q. 2017. Tracing the Phenomenon Concept and Research of Shrinking Cities. *Urban Planning International*, 32(3), 50–58. //高舒琦, 2017. 收缩城市的现象,概念与研究溯源. 国际城市规划, 32(3), 50–58.

Gao, Z., Yin, N. W., Tong, X. Y., Li, D. X. & Gu, J. 2019. Shrinkage under Urban Growth: A Case Study of Wuhan City. *Tropical Geography*, 39(1), 29–36. //高喆, 尹宁玮, 童馨仪, 李东欣, 顾江, 2019. 城镇增长下的收缩:以武汉为例. 热带地理, 39(1), 29–36.

Haase, A., Rink, D., Grossmann, K., Bernt, M. & Mykhnenko, V. 2014. Conceptualizing Urban Shrinkage. *Environment and Planning A*, 46(7), 1519–1534.

Harvey, D. 1982. *The Limits to Capital*. Oxford: Blackwell.

Hollander, J. B., Pallagst, K., Schwarz, T. & Popper, F. J. 2009. Planning Shrinking Cities. *Social Science Electronic Publishing*, 72(4), 223–232.

Hollander, J. B. & Nemeth, J. 2011. The Bounds of Smart Decline: A Foundational Theory for Planning Shrinking Cities. *Housing Policy Debate*, 21, 349–367.

Li, X., Du, Z. W. & Li, X. F. 2015. The Spatial Distribution and Mechanism of City Shrinkage in the Pearl River Delta. *Modern Urban Research*, 9, 36–43. //李郇, 杜志威, 李先峰, 2015. 珠江三角洲城镇收缩的空间分布与机制. 现代城市研究, 9, 36–43.

Li, X. et al. 2017. Academic Debates upon Shrinking Cities in China for Sustainable Development. *Geographical Research*, 36(10), 1997–2016. //李郇, 等, 2017. 局部收缩: 后增长时代下的城市可持续发展争鸣. 地理研究, 36(10), 1997–2016.

Li, Z. & Long, Y. 2018. An Analysis on Variation of Quality of Street Space in Shrinking Cities Based on Dynamic Street View Pictures Recognition: A Case Study of Qiqihar. *Urbanism and Architecture*, 6, 21–25. //李智, 龙瀛, 2018. 基于动态街景图片识别的收缩城市街道空间品质变化分析—以齐齐哈尔为例. 城市建筑, 6: 21–25.

Lin, X. B., Yang, J. W. Zhang, X. C. & Chao, H. 2017. Measuring Shrinking Cities and Influential Factors in Urban China: Perspective of Population and Economy. *Human Geography*, 32(1), 88–95. //林雄斌, 杨家文, 张衔春, 晁恒, 2017. 我国城市收缩测度与影响因素分析—基于人口与经济变化的视角. 人文地理, 32(1): 88–95.

Liu, H. 2016. Research on Quantitative Calculation Method of Shrinking City. *Modern Urban Research*, 2, 17–22. //刘合林, 2016. 收缩城市量化计算方法进展. 现代城市研究, 2, 17–22.

Liu, C. Y. & Yang, P. F. 2017. A Comparative Study on the Motivation Mechanism and Performance Characteristics of Chinese and Foreign Shrinking Cities. *Modern Urban Research*, 3, 64–71. //刘春阳, 杨培峰, 2017. 中外收缩城市动因机制及表现特征比较研究. 现代城市研究, 3, 64–71.

Liu, Y. B. & Zhang, X. L. 2017. A Study on the Shrinkage of Wuhan Metropolitan Area. *Planners*, 33(1), 18–25. //刘玉博, 张学良, 2017. 武汉城市圈城市收缩现象研究. 规划师, 33(1), 18–25.

Long, Y., Wu, K. & Wang, J. H. 2015. Shrinking Cities in China. *Modern Urban Research*, 9, 14–19. //龙瀛, 吴康, 王江浩, 2015. 中国收缩城市及其研究框架. 现代城市研究, 9, 14–19.

Martinez-Fernandez, C., Audirac, I., Fol, S. & Cunningham-Sabot, E. 2012. Shrinking Cities: Urban Challenges of Globalization. *International Journal of Urban and Regional Research*, 36(2), 213–225.

National Bureau of Statistics of China. 2008-2014. *China Statistical Yearbook*. Beijing: China State Statistical Press. //中国统计局, 2008–2014. 中国统计年鉴. 北京: 中国统计出版社.

Oswalt, P. 2006. *Shrinking Cities*. Berlin: Hatje Cantz.

Rieniets, T. 2009. Shrinking Cities: Causes and Effects of Urban Population Losses in the Twentieth Century. *Nature and Culture*, 4(3), 231–254.

Schilling, J. & Logan, J. 2008. Greening the Rust Belt: A Green Infrastructure Model for Right Sizing America's Shrinking Cities. *Journal of the American Planning Association*, 74(4), 451–466.

Smith, N. 1979. Toward a Theory of Gentrification: A Back to the City Movement by Capital, not People. *Journal of the American Planning Association*, 45(4), 538–548.

Smith, N. 2010. *Uneven Development: Nature, Capital and the Production of Space*. Athens: University of Georgia Press.

Wiechmann, T. 2008. Errors Expected-aligning Urban Strategy with Demographic Uncertainty in Shrinking Cities. *International Planning Studies*, 13(4), 431–446.

Wu, K., Long, Y. & Yang, Y. 2015. Urban Shrinkage in the Beijing-Tianjin-Hebei Region and Yangtze River Delta: Pattern, Trajectory and Factors. *Modern Urban Research*, 9, 26–35. //吴康, 龙瀛, 杨宇, 2015. 京津冀与长江三角洲的局部收缩: 格局、类型与影响因素识别. 现代城市研究, 9, 26–35.

Yang, Z. & Dunford, M. 2017. City Shrinkage in China: Scalar Processes of Urban and Hukou Population Losses. *Regional Studies*, 52(8), 1111–1121.

Yang, D. F., Long, Y., Yang, W. S. & Sun, H. 2015. Losing Population with Expanding Space: Paradox of Urban Shrinkage in China. *Modern Urban Research*, 9, 20–25. //杨东峰, 龙瀛, 杨文诗, 孙晖, 2015. 人口流失与空间扩张:中国快速城市化进程中的城市收缩悖论. 现代城市研究, 9, 20–25.

Zhang, B. B. & Li, Z. G. 2017. Shrinking Cities: International Progresses and Implications for China. *City Planning Review*, 41(10), 103–108. //张贝贝, 李志刚, 2017. "收缩城市"研究的国际进展与启示. 城市规划, 41(10), 103–108.

Zhou, K. & Qian, F. F. 2015. Shrinking City: On Searching for Urban Development in Non-growing Scenarios. *Modern Urban Research*, 9, 1–13. //周恺, 钱芳芳, 2015. 收缩城市: 逆增长情景下的城市发展路径研究进展. *现代城市研究*, 9, 1–13.

Zhou, K., Qian, F. F. & Yan, Y. 2017. A Multi-scaled Analysis of the "Shrinking Map" of the Population in Hunan Province. *Geographical Research*, 36(2), 267–280. //周恺, 钱芳芳, 严妍, 2017. 湖南省多地理尺度下的人口 "收缩地图". *地理研究*, 36(2), 267–280.

8 Urban shrinkage in the double periphery

Insights from the Sino-Russian borderland

Ekaterina Mikhailova and Chung-Tong Wu

Introduction

The border regions between Russia and China, closed to visitors until the early 1990s, attract the attention of researchers from diverse disciplines interested in issues ranging from economic development, migration, environmental change to politics but, until recently, few have studied the phenomenon of urban shrinkage that characterizes most of these regions and cities. This chapter examines the border cities of Blagoveshchensk (Blago) in Amur Oblast, Russia, and Heihe in Heilongjiang province, China, to explore the following. What are the characteristics of urban shrinkage in border regions and border cities? Does the border exacerbate or mitigate urban shrinkage? What additional factors might impact on urban shrinkage in a border setting?

Blago and Heihe, separated by 570 m across the Amur or Heilongjiang River, which forms this section of the border between Russia and China, are geographically the closest pair of cities on the border between the two countries. The population of Blago was about 225,000 (Mikhailova and Wu, 2017, p. 516) and that of Heihe's urban center (Aihui district) 124,900 in 2014 (Mikhailova, Wu, and Chubarov, 2019, p. 293). However, the regions of which they are a part are vastly different in size: Amur Oblast had a population of 794,000 in 2014 and was 67.5% urban and the population of Heilongjiang was 38.31 million and 60% urban. Although geographically close, these cities are perpetually one hour apart because China adheres to one time zone across the country. This is only one of the notable differences that generate inconveniences and, at times, cautious interactions.

This chapter begins with a brief review of relevant studies of shrinking border regions and cities followed by a discussion of a theoretical framework germane to this study of shrinking border regions and cities. The third, fourth, and fifth sections present the narratives from the perspective of Blago and Heihe. The conclusion will summarize aspects for further investigation.

DOI: 10.4324/9780367815011-10

Literature review and theoretical framework

Studies of cross-border shrinking cities

Available studies of shrinking border cities and regions are chiefly studies of cities in the European Union (EU) and Central and Eastern European (CEE) countries, but studies of Chinese counterparts are also emerging. Several themes can be identified: the impact of economic structure and competitiveness associated with peripheral location; and a methodology to identify urban shrinkage and a variety of cross-border interactions. A mono-industry base and outlying border location are identified as causes of urban shrinkage in diverse postsocialist countries including South Bulgaria (Simeonova and Milkova, 2018), Lithuania (Ubarevičienė, Van Ham, and Burneika, 2016), Latvia (Puzulis and Kule, 2016), Lithuanian and Polish border regions (Bruneckiene and Sinkiene, 2015), former East German border cities (Schätzl, 2006) and the Northeast of China (Cheng, Song, and Liu, 2019). Others assess the economic competitiveness (Bruneckiene and Sinkiene, 2015) of border cities. A number of studies investigate how best to identify urban shrinkage in the postsocialist context, relying on macro-level data such as gross domestic product (GDP) and population change (Cheng et al., 2019; He and Yang, 2019; Puzulis and Kule, 2016; Wang, 1997), or including other data such as nighttime illumination (Liu et al., 2018).

Cross-border interactions are represented chiefly by studies between EU countries. Hoekveld's study of three declining border regions of the Netherlands and its neighbors (Hoekveld, 2012) and that of Chouraqui of the Forbach region of France and its German neighbors (Chouraqui, 2018) identify a range of interactions, including commuting for employment, residential migration (from Germany to the Netherlands and Germany to France), and potential negative commercial impacts (for France). An earlier study investigates initiatives of trying to attract Polish citizens to take up residence in East German border regions as a means of bolstering their population (Schätzl, 2006). Similar studies of the borderlands of Slovakia and Czechia (Halás, 2006) and the Polish-Ukrainian border regions (Smetkowski, Nemeth, and Eskelinen, 2017) also focus on the asymmetrical interactions and regional differences. These studies highlight interactions that can develop in the setting of a porous border (Schengen Area) but, importantly, they also identify differences due to regional conditions and context (Bürkner 2002). Furthermore, there are cautionary studies, such as Bürkner's study of Polish and German border towns, about cultural obstacles to interactions, but others point to national level obstacles in cases such as China and its northern neighbors (Cheng et al., 2019). Several studies, including research on Lithuania (Ubarevičienė et al., 2016) and China (Gao and Long, 2017; Liu et al., 2018), identify the phenomenon of shrinkage at the rural or regional periphery and concentration at the urban center.

Core-periphery and peripheralization

We posit border regions and border cities require an analytical framework beyond those applied to shrinking cities. The conventional narrative of urban shrinkage is usually framed in terms of national and regional economic shifts leading to deindustrialization, loss of employment opportunities, and consequent population decline (Hollander et al., 2009; Oswalt and Rieniets, 2006; UN Habitat, 2008). For resource-based cities, the main causes of urban shrinkage are often exhaustion of the resource and/or increasingly uneconomic operations due to competition, additional environmental regulations, and rising costs (Martinez-Fermandez et al., 2012). Climatic preferences and other social attractions also contribute to population shifts (Beauregard, 2009; Weaver et al., 2016). While many, if not all the above, apply to border regions, there are additional contributing factors, including national security, foreign policy, and national political processes that lead to neglect by the central government. A multi-scaler (local, border regional, provincial, and national) and multi-actor (local, provincial, special commissions, and national) analysis is required to comprehend the multiple facets of cross-border shrinking cities and regions. Contextualizing the milieu that frames or inhibits the development possibilities of border cities is also necessary.

The core-periphery model (Friedmann, 1973; Wallerstein, 1974), and the concepts developed by the peripheralization discourse, provide the analytical framework for this chapter. More recent applications of the core-periphery model include the notion of "double periphery" applying to border regions (Wu, 2016) and to the US urban context (Silverman, 2018). The "double periphery" notion argues that it is not merely location but the social, economic, and political processes that relegate these regions to the margin. Silverman argues that global cities' analysis and the analytical framework of dependence are the keys to understanding the peripheral aspects of shrinking US cities (Silverman, 2018). While Silverman applies the term "periphery" to non-global cities, we use the term "periphery" in both a literal and metaphorical sense. Border regions are on the geographic periphery of nations but often also in political, social, and economic terms. Bordering another country or countries raises national security aspects that add to the complexity of border regions. The presence of military garrisons for border guard duties and immigration and customs officials to manage border crossings can constrain local autonomy. This formulation aligns with the peripheralization discourse, which emphasizes the processes of centralization (Lang, 2012, 2015), multidimensional aspects beyond location, accessibility, and density and a multi-scaler perspective (Eder, 2019; Kuhn, 2015). Friedmann's four processes of innovation: diffusion, control (decision-making), migration, and investment (Friedmann, 1973) are reflected in the economic, demographic, and political dimensions identified by Kuhn and innovation capacity investigated by Kuhn and Eder. Whereas the core-periphery model focuses on the formation of dependency relationships, the peripheralization concept emphasizes the dynamic process that allows regions to change their status over time.

The core-periphery model and the peripheralization discourse both posit that the development process is characterized by dependent relations between the core region and its periphery (Lang, 2015). The core region commandeers resources, finance, and talent from the periphery resulting in uneven development with the periphery that is mired in a downward spiral due to a combination of "cumulative causation" and "backwash" effects. Since key decisions are made in the core, the political processes reinforce the unequal pattern. In this conceptualization, the periphery can be the border region, but not necessarily so. The crucial feature is that these regions are dependent on the core. In this study, the periphery is the border region that is politically and economically dependent on the core region of its country. We further posit that the peripheral status of border regions in Russia and China, two geographically large countries, contribute to a path-dependent condition in which the border regions that were neglected in the past are further marginalized as the respective countries transitioned from a planned economy to a market-oriented economy based on encouraging growth in specific regions—coastal regions of China, and for Russia, Moscow, St. Petersburg and regions exporting natural resources.

Urban shrinkage at the Russo-Chinese borderland: case study cities

Blago in a shrinking region(s)

Blago is an administrative center of Amur Oblast, 1 of 11 federal regions comprising the Russian Far East (RFE). Stretching from Lake Baikal to the Pacific Ocean, RFE covers over 40% of Russia's territory but is inhabited by only 5.5% of Russia's population.[1] From 1990 to 2018, RFE experienced a severe demographic decline due to the outmigration of over 1.8 million. Even though the intensity of outmigration has slowed since 2010, the trend persists. In 2016, RFE's outmigration was a record low with 12,000 people moving out, half of the 2015 level.

RFE's population outflow gravitates toward the European part of Russia—a trend characterized as "the western migration drift". The East-West vector of outmigration was present already in the 1960s when the Soviet labor market was experiencing deficiency in labor resources combined with improved working conditions in the more attractive regions. With the start of the economic transition in the 1990s, "the western migration drift" and centripetal migration intensified as people rushed to regions and cities "where the economic collapse was less felt, the market developed faster and more jobs were available" (Zaionchkovskaya, 2013, p. 10).

However, within the vast territory of RFE there are variations. Within RFE, Amur Oblast represents the middle ground between the prosperous

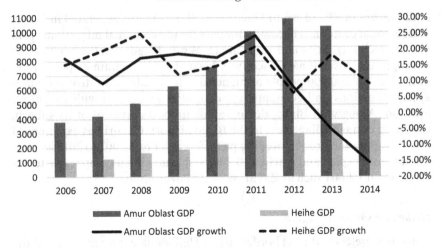

Figure 8.1 Amur Oblast and Heihe GDP per capita (US$) and GDP growth rates 2006–2014.

*GDP/capita has been converted from national currencies to USD based on OECD exchange rates (available at https://data.oecd.org/conversion/exchange-rates.htm).

and economically depressed regions. The monthly wage in Amur Oblast is generally equal to the national average, which is 40–50% less than Moscow. Since the 2014-ruble depreciation, the US dollars (USD) equivalent of nominal wages in Amur Oblast has dropped by one-third from $960 in 2013 to $675 in 2018. Between 2005 and 2017, the GDP per capita in Amur Oblast fluctuated between 65% and 85% of the Russian average. The USD equivalent of Amur Oblast GDP dropped by 45% from 2012 to 2016 reaching $5,100, the same as in 2007 (Figure 8.1).

Over the course of the last 30 years (1989–2019), Amur Oblast lost one-fourth of its population and shrank from 1,050,000 to 795,000 inhabitants, chiefly due to outmigration. While annual outflow of several thousands of people has become common for the region, there were several major peaks of outmigration: 19,000 in 1992; 23,000 in 1994; 20,000 in 2001; and 34,000 in 2009. The withdrawal of state subsidies from RFE, the consequent collapse of city-forming enterprises, the decline of coal, forest, and metal industries and the termination of the Baikal-Amur Mainline construction project in 1991 were the most significant economic factors that hastened population outflow from Amur Oblast. The other major cause was redeployment or demobilization of military units stationed in the region. Since 2010 the regional population loss has subsided to about 3,000–4,000 annually.

Urban shrinkage similar to that in Amur Oblast occurs in seven out of ten cities and is accompanied by the concentration of population and economic activity in certain nodal settlements, notably Blago. In 2015 about 28% of the Amur Oblast population lived in Blago, 10% more than in 1989. In addition, about 53% of Amur Oblast's GDP in 2015 was generated in Blago. The regional center has become a magnet for intraregional migrations from nearby declining cities and rural areas. While Blago is not immune from outmigration flows, it has, since 1987, been able to maintain its population at the level of 200,000 people or higher due to its administrative and managerial functions and economic opportunities associated with its border location. Along with Blago, demographic growth is typical of its suburb, Chigiri, a cottage village next to the Blago airport. Chigiri has experienced steady population growth over the past 15 years. From 2002 to 2018 its population quadrupled and reached 10,200 people making it comparable to cities of Amur Oblast such as Zavitinsk and Skovorodino.

Heihe in a shrinking region

Heihe is a prefectural-level border city of Heilongjiang province, one of the three provinces often referred to as the Northeast "rust belt" of China. The province recorded an outmigration of 2.05 million between 2010 and 2015 and a total fertility rate that is among the lowest in China—around 0.73 since 2010, a combination that led some to raise national security concerns because it has a long border with Russia (Jia, 2012; Song, Sha, and He, 2016). At the founding of the People's Republic of China (PRC), with help from the Soviet Union, heavy industries developed rapidly in the Northeast of China. Its celebrated heavy industries and energy base were touted as a national model during the Cultural Revolution. From the early 1980s, the region's fortune started to turn as its heavy industries became uncompetitive in a more market-oriented context and the resource base dwindled or became less viable as environmental regulations and economic conditions changed. This "rust belt" legacy is the subject of several national programs of economic revival promulgated first in 2003, then repeated in 2009, 2013, and again in 2016. The necessity for multiple programs attests to the persistent problems and the limited success of implemented policies and programs (Chung, Lai, and Joo, 2009). Heilongjiang's relative GDP per capita steadily declined from 5th in the nation in 1978 to 25th out of 31 in 2017. The minimum monthly wage set by the Heilongjiang government was just under 68% of that of Shanghai in 2019 (RMB 1,680 vs RMB 2,408), the latter being the highest in the nation. Heilongjiang is still heavily reliant on oil production and coal mining—the energy sector accounted for 70% of its industrial output and seven of its 13 prefecture-level cities are reliant on coal mining, oil, and timber production (Yu, 2015). The border regions of the province, in which Heihe is located, have even lower economic growth than the province—estimated to be just under 91% of the province and less than 77% of the national average in 2009 (Jia, 2012).

Institutional rigidity is also regarded as a contributing factor. Heilongjiang has the largest percentage of its agriculture and industries under state-owned enterprises, which are acknowledged to be slow to reform and uncompetitive

Table 8.1 Blago and Heihe indices (percentage change)

	2007	2008	2009	2010	2011	2012	2013	2014	2015	2016	2017	2018
Heihe (population)	0.28	–0.04	0.18	–0.05	–0.15	–0.15	–0.75	–1.06	–0.19	–0.4	–1.41	–0.75
Blago (population)	–1.46	–0.82	–0.34	3.79	–0.01	0.60	0.84	–1.31	1.87	0.06	0.04	0.30
Exports*		–2.5	–59	31.1	1.5	18.6	34.0	29.2	–76.7	–27.7	–2.6	–29.5
Imports*		20.9	–25.4	38.1	360	10.4	–33	–7.1	–38.9	–11.9	28.2	80.4

Sources: Exports and imports (provincial trade with Russia), Heihe Statistical Bureau, Amur Oblast Statistical Bureau (various years), (Song, 2019).

Note
* Trade between Russia and Heilongjiang Province.

in the new economy (Hancock, 2019; Liao and Wei, 2016). In the period 2000–2010, stagnant regions outnumbered the more vibrant regions in Heilongjiang (Wu and Zhuang, 2014). Heihe, therefore, exists in a region of a slowing industrial economy, with an agricultural sector reliant on national subsidies, a forestry sector with increasing restrictions, comparative low wages, and an institutional setting slow to reform. It is also located in a border region that is even more backward economically—the periphery of the periphery.

Heihe, already classified as a stagnant region in 2000 (Wu and Zhuang, 2014), has experienced fluctuations of its population (Table 8.1), but in 2003 its population started on a slow and persistent downward trend. Modest rises in population in the late 1990s were chiefly due to the initial boom of tourism, chiefly day tourism from across the border when China relaxed its visa requirements first in 1994 (restricted to visits to the tax-free trade zone on Heihe island) and then expanded in 2000 to allow Russian tourists access to the entire urban district. Adventurous traders and entrepreneurs from across the country came to exploit opportunities in Heihe, but the adverse ruble exchange rate since 2014 has led to a downturn in Russian tourism and associated trade, resulting in the departure of many immigrants. Conversely, Blago is experiencing a boom in cross-border tourism as Chinese day-trippers cross the border to shop for food and other consumer items, attracted by the low prices due to favorable currency exchange rates and the perceived high quality and safety of food items from Russia.

Trade hubs and cross-border interactions

Opening the border and cross-border interactions

While the literature on border regions and cities is replete with studies of traffic, people and goods flow across borders, there are also findings on cultural and historic factors that impact the size and direction of these flows. Transborder flows have developed between Blago and Heihe following the opening of their borders in the early 1990s, but their unique location on the border just across from each other has been perceived and used differently. Convinced that its border location is advantageous, successive Heihe mayors have been inspired to propose numerous schemes to realize its self-proclaimed goal of being a "window to Europe" (via Russia), a base for cross-border investment and a trade hub for exports to Russia and beyond.[2] Opening the border and easing visa conditions for Russian visitors generated a boom in border trade and tourism, but these initiatives were never reciprocated by Russia and were subject to the vagaries of currency exchange and economic conditions across the border. In Blago, location on the border stimulated dual perception of the city's primary purpose. The ideas of "fortress" or "national outpost" remain strong (Billé, 2012). Regional civil servants conduct themselves in ways to demonstrate their loyalty to the Russian state by making no concessions to the Chinese and maintaining the status of an equal partner at whatever joint activity eventuates (Mikhailova et al., 2019). Other Blago residents relate to the idea of a cross-cultural contact zone and connect their future with the increasing interactions with China, however, not without reservations.

Since the mid-2000s, Heihe has touted the virtues of a "twin city" ("two countries, one city") to Blago. Heihe's eagerness to foster twinning is at best seen by some Amur Oblast officials as being too pushy, insensitive to, or even dismissive of, how Amur Oblast may regard the proposed twinning relationship. At worse, Heihe is perceived as an upstart trying to show off to a neighbor that used to have much stronger political and economic prowess. In our previous exploration of the limited success of the twin city initiative (Mikhailova and Wu, 2017), we argued that in spite of trade, consumption, and investment ties, the twin city initiative can only be considered "ersatz" because few initiatives beyond limited economic ones have been implemented.

The vagaries of national policies meant that these two cities, less than one kilometer apart, are permanently a one-hour time zone apart such that synchronizing the office hours of border custom offices was not resolved until 2014. Other sensitive issues that have been subject to frequent intervention include cross-border passenger traffic; the tax-free quota for returning residents; the limits of the visa-free zone on the Chinese side of the border and the requirements for the visa-free travel—all elements that have direct and significant impacts on the potential of tourism and shuttle trade. Multiple interventions in 2000, 2002, 2003, 2006, and 2009 impacted on where individuals could stay, for how long and what tax-free amounts (by weight of goods or value) they could bring back to Russia (Zhuravskaya, 2011). These measures created confusion, imposed inconveniences for the travelers, impacted on cross-border trade and tourism and created opportunities for corruption (Ryzhova, 2008).

Other examples of interference by central government agencies abound. For example, in 2008, the Russian government introduced a new regulation to identify sailors without considering the river fleet involved in international water transportation (Mikhailova, 2018). Consequently, after so-called "sailors' passports" expired in January 2014, the Blago ferry crew had no valid documents to enter the port of Heihe. To avoid ferry border crossing on the Amur from shutting down, Blago and Heihe ports made a temporary bilateral arrangement to permit Russian sailors to enter Heihe based on their foreign travel passports and the muster roll until the issue of a proper seafaring identity document was solved.

Even activities that have obvious benefits to Blago have, from time to time, been inhibited by national regulations. From early 1990 until today, Amur Oblast is the only Russian region exporting energy to China. Energy export started in 1992 with the launch of the first Blago-Heihe power line (110 kW) and has gradually increased with the construction of new power lines in 1997 (110 kW), 2006 (220 kW), and 2011 (500 kW). From the 1990s to the early 2000s, energy export to China was a means of acquiring limited or unavailable goods at home, a crucial source of regional income. Initially, energy from Amur Oblast to China was exported on barter terms (Gordienko, 2019). In 2007, however, it was halted due to the increase in export tariffs imposed by the Russian Federal Tariff Service. That federal intervention distorted the market equilibrium between the excess energy-generating capacities of Amur Oblast and energy-deficient Heilongjiang province making it too expensive for Chinese buyers. Until 2009 when energy export was resumed, Amur Oblast hydroelectric power plants had to discharge water to let their generators idle.

These examples testify to the vulnerability of cross-border activities to interference by multiple national actors who have little regard for any local inconveniences or even mutual economic loss. Such a border mindset fostered asymmetric policies implemented without regard to their impacts on local livelihood. Furthermore, they illustrate the dependency relationships between these border cities and their respective central governments and the lack of autonomy as identified by the core-periphery model and the peripheralization discourse.

Growth paradigm and urban shrinkage

While persistent depopulation is acknowledged both in RFE and in Amur Oblast, the degree of acceptance of demographic shrinkage differs from one official document to another. Some regional and macro-regional documents, such as the Amur Oblast Strategy for socio-economic development until 2025, demonstrate "observation without acceptance" (Pallagast, Fleschurz, and Said, 2017, p. 17). Others show a "certain acceptance" (Ibid., p. 17) of urban shrinkage (see the Amur Oblast demographic forecast). Affirmations for growth overshadow this patchy acceptance of shrinkage.

At the national level, the growth orientation is in full sway. According to the Strategy for RFE Demographic Policy of 20 June 2017, the macro-regional demographic agenda includes two tasks: firstly, to stabilize RFE's population and, secondly, to reach a forecast of short and long-term regional demographic growth of 802,000 inhabitants by 2020 and 805,000 inhabitants by 2025.

At the regional level, the demographic situation is often perceived more realistically. The regionally prepared Amur Oblast demographic forecast expected an 8% population decline within the next 18 years—a loss of around 65,000 people by 2035.[3] However, it is hardly possible to adopt regional strategies that run counter to federal strategies. Consequently, the Amur Oblast Strategy for socio-economic development until 2025, a publicly available document obligatory for each Russian region, follows the federal vision of demographic growth. The Strategy predicts a change in migration rate from −46.7 in 2015 to +34.2 in 2025 (for every 10,000 inhabitants) without explanation. The same unrealistic predictions are reflected in economic indicators. The Strategy forecasts a 50% increase in the average monthly wage, a two-fold growth in GDP, and a six-fold growth in industrial output by Amur Oblast between 2020 and 2025 (Government of Amur Oblast, 2017, pp. 118–119).

Unsurprisingly, within a national context of growth at any cost, the narrative in Heihe and Heilongjiang has never deviated from the pursuit of economic growth. The unrealistic goal of becoming another Special Economic Zone (SEZ) like Shenzhen, an idea promoted by the then General Secretary Hu in 1982, was never going to be realized because none of the conditions that made Shenzhen possible are available in Heihe and its region. Nevertheless, an expectation of economic growth never diminished even under the "new normal". The 2016 Heilongjiang Provincial plan for the border regions continues to rely on promoting economic growth (Heilongjiang Development and Reform Commission, 2016) and only acknowledges the out-migration of skilled labor in one sentence in the 15-page document. This is a classic case of denial in the face of mounting evidence (Pallagast et al., 2017), but hopefully, the 2019 April State Council statement on shrinking cities may help to usher in changes.

Understandably, Heihe officials adhered to the national growth discourse without fail for the sake of obtaining additional funding and for personal advancement prospects. Even in the years when regional and urban shrinkage became evident, there was no official acknowledgment of a problem nor any discernible efforts to counter these trends because that would have been regarded as contrary to the narrative of growth. Despite deepening urban shrinkage in the province and in the border region, this aspect was completely ignored in the annual work reports submitted by successive mayors even though the downward population was noted in the annual economic and social development report.

Peripheralization, core-periphery relations, and urban shrinkage

The above section demonstrates that interactions across the border developed when the borders became more permeable, but the Chinese officials have a more grandiose view of border development and the trade hub is at best seen as a stepping-stone to Heihe's role as a "window to Europe". Even the trade hub development was hampered by practical issues of transportation and cultural differences but most importantly by numerous interventions of central government agencies on both sides of the border. Instead of the type of interactions reported in cross-border literature, asymmetrical institutions, and practices that hamper interactions developed.

While the fitful development of trade and tourism-related activities between Blago and Heihe is hampered by specific bureaucratic interventions, the development of core-periphery relations is manifested in the process of peripheralization starting with their narratives of cross-border interactions. Heihe, far from the coastal regions of China, was relegated to the periphery of China's reform policies. The opening of the border, more than a decade after reforms were first implemented in the rest of the country, barely altered the processes that reinforced the core-periphery relations well underway in the rest of the country (Lu and Wei, 2007).

Ambivalent cross-border narrative

Over the last several decades, Heihe, being a part of the border region, has been included in numerous Chinese attempts to establish co-development projects with Russia, including the ambitious programs under the banner of Dongbei (Northeast) Russia Far East integration which saw several rounds of agreements signed in 2005, 2007, and 2009 (People's Republic of China, 2009). The Chinese seemed indifferent or oblivious of how Amur Oblast or Russian officials might perceive these proposals. Time and again the Chinese were frustrated by the lack of progress with 200 or more projects under these agreements. As Christofferson observes, since 2015, it seems to have merely placed the "decades-old ideas and incorporated them into a new framework" of the Silk Road Economic Belt or the Belt and Road Initiative (Christofferson, 2016).

The Chinese national government's desire to cast the development of Heilongjiang and the Northeast provinces within the ambit of its infrastructure diplomacy initiatives (Jia and Bennett, 2019) is consistent with its growth paradigm and its geo-political ambitions, but none of these have taken into account the nature of the shrinking region and its shrinking cities. These national initiatives reflect the same approach applied in the Northwest region of China using infrastructure projects for the purposes of "territorial integration and to enhance its representational function as a land gateway to Central Asia and beyond…" (Joniak-Lüthi and Bulag, 2016, p. 5). In the case of Heilongjiang, these initiatives met with a reluctant neighbor who is wary of China's intentions, similar to situations identified in the EU context (Bürkner 2002). The programs that were tried in Blago and Heihe never had any real chances of success because the overall context involved multi-scaler perspectives and multi-actors making policies that weakened the potentials of the regions and the cities, rather than strengthening them.

Border region mindset

Until the early 1990s, Blago and Heihe, both cities in border regions that have experienced numerous conflicts or were proximate to conflicts, were closed to visitors. While the change of their status heralded brief periods of increasing and intense trade and investment links across the border, the lack of autonomy at the local level on both sides of the border hindered many early initiatives. These included a proposal to build a road and rail bridge linking the two cities, the establishment of a mutual trade zone, proposals for a winter pontoon bridge, and city twinning. These initiatives involved multiple central and provincial government agencies' approvals and financial assistance—not the least because all these involved crossing the national boundary literally or figuratively.

The torturous path of the proposal to build a bridge linking the two cities is the most telling example of the local governments' reliance on national, provincial, and even bilateral approvals and funding to realize the proposal. This initiative embodied the reliance on grandiose projects, that is, a multi-scaler approach and multi-actors. The idea of a bridge was first mooted by the then Amur Oblast administration in 1988 and was immediately embraced by the Chinese. In 1996 the agreement was signed for a rail and road bridge construction. This proposal was the key physical element that would promote Heihe's role as a metaphoric bridge to Europe via Russia with the potential to open trade and investment for both NE China and the RFE. Despite numerous subsequent negotiations and agreements, the pact to build the bridge was not finalized until 2015 and the physical link of a road bridge was finally completed in late 2019, with opening repeatedly delayed from mid-2021 to 2022 due partly to the global pandemic and the respective governments' off and on border closures to try to stop the spread of the virus. Indeed, due to the uncertainties associated with the COVID pandemic, there is no assurance when through traffic will commence.

Without a physical link, transport between Heihe and Blago relied on ferries for passengers and cargo ships/barges during the summer months, hovercrafts during the winter, and, since 2011, a pontoon bridge on the frozen river. Even the winter pontoon bridge took several years to negotiate, starting in 2007 and finally realized in 2011. The limited cargo capacity in the winter does not facilitate trade during several months of the year. The impost on individuals, enterprises, and entities engaged in tourism, trade, and commerce was plain to observe but still took almost three decades for a resolution. The few relatively stable cross-border activities that emerged almost simultaneously with the opening of the border were electricity export, ferry border crossing, and shuttle trade, but even these met a variety of obstacles, as explained in the previous section.

Imposed models

A variety of national and foreign policy initiatives impact on development possibilities in Blago and Heihe. For Heihe, programs aiming at economic revival are focused on industrial parks for export processing that are supposed to attract investors from Russia or elsewhere. There are currently five industrial zones in Heihe. These projects have been implemented despite two fundamental issues—the lack of year-round direct transport between Heihe and Blago (dependent on the situation with the COVID-19 pandemic, the bridge may be in service for freight by mid-2022 and for passenger traffic by 2023 – see Grigorieva, 2022) and the lack of energy or the need to import energy from Blago/Amur Oblast for the industrial parks to function. These projects were partly based on the national programs of Northeast Revival, starting in 2003, 2009, and 2013. By adhering to these programs, extra funds were made available. Local officials, therefore, were glad to embrace these programs whether they believed they had any chance of success or what other conditions were necessary for their success (Ivanov, 2013).

However, in August 2019, the State Council announced the establishment of new "Pilot Free Trade Zones" (PFTZ) in 6 provinces, including Heilongjiang (State Council People's Republic of China, 2019) in addition to the 11 similar zones already in operation in the rest of China. In the Heilongjiang configuration, the PFTZ is split into three locations, one each in Harbin, Heihe, and Suifenhe. The zone is typical of the "zoning for growth" approach through designating an area for tax concessions, policy reforms, and attraction of foreign direct investment (FDI) and domestic investments, with features akin to the SEZs, like Shenzhen. Indeed, Harbin (the provincial capital) is regarded as the key zone with links to east and west (via Mongolia to Chita/Transbaikalia, Russia and through Suifenhe to Vladivostok/Primorye, Russia) and to north and south via Tongjiang railroad bridge or the road bridge of Heihe (Song, 2019). The Heihe zone is 20 km^2 (inclusive of the bridge zone) and expected to focus on energy, green food processing, logistics, tourism, health, and financial industries (State Council People's Republic of China, 2019 Document 16). Unlike the 1997 proposal to establish a Mutual Trade Zone, which invited Blago to establish a similar area across the river and mirror the benefits offered in the Heihe zone (Cheng and Xu, 2006), this program does not mention that possibility. One could regard this initiative as the Central government finally deciding it is time to remedy the lack of global reach of the region. However, is this the appropriate project at this juncture?

Similarly, in Russia, one of the most significant recent territorial development tools launched by federal authorities is the "priority development areas" (PDAs) program—economic zones with a favorable taxation regime. Since the implementation of PDAs in late 2014, 20 zones have been established in RFE including three in Amur Oblast. These three PDAs in Amur Oblast are expected to create 4,750 new jobs and attract about 16.6 million USD investments to the region. As of January 2020, twelve of the RFE's PDAs have foreign investment. Seven of them, including one in Amur Oblast, have Chinese investments. Each PDA in Amur Oblast has a distinct economic profile. Launched in 2015, the Priamurskaya PDA is located right next to the newly constructed Amur River/Heilongjiang bridge, 20 km from Blago. Its specialization includes transport services, logistics and storage, petroleum, and fertilizer production, overlapping those proposed for the Heihe PFTZ.

Paradoxically, at the juncture when a physical link crossing the border becomes a reality, making possible a new era of cross-border cooperation, both sides have decided to implement almost the same kind of economic strategy in direct competition with each other. In the case of Heihe, it is being burdened with a model of development 40 years after it has been tried with varying degrees of success elsewhere in the country. The imposed models, implemented without regard to whether these strategies fit the current circumstances of the regions, again exemplify the dependency relationships between Heihe and Blago with their respective central governments. This approach reinforces the border, affirms the growth narrative, and ignores the settings of shrinking regions.

Conclusion

Blago and Heihe demonstrate that the core-periphery dependent relationships are not merely the result of being geographically far from the center/core; rather, they are the consequences of being on the periphery of the national growth discourse. In the postsocialist era, both centralized regimes continue to implement top-down regional development policies that have led to mass-outmigration from these border regions. Furthermore, from time to time, the center imposes policies that constrain local initiatives in their attempts to kindle economic development. The characteristics of the dependent relationship that impacts on these border regions are: (1) national priorities for these regions seldom go beyond announcements; (2) proposals and programs that apply to them are generally a catch up subject to whatever national policies may be at the time; (3) few of the proposed "solutions" take into account the circumstances of the regions since they are considered to be "places that don't matter" (Rodriguez-Pose, 2018); and (4) there is a lack of a national approach toward removing or resolving the historic burdens that could hinder closer collaborations across the border to promote partnerships for mutual benefit. Even as the decades in the making of the physical link, the road bridge, is about to become a reality, these border regions have imposed on them programs found to be successful elsewhere several decades ago, but which do not take into account the current situation at the border regions or the prevailing global context. After decades of being on the fringe of the national development paradigm, they are now burdened with almost identical and competing developments located just kilometers apart. Neither side has identified innovative programs that might exploit the opportunities presented by their cross-border location or that would fit the conditions of the shrinking regions.

While the case studies show that trade hubs in shrinking regions could become intraregional migration magnets due to economic opportunities derived from their border locations, their development is subject to diverse factors such as national decisions and global events that are beyond the control of the cities. Such trade hub roles are short-term and unsustainable and could not be relied upon to address population shrinkage. Population shrinkage remains largely ignored and, where acknowledged, assumed to be solved by growth-oriented strategies. This is the case in China even as the official mantra is supposed to be that of a "new normal" where economic growth is not the only goal. That none of the imposed projects acknowledge the reality of population shrinkage in the regions reinforces Eder's call to take into account regional "specificities" (Eder, 2019). Blago and Heihe illustrate the value of the core-periphery model and the peripheralization discourse to identify the processes of peripheralization imposed by increasingly centralized regimes on these cross-border cities and regions. Additional case studies of shrinking border cities and regions will enhance our understanding. We propose three specific aspects for further research.

Firstly, Blago and Heihe pinpoint an issue of institutional rigidity within the context of centralized regimes. In the case of Heihe, slow to reform state-owned enterprises were identified as part of the reasons for the lack of economic progress in the border region leading to interventions by the central government. In turn, this further debilitated local initiatives leading to a vicious cycle of increasing dependency on the center. Although the issue of backwash effects was identified in the core-periphery model, there is little specific attention given to institutional rigidity deepening dependency.

Secondly, the cases of Blago and Heihe challenge us to revise our understanding of local responses to urban shrinkage. These cases highlight the necessity to consider whether local and regional officials have the decision space to embrace the issue and present policies to deal with them. In these case studies, it is obvious this is not feasible in the context of highly centralized regimes. How to promote programs that can realistically manage urban shrinkage in the context of limited local autonomy remains a challenge to researchers and policymakers.

Thirdly, these cases underline the need for more studies of cross-border shrinking cities and regions to broaden our understanding of shrinking cities and regions in diverse settings. Whereas Blago and Heihe have specific contextual features even in a postsocialist context, they draw attention to the role of regional inequality and core-periphery relations on urban shrinkage. Additional case studies will broaden our understanding of cross-border shrinking cities and regions.

Notes

1 On 3 November 2018, the Presidential Decree No. 632 expanded RFE's territory by incorporating the Republic of Buryatia and Transbaikal Krai. This change of administrative boundaries restored the number of RFE's residents to its historical maximum reached in the final year of USSR's existence—8 million inhabitants.

2 These were the slogans on the website of Heihe until late 2019.
3 Unpublished document.

References

Beauregard, R. A. 2009. Urban Population Loss in Historical Perspective: United States 1820–2000. *Environment and Planning A*, 41(3), 514–528.

Billé, F. 2012. On ideas of the border in the Russian and Chinese social imaginaries. In Bille F., Delapace G. & Humphrey C. (eds.) *Frontier Encounters: Knowledge and Practice at the Russian, Chinese and Mongolian Border* (pp. 19–32). Cambridge: Open Book.

Bruneckiene, J. & Sinkiene, J. 2015. The Economic Competitiveness of Lithuanian-Polish Border Region's Cities: The Specific of Urban Shrinkage. *Equilibrium*, 10(4), 133–149.

Bürkner, H. J. 2002. Border Milieux, Transboundary Communication and Local Conflict Dynamics in German-Polish Border Towns: The Case of Guben and Gubin. *Die Erde*, 133(1), 69–81.

Cheng, Y., Song, T. & Liu, H. M. 2019. China's Border Shrinking Cities: Pattern, Types and Impact Factors. *Beijing Planning Review*, (3), 48–52. //程艺, 宋涛, 刘海猛, 2019. 我国边境收缩城市: 格局、类型与影响因素. 北京规划建设, 3, 48–52.

Cheng, R. & Xu, Z. Y. 2006. Sino-Russian Border Residents Mutual Trade Development from the Perspective of Heihe and Blagoveshchensk. *Russia and Central Asia Markets*, 1, 32–35. //成榕, 徐志尧, 2006. 从黑河—布拉戈维申斯克看中俄边民互市贸易的发展轨迹. 俄罗斯中亚东欧市场, 1, 32–35.

Chouraqui, J. 2018. The cross-border location of a medium-sized shrinking city, at once a potential asset and a factor of fragility: the case of Forbach. https://agence-cohesion-territoires.gouv.fr/sites/default/files/2020-09/medium-sized-cities_cget-4.pdf [Accessed 26 August 2019].

Christofferson, G. 2016. The Russian Far East and Heilongjiang in China's Sild Road Economic Belt [Online]. *Asia Dialogue*. https://theasiadialogue.com/2016/04/25/the-russian-far-east-and-heilongjiang-in-chinas-silk-road-economic-belt/ [Accessed 8 Sept. 2019].

Chung, J. H., Lai, H. & Joo, J. H. 2009. Assessing the "Revive the Northeast" (*zhenxing dongbei*) Programme: Origins, Policies and Implementation. *The China Quarterly*, 197, 108–125.

Eder, J. 2019. Peripheralization and Knowledge Bases in Austria: Towards a New Regional Typology. *European Planning Studies*, 27(1), 42–67.

Friedmann, J. 1973. *Urbanization, Planning and National Development*. Beverly Hills: Sage.

Gao, S.Q. & Long, Y. 2017. Distinguishing and Planning Shrinking Cities in Northeast China. *Planners*, 33(1), 26–32. //高舒琦, 龙瀛, 2017. 东北地区收缩城市的识别分析及规划应对. 规划师, 33(1), 26–32.

Gordienko, O. 2019. "Chinese" tariff: at what price do they sell electricity in China and how much do people in Heihe pay for electricity [in Russ.], *Amur Pravda Newspaper*. Available at: https://ampravda.ru/2019/11/28/092536.html [Accessed 12 January 2020].

Government of Amur Oblast. 2017. Strategy for socio-economic development until 2025 of July 13, 2012 (with amendments of November 8, 2017) [In Russ.]. Available at: https://www.amurobl.ru/upload/iblock/315/3159ba0160ccd60dbddf3d329d171102.pdf [Accessed 10 January 2020].

Grigorieva, I. 2022. Possible opening date of road bridge between Blagoveshchensk and Heihe announced. Amurskaya Pravda [In Russ.]. Available at: https://ampravda.ru/2022/01/11/109759.html [Accessed 12 January 2022].

Halás, M. 2006. Theoretical preconditions versus the real existence of cross-border relations in the Slovak-Czech borderland. In Komornicki T. & Czapiewski K. (eds.) *EUROPA XXI–15 Regional Periphery in Central and Eastern Europe*, Vol. 15 (pp. 63–76). Warszawa: Stanislaw Leszczycki Institute of Geography and Spatial Organization, PAS.

Hancock, T. 26 April 2019. China's shrinking cities: "most of my classmates have left". *Financial Times*. https://www.ft.com/content/bd1e1bf2-661f-11e9-a79d-04f35047d62 [Accessed 15 Sept. 2019].

He, Y. & Yang, L. 2019. Research on the spatial distribution and influencing factors of shrinking cities in Jilin Province. In Long Y. & Gao S. Q. (eds.) *Shrinking Cities in China: The Other Facet of Urbanization*. Singapore: Springer.

Heilongjiang Development and Reform Commission, 2016. *Heilongjiang Promote Wealth of Border Region Citizens Movement Plan (2016–2020)*. Harbin, Heilongjiang. //黑龙江省发展和改革委员会, 2016. *黑龙江省兴边富民行动规划 (2016-2020年)*. 黑龙江哈尔滨.

Hoekveld, J. J. 2012. Time-space Relations and the Differences Between Shrinking Regions. *Built Environment*, 38(2), 179–195.

Hollander, J. B. et al. 2009. Planning Shrinking Cities. *Progress in Planning*, 72(4), 223–232.

Ivanov, S. A. 2013. Economic and symbolic capital at the border of globalizing China: the case of Heilongjiang province. In Sevastianov, S. V., Kireev, A. A. & Richardson, P. (eds.) *Borders and Transborder Processes in Eurasia* (pp.188–203). Vladivostok: Dalnauka.

Jia, Y. M. 2012. A Study of Population Security and Economic Social Development in Border Areas. *Population Journal (Renkou Xuekan)*, (5), 22–29. //贾玉梅, 2012. 边境地区人口安全与经济社会发展研究—以黑龙江省边境地区为例. 人口学刊, (5), 22–29.

Jia, F. & Bennett, M. M. 2019. Chinese Infrastructure Diplomacy in Russia: The Geopolitics of Project Type, Location, and Scale. *Eurasian Geography and Economics*, 59(3–4), 340–377.

Joniak-Lüthi, A. & Bulag, U. E. 2016. Introduction: Spatial Transformations in China's Northwestern Borderlands. *Inner Asia*, 18(1), 1–14.

Kuhn, M. 2015. Peripheralization: Theoretical Concepts Explaining Socio-Spatial Inequalities. *European Planning Studies*, 23(2), 367–378.

Lang, T. 2012. Shrinkage, Metropolization and Peripheralization in East Germany. *European Planning Studies*, 20(10), 1747–1754.

Lang, T. 2015. Socio-economic and Political Responses to Regional Polarisation and Socio-Spatial Peripheralisation in Central and Eastern Europe: A Research Agenda. *Hungarian Geographical Bulletin*, 64(3), 171–185.

Liao, F. H. & Wei, Y. D. 2016. *Sixty Years of Regional Inequality in China: Trends, Scales and Mechanisms*. Santiago, Chile: RIMISP.

Liu, F. B. et al. 2018. The Research on the Quantitative Identification and Cause Analysis of Urban Shrinkage from Different Dimensions and Scales: A Case Study of Northeast China during Transformation Period. *Modern Urban Research*, 7, 37–46. //刘风豹,等, 2018. 城市收缩多维度、多尺度量化识别及成因研究—以转型期中国东北地区为例. 现代城市研究, 7: 37–46.

Lu, L. & Wei, Y. D. 2007. Domesticating Globalisation, New Economic Spaces and Regional Polarisation in Guangdong Province, China. *Tijdschrift Voor Economische En Sociale Geografie*, 98(2), 225–244.

Martinez-Fermandez, C., Wu, C. T., Schatz, L. K., Nobuhisa, T. & Vargas-Hernandez, J. G. 2012. The Shrinking Mining City: Urban Dynamics and Contested Territory. *International Journal of Urban and Regional Research*, 36(2), 245–260.

Mikhailova E. 2018. Collaborative Problem Solving in the Cross-Border Context: Learning from Paired Local Communities along the Russian Border. *Journal of Borderlands Studies*, 33(3), 445–464.

Mikhailova, E. & Wu, C. T. 2017. Ersatz Twin City Formation? The Case of Blagoveshchensk and Heihe. *Journal of Borderlands Studies*, 32(4), 513–533.

Mikhailova, E., Wu, C. T. & Chubarov, I. 2019. Blagoveshchensk and Heihe: (un)contested twin cities on the Sino-Russian border? In Garrard J. & Mikhailova E. (eds.) *Twin Cities: Urban Communities, Border and Relationships over Time* (pp. 288–300). New York and London: Routledge.

Oswalt, P. & Rieniets, T. (eds.) 2006. *Atlas of Shrinking Cities*. Ostfildern, Germany: Hatje Cantz Publishers.

Pallagast, K., Fleschurz, R. & Said, S. 2017. What Drives Planning in a Shrinking City? Tales from Two German and Two American Cases. *Town Planning Review*, 88(1), 15–28.

People's Republic of China. 2009. *People's Republic of China North East Region and Russia Federation Far East and East Siberia Regional Cooperation Plan Outline 2009–2018*. In Development and Reform Commission (eds.) Beijing. http://www.chinaeast.gov.cn/2010-06/03/c_13331199_2.htm. //中华人民共和国, 2009.中华人民共和国东北地区与俄罗斯联邦远东及东西伯利亚地区合作规划纲要(2009-2018年).

Puzulis, A. & Kule, L. 2016. Shrinking of Rural Territories in Latvia. *European Integration Studies*, (10), 90–105.

Rodriguez-Pose, A. 2018. The Revenge of the Places that Don't Matter (and What to Do About It). *Cambridge Journal of Regions, Economy and Society*, 11(1), 189–209.

Ryzhova, N. 2008. Informal Economy of Translocations: The Case of the Twin City of Blagoveshensk-Heihe. *Inner Asia*, 10(2), 323–351.

Schätzl, L. 2006. Impacts of European unification on shrinking border cities of East Germany. Paper presented at the ENHR conference *Housing in An Expanding Europe: Theory, Policy, Participation and Implementation*. Ljubljana: Solvenia.

Silverman, R. M. 2018. Rethinking Shrinking Cities: Peripheral Dual Cities Have Arrived. *Journal of Urban Affairs*. 1–18. doi:10.1080/07352166.2018.1448226.

Simeonova, V. & Milkova, K. 2018. Deindustrialization and Urban Decline in the Border Mountainous Regions in the Context of Regional Development (after the Example of the Rhodopes and Strandzha-Sakar). *ПуБ ЛИ ЧНИ ПОЛИТИКИ.bg*, 9(4), 36–48.

Smetkowski, M., Nemeth, S. & Eskelinen, H. 2017. Cross-border Shopping at the EU's Eastern Edge: The Cases of Finnish-Russian and Polish-Ukrainian Border Regions. *Europa Regional*, 24(1/2), 50–64.

Song, L. 2019. Strategic Priorities of Cooperation Between Heilongjiang Province and Russia. *R-economy*, 5(1), 13–18.

Song, C., Sha, J. H. & He, G. Y. 2016. Relation Between Population Migration and Economic Growth in Northeastern China Based on an Economic Growth Model with Endogenic Population Migration. *Resources and Industries*, 18(2), 129–137. // 宋慈, 沙景华, 何更宇, 2016. 东北人口迁移与经济增长关系研究—基于内生人口迁移的经济增长模型. 资源与产业, 18(2), 129–137.

State Council People's Republic of China. 2019. *State Council Agrees to Establish Six New Pilot Free Trade Zones*. Beijing: State Council. http://english.www.gov.cn/policies/latestreleases/201908/26/content_WS5d63aca2c6d0c6695ff7f4d4.html

Ubarevičienė, R., Van Ham, M. & Burneika, D. 2016. Shrinking Regions in a Shrinking Country: The Geography of Population Decline in Lithuania 2001–2011. *Urban Studies Research*, 1–18. Article ID5395379.

UN Habitat. 2008. *State of the World's Cities 2008/2009: Harmonious Cities*. London: Earthscan.

Wallerstein, I. 1974. *The Modern World System*. New York: Academic Press.

Wang, Z. G. 1997. *Bianjing Diqu Xianyu Jingji Fazhan Yanjiu (Research into the Development of Border Regions and County Districts)*. Harbin: Heilongjiang Jiaoyu Chubanshe (Heilongjiang Education Publisher).

Weaver, R. et al. 2016. *Shrinking Cities: Understanding Urban Decline in the United States*. New York: Routledge.

Wu, C. T. 2016. Periphery, borders and regional development. In Prangan H. et al. (eds.) *Insurgencies and Revolutions: Reflections on John Friedmann's Contributions to Planning Theory and Practice* (pp. 73–83). New York: Routledge.

Wu, X. & Zhuang, H. 2014. Study on the Coordinative Development of Population Distribution and Pattern of the Economy in Heilongjiang Province. *Areal Research and Development*, 33(1), 164–169. //吴相利, 庄海燕, 2014. 黑龙江省人口分布与经济格局协调发展研究. *地域研究与开发*, 33(1), 164–169.

Yu, L. T. 12 November 2015. In for the Long Haul. *Beijing Review*, (46), 38–39.

Zaionchkovskaya, Zh. A. 2013. Federal Districts on the Migration Map of Russia. *Regional Research of Russia*, 3(4), 328–334.

Zhuravskaya, T. 2011. "Seryy" import na rossiysko-kitayskoy granitse: chto novogo? ("Grey" Importation over the Russian-Chinese Border: What's New?). *Economic Sociology*, 12(5), 54–71.

Part III
Russia

9　Introduction to the Russia section

Maria Gunko

Patterns of Russian urbanization in the 20th century

The 20th century in Russia may be referred to as the "urban age", not only due to the increase in the number of cities and share of urban population but also due to the profound transformations in the way of life which followed urbanization. Soviet industrialization, in which cities served as "departments of a national industrial corporation" (Vizgalov, 2015:267), led to a period of rapid urbanization between the 1930s and the 1950s. New cities and towns were established for the purposes of natural resources extraction, logistics, and manufacturing including in areas where no permanent settlement have ever been before (e.g., Russian Far North and Far East). The story of many such cities and towns is linked to the GULag system[1] that used the prisoner workforce in mining and construction activities (Barenberg, 2014). Thus, their foundation was a result of unsustainable economic situation in the Soviet Union when rapid industrialization required a cheap workforce, the demand that was satisfied with forced labor (ibid). Often cities were planned against all odds in impossible extreme conditions, realizing ambitious ideas of the state without counting economic, environmental, or human cost (e.g., Bolotova, 2014; Jull and Cho, 2017). Along newly emerging cities, many former rural settlements in various parts of the country were assigned urban status. In addition to pragmatic rationality of city-building for allocation of productive forces in the early 20th century, the Soviet state also assigned cities ideological meaning. They were viewed as "fort posts" of communist agenda-setting manifesting the power of the ruling party (Bruno, 2016; Paperny, 2016). In the mid-20th century, urbanization in the Soviet Union slightly slowed its pace and ideological component of city-formation weakened. Nevertheless, new cities in Soviet Russia were established until the late 1980s. Overall, according to population censuses, the level of urbanization in Russia rose steadily throughout the 20th century—from 15% in 1897 to 51% in 1959, and to 73% in 1989 (Polyan et al., 2001). Since then, it remains stable—roughly 74% at the end of 2019 (Rosstat).

Rural to urban migration is an important mechanism of urbanization. However, the nature of migration in the Soviet Union was controversial.

DOI: 10.4324/9780367815011-12

Official sources state that in the period 1926–1979, rural to urban migrations contributed to a 55% increase in urban population (Korel, 1982). At the same time, it is widely acknowledged that the period 1930–mid-1950s in the Soviet Union was the period of mass repressions including forced migrations and deportations (e.g., Hagenloh, 2009; Polyan, 2001; Barenberg, 2014). Only in the late 1950s, after the death of Josef Stalin and subsequent alleviation of the repressive machine, voluntary migration set in. Still, it was limited by the institute of *propiska* (registration at the place of permanent residence) that restricted free movement and choice of residency. Furthermore, rural dwellers, more specifically workers of state collective farms (*kolkhoz* and *sovkhoz*) were deprived of passports until 1974, which meant that their relocation to the city was associated with numerous hurdles. Thus, rural to urban migration was most intensive in the period 1970–1980 determining rural depopulation.

Though rural shrinkage was already gaining momentum in Russia in the second half of the 20th century, urban shrinkage was far less common being restricted to old-industrial regions with aged population structure (Batunova, Gunko, and Medvdev, 2021). The abrupt collapse of state socialism in Russia and subsequent turn toward (neoliberal) capitalism reinforced negative demographic tendencies in the country, as well as contributed to considerable intensification of migration flows. As noted by Gaddy and Ickes (2013), the Soviet state ignored the cost of cold and remoteness due to the artificially low prices for energy and other inputs making much of its cities dependent on state support. Therefore, when the priorities of the new Russian state changed such cities began to rapidly lose population. As of 2019, about 70% of Russian cities lost population in comparison to 1989, the year of the last Soviet population census (Rosstat).

Structural legacies, such as the location and distribution of industries continue to shape urban development and government policies in Russia (Gaddy and Ickes, 2013; Crowley, 2016); though, it is the company towns (*monogorods*),[2] where socialist legacies remain particularly strong. The collision of the specific urban structures and infrastructures created as a result of Soviet planning and neoliberal rationality uncovered that under conditions of shrinkage, they are very difficult to adapt to new conditions, requirements, and technologies (Collier, 2011). Moreover, hectic privatization over land and real estate, as well as the redistribution of ownership rights between various tiers of governance soon after the collapse of state socialism in Russia have left numerous gaps and complex property issues that impede managing the effects of urban shrinkage on the local level.

Overview of current demography trends in Russia on the national level

Demographic situation in Russia, as well as reasons for urban shrinkage will be in detail discussed in chapters within this section (especially Chapters 10 and 11). Here I would like to briefly outline the main trends.

Along with many other countries of the world, Russia has been affected by the decrease in birth rates already in the Soviet era because of the demographic transitions. Moreover, the adverse events in the country's past in the first half of the 20th century—wars, several waves of famine and repressions–have altered the population structure. Whereby decrease in birth rates periodically repeats itself following the periodic decrease in the number of women of fertile age. Along with this, Russia is characterized by a considerably high mortality rate in comparison to the EU countries and a significant gap in life expectancy between men and women (Eberstadt, 2010; Vishnevsky and Shchur, 2019). At the same time, the Russian migration balance remained positive throughout the whole post-Soviet period including at the expense of repatriates from the former Soviet Republics. Nevertheless, in-migration did not compensate for acute internal demographic crises in Russia at the wake of postsocialist transition—the so-called "Russian cross" with mortality significantly outstripping fertility (Vishnevsky, 2000). Overall, between 1989 and 2009, Russia has lost around 4% of its population (declining from 147.4 m to 141.9 m).

Since the beginning of the 2000s, the demographic trends started to slightly change in a more positive direction. An increase in birth rates (in part due to the big cohort of women who entered a fertile age and in part due to improvement of the overall socio-economic situation in the country) was observed along with declining mortality thanks to improvement of economic situation, healthcare delivery, and rising popularity of healthier lifestyle. By 2010, the country managed to achieve population growth which continued until 2018, inclusive. The population changed from 141.9 m to 146.8 m, increasing by slightly over 3%. Policymakers attributed this effect to the state pro-natalist demographic policy commenced in Russia since the late-2000s (Yakovlev, 2020; also, see Chapter 13 in this volume). Though scholars have not come to a single opinion on the effects of this policy, largely arguing that the rising birth rates are rather associated with the specifics of the national population structure (Novaya Gazeta, 2020).

Since 2019, a sharp slowdown in the pace of in-migration to Russia is evident alongside natural population decline, both in part due to increased political and economic uncertainty (RBC, 2020). The Covid-19 pandemic further reinforced these trends, alongside its contribution to excess in the total death counts. In 2020, this excess was around 360,000 people in comparison to the same period of 2019 (Kobak, 2021). According to the first draft of the "Unified plan for achieving national development goals of the Russian Federation until 2024 and for the planning period until 2030", it is expected that in 2021 Russia's population will decrease by 290,000 people, in 2022 by 238,100, in 2023 by 189,100, and in 2024 by 165,300 people (RBC, 2020). Thus, the total population of the country, according to the above document, will decrease by more than 1.2 million people in 2020–2024 and the return to growth is expected only by 2030. The overall trend toward depopulation will without a doubt affect all regions and cities of the country. But given spatial polarization within Russia, some places (e.g., rural areas, small cities in general and especially those in the Far North and Far East) will be more vulnerable to shrinkage than others.

Academic research and policy responses to urban shrinkage

Against adverse demographic trends in Russia, it would be logical to assume that urban shrinkage has gained much attention within both scholarly and policymaking debates. It is true that demographic issues have been in the spotlight of both discussion; however, urban shrinkage *per se* has not been discussed widely. The knowledge about this phenomenon in the Russian context is still lacking both in theoretical and in practical terms.

One of the first studies of urban shrinkage in the Russian context was the research of the team lead by Philipp Oswald that highlighted the case of old-industrial Ivanovo—declining center of the textile industry—which became a canonical example of a Russian shrinking city (Oswald, 2004). Further studies of shrinking cities in Russia started to emerge in the mid-2010s (e.g., Antonov et al., 2014; Cottineau, 2016; Batunova, 2017; Batunova and Gunko, 2018; Gunko, Eremenko, and Batunova, 2020). However, there is still no comprehensive definition of urban shrinkage/shrinking city in Russia, as well as a lack of understanding of shrinkage's consequences for the built environment and urban economy. Research of urban planning and policymaking under conditions of shrinkage is also negligible. This is not least due to the traditional for the Russian academia weak connection between urban geography and planning studies.

The minor attention to shrinking cities in the Russian context within scholarly research in part defines scarcity of planning and policy responses. The Russian state pursues a pro-growth agenda; whereby countering depopulation is the sole possible strategy at all levels of governance (Batunova and Gunko, 2018; also, see Chapter 11 in this volume). In terms of urban planning and managing the built environment, it seems much more preoccupied with insignificant urban maintenance, urban design, and landscaping projects (*blagoustroystvo*)[3] which require modest funding and provide immediate political gains (Zupan and Gunko, 2019), unlike comprehensive renewal projects for shrinking cities that suffer from underfunding, progressive vacancies, and deteriorating infrastructures.

Clémentine Cottineau (2016), who extensively researched the Russian urban system, claims that it seems to be the most shrinking urban system in the world. Shrinking cities are located through the vast territory of the country—east to west and north to south—in various geographic conditions and cultural contexts. Thus, Russia is an instructive case for researching urban shrinkage and its implications. Though the national governing metanarrative is in countering shrinkage, there are cases when local authorities and actors come up with creative approaches toward planning for shrinkage both formally and informally. However, it is hard to find such cases as official planning documents tend to reflect only the national discourses and indicators (see Chapters 11 and 14 in this volume).

Russian administrative-territorial division and planning system

Before delving into urban shrinkage in Russia, a brief explanation of the Russian administrative-territorial division and planning systems is required to properly understand the issues at hands.

Administrative-territorial division

Currently, there are three levels of governance in Russia according to the Constitution–national, subnational (regional), and municipal (local). Until recently, only national and subnational levels were acknowledged as state levels of governance, while the local (municipal) level was *de jure* an independent level represented by local self-governance bodies. Despite the proclaimed autonomy, the local level has always been dependent on higher levels of governance due to the redistributive fiscal policy. With the adoption of the new Constitution of the Russian Federation in summer 2020, the local level was firmly integrated into the so-called Russian state "power vertical" (Gelman and Ryzhenkov, 2011).

Subnational level in Russia is presented by a variety of regions, officially 85 in total as of 2020—*oblast (46), krai (9), avtonomniy okrug (4), avtonomnaya oblast (1), Respublika (16)*, and cities of federal significance (3). Essentially there is no difference between the first four, the title remains a tribute to the historical tradition. Republics (*Respubliks)* were formed based on ethnic principle and currently have more autonomy from the national government, including having an own Constitution and the right to establish own state language (the language of the ethnic group) in addition to Russian. Cities of federal significance include Moscow, Saint-Petersburg, and Sevastopol. The latter is located in Crimea—territory annexed by Russia from Uktraine in March 2014. Regions are aggregated into eight Federal districts, which are non-constitutional administrative-territorial units. The composition of Federal Districts changes over time due to the changing priorities of the state.

Given that there are no established translations for Russian regions, in various chapters of this section, authors may choose different forms. Here, I provide an example of Tverskaya oblast [Тверская область], which can also be referred to as Tver oblast or Tver region after the regional center Tver. Tver' may also be spelled with ' indicating the soft sigh [ь].

The local or municipal level is subdivided into two tiers. First of them is represented by a mixed rural-urban county (*munitsipalny rayon)* and an urban county (*gorodskoy okrug)*. Both are directly subordinated to the regional government. Second tier is represented by settlements (*poselenie)* which can be urban (*gorodskoe)* or rural (*selskoe)*.

Statistical data on Russian administrative-territorial units are available at all constitutional levels of governance. Though the municipal level (especially its second tier) is less covered in statistical accounts.

Planning system

Two types of planning subsystems concurrently exist in Russia with respective set of planning documents. Firstly, strategic planning subsystem concerned with general aims and development tasks of the territorial entity in the economic and social spheres. Secondly, spatial planning subsystem concerned with the built-up environment (including housing, infrastructures, green spaces), zoning, and land use. Both types of planning are carried out at all levels of governance: national level, regional level, and two municipal tiers—counties and settlements. Strategic and spatial planning are implemented by different governmental structures and have little in common, as laws that regulate planning subsystems are poorly coordinated (Grishina et al., 2016).

According to the national law, spatial planning should ensure the territorial component of strategic planning's aims and tasks. In reality, this sequence is rarely achieved. The development of planning regulations in post-Soviet Russia has been slow and uneven. Spatial planning at all levels of governance became mandatory with the adoption of the Urban Planning Code in 2004 (Federal Law no. 190-FZ from 29 December 2004), while the Federal Law on Strategic Planning was approved only in 2014 (Federal Law no. 172-FZ from 28 June 2014). The Urban Planning Code states that the development of spatial planning documents is obligatory for every administrative-territorial unit of the Russian Federation from national to municipal levels. At the same time, according to the Federal Law on Strategic Planning, the development of a socio-economic strategy at the municipal level is not obligatory and depends on the local policymakers' decision.

Structure of the section

Given the scarcity of research on urban shrinkage in Russia both within the country (in Russian) and internationally, the present volume is unique in terms of the issue coverage bringing together both national overviews and highlighting specific regional and local cases. The cases covered in this section are illustrated in Figure 9.1. The section starts with the present introduction (Chapter 9) aimed at explaining the broad patterns of historical background, contemporary state of urbanization, and demographic situation in the country. It also provides an explanation to some commonly mentioned Russian terms, as well as describes the territorial-administrative division and planning in Russia.

Chapter 10 by Kseniya Averkieva and Vera Efremova provides an empirically rich overview of the geography and functional typology of Russian

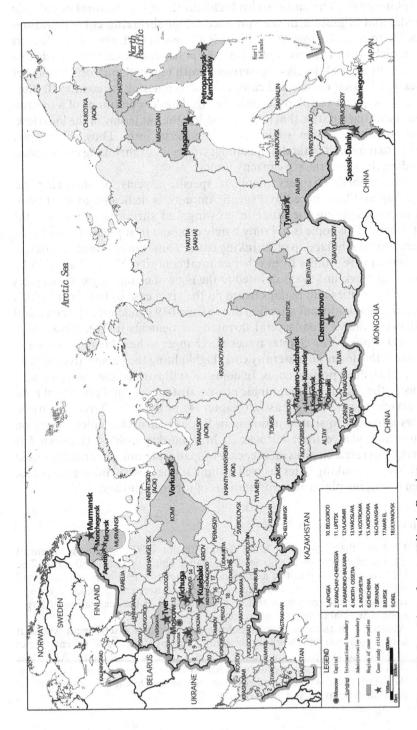

Figure 9.1 Location of case studies in Russia.

shrinking cities. The authors also highlight the role of natural population decline and migration in the "production" of shrinking cities in various Russian regions. Chapter 11 by Elena Batunova and Maria Gunko delivers a comprehensive analysis of planning and policy responses to urban shrinkage in Russia at all levels of governance with regional and local responses researched based on selected cases in various parts of the country. It illustrates how urban shrinkage is (mis)managed under conditions of a centralized authoritarian state that *de facto* restricts the autonomy of the local level where urban shrinkage is manifested in material form. Thus, this chapter argues that for understanding planning strategies of shrinking cities understanding their governance capacity is vital.

Further, three chapters highlight specific aspects of shrinking cities. Chapter 12, authored by Evgenii Antonov, is dedicated to local labor markets, comparing the latter in growing and shrinking cities over the last decade. This topic is not only barely explored in Russia but also in the international scholarship of shrinking cities. Thus, the obtained results call for comparative studies from other cultural contexts. Next, Chapter 13 by Anastasia Novkunskaya is devoted to the issue of changing public services in shrinking cities and regions based on the vivid case of healthcare which has attracted significant attention due to Covid-19 pandemic. The empirical evidence is drawn from several shrinking settlements in Tverskaya oblast (Central Russia). The chapter traces the changes to healthcare provision in Russia with a focus on maternity care, highlighting the perspectives of families and medical professionals. In doing so, it illustrated the contradictions between the national demographic pro-natalist policies and policies of public services restructuring. Lastly, Chapter 14, authored by Daria Chigareva, draws attention to spatial planning in Russian shrinking cities. Based on empirical evidence from Kirovsk in Murmanskaya oblast (Russian Far North), this research gives an overview of spatial planning instruments used in Russian shrinking cities, as well as limitations that shrinking cities face when managing physical manifestations of urban shrinkage.

Notes

1 GULag—the acronym of Main Administration of Camps, the government agency in charge of the Soviet forced-labor camp system (1930–1950). The GULag had both punitive and economic functions. Prisoners were assigned distinct economic tasks, including the exploration and mining of natural resources, colonization of remote areas, as well as the realization of large-scale industrial construction projects.

2 According to the official terminology, *monogorod* is a settlement where at least 20% of residents work for the same company ("city-forming" enterprise) or group of companies operating within a technological chain. As a rule, the company or group accounts for 50% or more of gross output. Since 2014, the list of officially recognized *monogorods* is defined by the Government of the Russian Federation and updated annually.

3 *Blagoustroystvo* (literally "goodness creation") is a notion that does not have a direct equivalent in English. It pertains to measures of cityscape maintenance, landscaping, and urban design aimed at ensuring and improving the comfort of living conditions, as well as upgrading the sanitary and aesthetic state of the territory.

References

Antonov, E., Denisov, E., Efremova, V. & Faddeev, A. 2014. Sovremennie problemy razvitiya ubyvaushchikh gorodov na severo-vostoke respubliki Komi [Contemporary Issues of Development in Shrinking Cities of the North-Eastern Part of Komi Republic]. *Vestnik Moskovskogo Universiteta: Seriya 5: Geografiya*, (2), 55–61.

Barenberg, A. 2014. *Gulag Town, Company Town: Forced Labor and Its Legacy in Vorkuta*. New Haven, CT: Yale University Press.

Batunova, E. 2017. Uchet depopulyatsionnykh protsessov v dokumentakh territorialnogo planirovaniya gorodov uga Rossii [Accounting for Depopulation Processes in Spatial Planning Documents of Cities in Southern Russia]. *Regionalnie issledovaniya*, (1), 64–72.

Batunova, E. & Gunko, M. 2018. Urban Shrinkage: An Unspoken Challenge of Spatial Planning in Russian Small and Medium-Sized Cities. *European Planning Studies*, 26(8), 1580–1597.

Batunova, E., Gunko, M. & Medvedev, A. 2021. Neupravlyaemoe prostranstvo: planirovanie i politika v usloviyakh depopulyatsii na primere Ivanovskoy oblasti [Mismanaged Space: Planning and Policymaking in the Context of Depopulation in Ivanovo Region]. *Vestnik Sankt-Pterburgskogo Unversiteta. Seriya Nauki o Zemle*, 66(3), 440–459.

Bolotova, A. 2014. *Conquering Nature and Engaging with the Environment in the Russian Industrialized North*. Rovaniemi: University of Lapland, Acta Universitatis Lapponiensis.

Bruno, A. 2016. *The Nature of Soviet Power*. Cambridge: Cambridge University Press.

Collier, S. 2011. *Post-Soviet Social: Neoliberalism, Social Modernity, Biopolitics*. Princeton, NJ: Princeton University Press.

Crowley, S. 2016. Monotowns and the Political Economy of Industrial Restructuring in Russia. *Post-Soviet Affairs*, 32(5), 397–422.

Cottineau, C. 2016. A Multilevel Portrait of Shrinking Urban Russia. *Espace Populations Sociétés*, 2015/3–2016/1. https://journals.openedition.org/eps/6123

Eberstadt, N. 2010. *Russia's Peacetime Demographic Crisis: Dimensions, Causes, Implications*. Seattle: National Bureau of Asian Research.

Gaddy, C. & Ickes, B. 2013. *Bear's Traps of Russia's Road to Modernization*. London and New York: Routledge.

Gelman, V. & Ryzhenkov, S. 2011. Local Regimes, Sub-National Governance and the "Power Vertical" in Contemporary Russia. *Europe-Asia Studies*, 63(3), 449–465.

Grishina, I., Pelyasov, A., Zamyatina, N., Gunko, M., Mikhailov, D., Putilova, E., Grishin, A., Kulikova, U. & Nesterkina, E. 2016. *Nauchno-metodicheskoe obespechenie dokumentov strategicheskogo planirovaniya na munitsipalnom urovne* [Scientific and Methodological Support for the Development of Strategic Planning Documents at the Municipal Level]. Report on Applied Economic Research (Theme П302-27-16). Moscow: SOPS.

Gunko, M., Eremenko, U. & Batunova, E. 2020. Strategii planirovaniya v usloviyah gorodskogo szhatiya v Rossii: issledovanie malyh i srednih gorodov [Planning Strategies in the Context of Urban Shrinkage in Russia: Research of Small and Medium-Sized Cities]. *Mir Rossii: Sociologiya, Etnologiya*, 29(3), 121–141.

Hagenloh, P. 2009. *Stalin's Police: Public Order and Mass Repression in the USSR, 1926–1941*. Washington, D.C.: Woodrow Wilson Center Press.

Jull, M. & Cho, L. S. 2017. Architecture and urbanism of Arctic cities: Case study of Resolute Bay and Norilsk. In Orttung, R. W. & Laruelle, M. (eds.) *Urban Sustainability in the Arctic: Visions, Contexts, and Challenges* (pp. 331–339). Washington DC: Institute for European, Russian and Eurasian Studies.

Kobak, D. 2021. Excess Mortality Reveals Covid's True Toll in Russia. *Significance*, 18(1), 16–19. https://doi.org/10.1111/1740-9713.01486

Korel, L. 1982. *Premeshchenie mezhdu gorodom i selom v usloviyakh urbanizatsii [Urban to Rural Mobility in the Course of Urbanization]*. Novosibirsk: Nauka.

Novaya Gazeta. 2020. "Nadezhd na reshenie problemy rozhdaemosti v Rossii net". Demograf Anatoliy Vishnevsky—o tom, pochemu prezidentskie meri ne pomogut "priumnozhyt"rossiyskoenaselenie["Thereisnohopeinsolvingtheissueoffertilityin Russia". DemographerAnatoliyVishnevskyonwhypresident'smeasureswillnothelp multiply Russian population]. *Novaya Gazeta*, January 17. https://novayagazeta.ru/articles/2020/01/16/83471-nadezhd-na-reshenie-problemy-rozhdaemosti-v-rossii-net

Oswald, P. 2004. Ivanovo. *Eine Stadt in Postsozialistische Transformation*. Working Paper, February 2004. http://www.shrinkingcities.com/fileadmin/shrink/downloads/pdfs/WP_Band_1_Ivanovo.pdf

Paperny, V. 2016. *Kultura dva [The Second Culture]*. 4th ed. Moscow: NLO.

Polyan, P. 2001. *Ne po svoey vole: istoriya i geografiya prinuditelnikh migratsiy v SSSR [Not by Choice: History and Geography of Forced Migration in USSR]*. Moscow: OGI.

Polyan, P., Nefedova, T. & Treivish, A. (eds.) 2001. *Gorod i derevnya v Rossii: 100 let peremen [City and Village in European Russia: 100 Years of Change]*. Moscow: OGI.

RBC. 2020. *Pravitelstvo rezko ukhudshilo prognoz po ubyli naseleniya Rossii [The Government Has Sharply Worsened Its Forecast for Russia's Population Decline]*. RBC, October 15. https://www.rbc.ru/economics/15/10/2020/5f8846b39a7947323dcb06c5

Vishnevsky, A. 2000. Podem smertnosti v Rossii v 90-e gody: fakt ili artefakt [The Rise of Mortality in the 1990s: Fact or Artifact?] *Mir Rossii: Sotsiologiya i Etnologiya*, 9(3), 153–160.

Vishnevsky, A. & Shchur, A. 2019. Smertnost i prodolzhytelnost zhyzni v Rossii za polveka [Mortality and Life Expectancy in Russia for Half a Century]. *Orgzdrav: novosti, mneniya, obochenie*, 5(2), 10–21.

Vizgalov, D. 2015. *Pust goroda zhyvut [Let Cities Live]*. Moscow: Sektor.

Yakovlev, E. 2020. Kak materinskiy kapital povliyal na rozhdaemost. *Vedomosti*, February 27. https://www.vedomosti.ru/opinion/articles/2020/02/27/823925-materinskii-kapital

Zupan, D. & Gunko, M. 2019. The "Comfortable City" Model: Researching Russian Urban Planning and Design Through Policy Mobilities. *Gorodskie Issledovaniya i Paktiki*, 4(3), 7–22.

10 Urban shrinkage in Russia

Concepts and causes of urban population loss in the post-Soviet period

Ksenia Averkieva and Vera Efremova

Introduction

The Russian urban system is rapidly polarizing with a few fast-growing large urban centers and a significant number of cities experiencing population loss, economic downturn, and cityscape deterioration. In international urban research and planning practice cities facing depopulation and structural crises are referred to as "shrinking" (see Haase et al. 2014).

Since the 2000s, cities with persistent and severe population loss became the focus of urban research. The expanding and dynamic academic debate on urban shrinkage includes the following focuses (Haase et al. 2014, 2016; Pallagst et al., 2017):

- Complexity and the interrelationship of factors that lead to population loss, as well as the consequences of depopulation at the urban level;
- Transformation of the built environment and infrastructure under conditions of urban shrinkage and the growing mismatch between supply and demand;
- Urban planning and policymaking in the context of shrinkage: strategies (accepting or countering), planning approaches and instruments aimed at adapting to shrinkage.

At the same time, to date the concept of urban shrinkage remains "fuzzy" having theoretical and methodological limits such as the problem of threshold definition, empirical contradictions, and insufficient understanding of urban development as a historically contingent process (Bernt 2016).

In Russian urban studies, the topic of urban shrinkage with its conceptualization and policy responses can be described as barely delineated despite the growing number of shrinking cities in the country (Batunova & Gunko 2018; Cottineau 2016). There is clearly a need for comprehensive and multi-dimensional research of urban shrinkage at different scales, as well as conceptualization of phenomenon within the specific cultural and geographic context.

DOI: 10.4324/9780367815011-13

This chapter provides a brief overview of research on, and identification of, shrinking cities in Russia. A shrinking city is defined as one with a population loss over 1% annually within periods of population censuses. We discuss urban shrinkage in Russia in terms of different time periods, city size and geographical location. This overview also includes a typology of shrinking Russian cities based on different indicators.

Methods and data

As was already noted in the introduction, international literature on urban shrinkage does not offer an established and widely agreed-upon quantitative criterion for a shrinking city. One of the most common definitions states that a shrinking city is a municipal unit with a minimum population threshold of 10,000 residents that has experienced population losses for more than two years and is undergoing an economic transformation with some symptoms of a structural crisis (Pallagst et al. 2009; Wiechmann 2008). In different countries, there may be a variety of definitions at the national level that include different indicators. For instance, in the United States, a shrinking city may be defined as having a population loss of more than 25% for 40 years from 1970 to 2010 (Weaver et al. 2017, pp. 7–8); in France, it should have an annual negative population change between 1975 and 2007 (Wolff et al. 2013); in Germany, several indicators are taken into account—migration and economic indicators for five years: population change, total net migration, the dynamics of the population of working age, the development of employees subject to social insurance contributions at the place of work, change in the unemployment rate and the development of the business tax base revenue per capita (BBSR 2018). Thus, the criteria for identifying shrinking cities are relative, being strongly reliant on the country. Despite differences between the contexts, in most studies and policy documents, population loss is identified as a key feature of urban shrinkage. Population change is associated with changes in the economic situation and taken together, these define the alteration of urban development in terms of agglomeration effects and productivity and the demand for consumer goods and services, housing, and infrastructure (Turok & Mykhnenko 2007).

The current analysis of shrinking cities in Russia is underpinned by considering demographic indicators. However, though available, such data has only relative temporal comparability. The methods of statistical accounting for economic and housing indicators changed during the 1990s and 2000s. Census dates (1959, 1970, 1979, 1989, 2002, 2010) are used as reference dates, and annual current population accounting data (population of cities on January 1, 2019) are used to illustrate the current population dynamics.

We consider annual population loss of over 1% as an indicator for delineating shrinking cities in Russia based on population change described with an exponential function:

$$P_t = P_0 e^{rt},$$

where P_t is the population after t years, P_0 is the initial population, and r is the continuous rate of change (Plane & Rogerson 1994). Using a logarithmic function to analyze population dynamics allows us to mitigate the value of the city size.

According to the proposed criteria, $\dfrac{P_t}{P_0} = 0.99$, $t = 1$

$$r = \ln\left(\frac{P_t}{P_0}\right) t^{-1} = -0.01005.$$

Cities with a population change of at least −0.01005 per year will shed a minimum of 1% annual population loss during the inter-census time frame. Along with cities that correspond to the r criterion, we also consider as shrinking cities where the absolute population loss exceeded 20,000 people over ten years.

Overall, our review comprised 864 cities (out of 1054 Russian cities). Firstly, we excluded cities with less than 10,000 people (according to the 2010 census) since their population number did not correspond to the threshold used in the most common definition of a shrinking city (Pallagst et al. 2009). Secondly, cities within Moscow and St. Petersburg capital regions with significant differences in population dynamics were not included. Thirdly, cities that had a status of a closed administrative-territorial unit[1] (*ZATO*) were excluded due to the lack of available official data.

Historical factors defining population decline in Russian cities

The features of the contemporary Russian urban network and dynamics of urban development were largely determined by state policies implemented during the Soviet period, as was the demographic change since the collapse of state socialism. The key pattern of urban development in Russia (and in the former USSR) was the rapid growth of cities because of industrial development partly due to the "conquest" of remote areas, often with extreme natural conditions (Bruno 2016). In the 1930s, cities associated with coal mining, metallurgy, and the timber industry were actively supported by the state, receiving additional development incentives during the Second World War. Between the 1930s and 1950s, extensive development of the Soviet Far North took place leading to the formation of large urban centers functioning around natural resources extraction. First massive flows of population to the Far North were associated with the Soviet repressive machine and comprised two types of forced migrants: the so-called special contingent (labor camp prisoners) and special settlers who were deportees of various social and ethnic backgrounds (for example, various ethnic groups of the North Caucasus, Koreans, Germans; *kulacks*—prosperous peasants from predominantly Southern regions of the USSR). Many Russian northern cities, including large ones, were originally founded between the late 1920s and early 1950s as campsites within the GULag.[2] Among them are Vorkuta, Inta, Magadan, Norilsk, and other cities as part of territorial branches of the

GULag (for example, Dalstroy, Sevvostlag, and Norillag). Later, these settlements received the city status and began to accept various population groups thus gradually acquiring new functions. After the death of Josef Stalin at the beginning of the 1950s and the gradual decline of the GULag system, the Soviet state started an ideological campaign actively using the rhetoric of "conquering nature" (Bolotova 2014) in order to promote voluntary migration to the North, Siberia, and the Far East. Furthermore, potential settlers were motivated to live and work in uncomfortable and even extreme natural conditions with various economic measures such as "northern" benefits and bonuses. This meant higher salaries (salaries with "northern" coefficients), easier access to housing and longer vacation periods. Since cities anticipated migration in-flow, they, on average, were provided with better infrastructure and higher quality housing. Overall, given the intensive forced and voluntary migration, Soviet cities of the North, Siberia, and the Far East experienced a drastic increase in population over a noticeably short period of time.

The downside of rapid and "artificial" urban growth during the Soviet era was the severe population out-flows from the North, Siberia and the Far East as soon as the comparative advantages of living in adverse conditions were lost. In the 1990s, a mass out-migration from many of the above regions was underpinned by a reduction of "northern" coefficients and abolition of subsidies resulting in lower wages on the backdrop of exceedingly high living costs. This process may be referred to as "stress migration". Between 1989 and 2010, Russia's most remote north-eastern region, Chukotka, lost over two-thirds of its population. In the 1990s, the rate of migration loss here was more than 2% annually. Since the mid-2000s, the population of the region started stabilizing. Similar trends were observed in Magadanskay oblast, Kamchatskiy krai, as well as in some cities of Sakha Republic (Yakutia). However, it is important to note that stress migration associated with a rapidly declining quality of life was not the only reason for population out-flow from the North, Siberia, and the Far East. Another reason was the initial attitude of those who settled in these regions—cities here were often perceived as transitional points where rapid earnings could be achieved with subsequent relocation to more comfortable regions in the center and south of the country.

Demographic and migration components of urban population change

The decline in urban population is simultaneously defined by many processes: out-migration; high mortality due to aging, as well as a large number of accidents and early deaths caused by lifestyle factors; low fertility associated with the age structure of the population and the reproductive behavior of individuals. From the beginning of the 1990s until the end of the 2000s, the average total fertility rate in Russia was less than two. This means it was impossible to ensure simple population reproduction when talking about population growth. Furthermore, fertility rates among the urban population were always lower than among the rural population. Given the rate of urbanization, around 70% in Russia, this situation threatened to result in a sharp decline in the population in most regions. Against this background, in the 1990s, many regions, especially in the European part of Russia, received immigrants from the former Soviet Republics (predominantly compatriots). Therefore, the extremely low fertility did not lead to catastrophic population decline. Currently, the fertility rate in Russia is, on average, more optimistic. The total fertility rate has been kept above two for several years (in 2013, it was already 2.2—almost reaching the level of simple reproduction), but in the cities, this figure is still lower—only 1.6.

Regarding the overall mortality rates, the worst situation may be observed in small and peripheral cities and districts of European Russia (predominantly within Central and North-Western Federal Districts) and the Urals where mortality is fueled by lifestyle factors and aging. In the cities of Siberia, mortality is on average lower due to the predominance of younger population cohorts. The same holds true for the Republics of the Caucasus and southern Siberia, where the age structure of the population plays a strong role. As a result, fertility and mortality trends contribute to a sharp polarization of Russian regions in terms of natural population movement: high rates of natural population loss are a distinctive feature of many cities in Central Russia, while natural population increase may be observed in the Caucasus and Siberia (especially its southern part).

The broad picture of internal migration movement in Russia may be characterized as a "western drift"—from regions and cities of the North and Far East to the central and southern regions of European Russia (Mkrtchyan 2011). These migrants comprise people of working age in search of better job opportunities, as well as individuals of pre-retirement age who seek comfortable residential arrangements. However, there are variations on the intraregional level. Migration, to some extent, reinforces the trend of natural population movement. In addition to natural population decline in small peripheral cities of Central Russia, there is out-migration of predominantly young population cohorts, and at the same time, the "youthful" resource cities in Siberia attract new residents. However, migration may also run contra to the trends of natural population movement. This is evident in Caucasian cities, which experience both natural population increase along with youth out-migration. Natural population decline in the regions of Central Russia is to some extent compensated for by in-migration.

Variety of shrinking cities in Russia based on demographic data

During the second half of the 20th century, Russia experienced periods of intensive urban growth, as well as periods when population dynamics varied significantly among cities. Further, we will discuss the core trends in accordance with population census timing.

The period 1959–1970 was marked by a considerable increase in urban population. Positive dynamics were typical for the vast majority of cities—regional and district centers, as well as industrial cities such as the resource cities of the north, centers of chemical and petrochemical industry, and cities with hydropower plants. Stable or negative population dynamics during this period was observed in a negligible number of cities such as old coal mining and metallurgical centers (for example, in the Urals).

In 1970–1979, the pace of population growth slowed, but most cities were still growing. New centers, mostly cities of the Volga region where large-scale construction industrial projects were realized, became leaders of population growth. The number of cities with negative population dynamics decreased due to intensified rural to urban migration. This may be partly attributed to the loosening of control over residency change for peasants (employees of state collective farms) who had been issued passports since 1974.

The period 1979–1989 was associated with a further slowing of urban population growth. The composition of the leading cities in terms of growth rates remained stable, including centers of the petrochemical industry in the Volga, Ural, and Siberian regions. The number of cities with population loss started to increase; among them, many were the old industrial centers of Central Russia, for example, cities associated with the Moscow coal basin (Tulskaya oblast) and textile cities (Ivanovskaya oblast). However, urban shrinkage was still not a pervasive phenomenon.

In post-soviet period, the trajectories of urban population dynamics changed significantly in comparison to those during the Soviet era (Figure 10.1). In 1989–2002, growth was limited to a small group of cities, including cities in ethnic republics—Tatarstan and Bashkortostan, republics of the North Caucasus and southern Siberia; and cities in economically successful regions—Belgorodskay oblast, Tyumenskay oblast. Cities experiencing population decline included a vast variety of types—old industrial cities of Central Russia, the Urals and Siberia; resource cities of the North and Far East; and company towns.

The period 2002–2010 was marked by further urban shrinkage. In many cities, population decline had become much more pronounced. The group of shrinking cities diversified further by including industrial centers with specialization in machinery and chemical production. Among other discussed periods, 2002–2010 is the one with the most severe and widespread urban shrinkage.

In 2010–2019, the country slowly recovered from the acute demographic crises and interregional migration flows slowed down. This resulted in an increase in the number of cities with positive population dynamics. However, growth was mainly confined to regional centers and large cities.

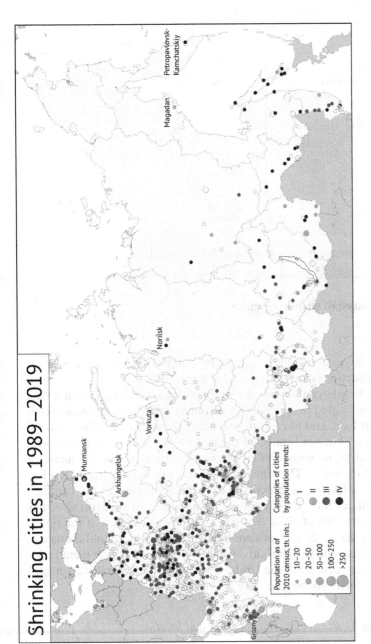

Figure 10.1 Russian shrinking cities in 1989–2019.

Source: Census data 1989, 2002, 2010; annual population account 2019 (Rosstat); Draft and design by V. Efemova and K. Averkieva.

Notes: Categories of cities by population trends: I—stable/growing cities over the whole period; II—cities stable/growing in 1990s, shrinking in 2000s and/or in 2010s; III—cities shrinking in 1990s, shrinking in 2000s and/or in 2010s; IV—cities shrinking in 1990s, stable/growing in 2000s nor in 2010s.

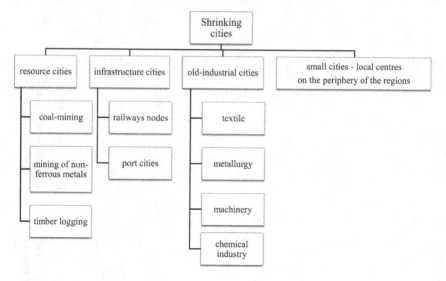

Figure 10.2 Functional typology of Russian shrinking cities.

Source: Own compilation. Draft & design: V. Efremova and K. Averkieva.

Typology of Russian shrinking cities

Functional typology

Summarizing this analysis, the following typology of shrinking cities based on economic functions can be identified (Figure 10.2).

The predominant type among shrinking cities is resource-based cities, especially those located in regions with extreme natural conditions (cities of Siberia, the Far East, and Far North, excluding those associated with the oil and gas industry) but not limited to them. Another type is infrastructural and logistic centers, composed of cities associated with the maintenance of various transport routes. They include port cities in the North and the Far East, cities on railway lines, especially in eastern Siberia and the Far East (Trans-Siberian railway, Baikal-Amur mainline). "Infrastructural" cities account for about 20% of cities that had been losing over 1% of the population annually. Roughly 15% of several shrinking cities are old industrial cities, mostly located in the Urals and Central Russia. These are metallurgical and machinery centers of the Urals, textile cities in Central Russia, and cities associated with timber industry in the north.

Shrinking cities by size groups and geographical location

Cities with a population loss of over 1% per year are predominantly small- and medium-sized cities (Table 10.1) located on the periphery of the regions. The population of cities in this group seldom exceeds 50,000 people (often

Table 10.1 Shrinking cities in Russia by population categories, 1959–2019

City size (2010) persons	Number of cities (2010)	1959– 1970	1970– 1979	1979– 1989	1989– 2002	2002– 2010	2010– 2019	2010– 2019%
Very small (10,000– 20,000)	249	12	10	3	80	113	99	40
Small (20,000– 50,000)	337	7	6	4	65	110	87	26
Medium-sized (50,000– 100,000)	131	–	1	4	14	18	13	10
Large (100,000– 250,000)	76	–	–	1	6	12	12	16
Major (over 250,000)	71	–	–		10	9	4	6
Total	864	19	17	12	175	262	215	98

Sources: Census data 1959, 1970, 1979, 1989, 2002, 2010; annual population account 2019 (Rosstat 2019); own compilation.

even less than 20,000 people). Among them, there are several company towns, however, this type of city is not pervasive. Generally, industrial production in these cities is represented (or was represented) by small enterprises specializing in processing local raw materials (food, forestry, construction materials manufacturing). In the 1990s and 2000s, many of these enterprises struggled to adapt to new conditions and some went bankrupt. In addition to the decline of the urban economy (which meant problems on the local labor market), people living in the small cities, especially young and middle-aged groups, were disgruntled by the low-quality urban environment, public services, and amenities, as well as limited educational and leisure opportunities. Overall, despite some in-flow of migrants predominantly from surrounding rural areas, Russian small cities are characterized by a negative migration balance and extremely high mortality rates.

The list of severely shrinking cities with populations between 10 and 20 thousand people duplicates the list of all-Russian leaders in relative population losses. Firstly, these are small cities of structurally weak regions in the Central and Volga Federal Districts, among them, there are many old cities with historical heritage. These cities are characterized by long-term population decline that had already begun in the Soviet era. Initially, it was caused by migration out-flow, which subsequently changed the age structure and led to aging manifested in significant mortality rates. Currently, both out-migration and natural population decline is observed here. Secondly, we must note small cities of the North and Far East, associated with natural resources extraction and logistics. These cities are located in extreme natural conditions, so the main reason for population loss is out-migration. Natural decline here is low due to the increased proportion of young population cohorts, especially those of working age. Lastly, small old industrial and structurally weak cities of the Urals (some are company towns with crises at the "city-forming" enterprise) are characterized by both out-migration and natural decline due to aging.

Among shrinking cities with populations of between 20 and 50 thousand people, the majority are mining cities of the Far East, Far North (Murmanskaya oblast, Komi Republic), and Permskyi krai. This group includes the shrinking cities of Central Russia; not only those on the periphery of regions but also ex-coal mining centers and centers of the chemical industry.

The number of medium-sized and large shrinking cities is relatively moderate. However, the effects of urban shrinkage here may be harder to manage due to the presence of complex urban infrastructure and the predominance of multi-apartment housing within large housing estates. The group of medium-sized shrinking cities consists of (ex)coal-mining centers, industrial cities of the North (Murmanskaya oblast) and old industrial centers of the Central and Volga Federal Districts. Among large shrinking cities are resource cities and centers of machinery and chemical industries. The most severe shrinkage has been in Grozny (Republic of Chechnya) due to military operations in the 1990s and 2000s[3] and Murmansk (Murmanskaya oblast)—the largest city of the Arctic—due to structural crises caused by the decline in state investment into logistics and military infrastructure. Furthermore, among large shrinking cities are Arkhangelsk (Arkhangelskay oblast)—large Northern port; Nizhny Tagil (Sverdlovskay oblast)—center of metallurgy and mechanical engineering; old industrial Taganrog (Rostovskaya oblast) and Komsomolsk-na-Amure (Khabarovskiy krai); textile center Ivanovo (Ivanovskaya oblast)—the so-called Russian "Manchester". Among shrinking cities are also those with a population over one million people—Perm (Permskiy krai) and Nizhny Novgorod (Nizhegorodskaya oblast) are regional centers facing economic decline due to the crisis of industrial production, especially machinery. Between 1989 and 2010, Nizhny Novgorod lost the largest number of inhabitants in Russia, over 187,000 people.

The geographic distribution of shrinking cities in Russia is assessed by location quotient Q_i:

$$Q_i = \frac{s_i / n_i}{S / N},$$

where s_i is the number of shrinking cities in macroregion i, n_i is the total number of cities in macroregion i, S is the total number of shrinking cities in Russia and N is the total number of cities in the dataset.

Table 10.2 Geographic distribution of shrinking cities in Russia, 1989–2019

Economic macroregions	Location quotient Q_i		
	1989–2002	*2002–2010*	*2010–2019*
Northern	2.23	1.85	1.56
Northwestern	1.25	1.29	1.29
Central	1.12	0.99	1.54
Volgo-Vyatka	0.78	1.57	1.12
Central Chernozem	0.23	0.76	0.65
North Caucasus	0.14	0.38	0.27
Volga	0.30	0.56	1.08
Ural	1.15	1.19	1.11
West Siberian	0.59	0.63	0.48
East Siberian	1.80	1.20	0.67
Far Eastern	2.71	1.74	1.27

Source: Census data 1989, 2002, 2010; annual population account 2019 (Rosstat); own compilation.

Changes in the location quotients in 1990s–2010s (Table 10.2) reveal that shrinking cities became more evenly distributed across the country. In 1989–2002, shrinking cities were foremostly limited to the North and Far East; however, recently, they may be found in all Russian regions (though they are less prominent in the south of European Russia, including in the North Caucasian Republics). During 2010–2019, the North and Far East maintained the status of leaders in the number and share of shrinking cities, though Central Russia has also become a notable region of urban shrinkage. Urban shrinkage has been activated in cities of the Volga region compared to the 1989–2002 period. In both Federal Districts, Central Russian and Volga, the rise of urban shrinkage was due to rapid depopulation in small and medium-sized cities. They remained stable between 1989 and 2002 owing to in-migration of compatriots from former Soviet republics; however, this flow weakened in the mid-2000s and population decline took place instead.

To summarize, many Russian cities are characterized by urban shrinkage, but Russian urban policymaking and agenda setting remain growth-oriented, drawing inspiration from major cities, which are economic engines and cores of urban agglomeration. At the same time, urban shrinkage is increasingly becoming a challenge to conventional planning and policymaking since shrinking cities are characterized by an oversupply of housing stock, amenities, social and technical infrastructure, as well as fragmentation of the cityscape. Adaptation of the built environment, infrastructure, public services, and amenities to changes driven by shrinkage requires significant financial and managerial resources often beyond local capacity. Therefore, urban shrinkage must be recognized at higher tiers of governance by the formulation of a new urban planning agenda.

Conclusions

The specifics of Russian urban shrinkage are determined by the country's history of urban development in the 20th century, including the establishment of new cities located in extreme or adverse environmental conditions and the role of various forms of forced migration (for example, GULag settlements). The changed priorities and opportunities of the Russian state after the collapse of state socialism led to a rapid migration out-flow from many cities in the North and Far East. This had a backdrop of severe demographic crises with lowering fertility and increasingly high mortality that aggravated stress migrations.

In the 1990s, the greatest depopulation was observed in cities located in extreme and adverse natural conditions of the North and Far East; while in the 2000s and 2010s, the population began to decline in various types of cities all over the country—resource cities, company towns, and in small towns on regional peripheries. Moreover, since the 2000s, spatial polarization in Russia deepened with urban shrinkage capturing more and more small and medium-sized cities. Though it would be wrong to conclude that urban shrinkage in Russia is limited solely to smaller cities, there are also many large shrinking cities.

When identifying shrinking cities at the national level, we argue that varying threshold criteria should be used for cities of different populations. High relative population decline in small cities can have "soft" manifestations for the urban economy, built environment, and infrastructure. On the contrary, in large cities, relative changes in numbers are less pronounced, but the high absolute values of population loss lead to pronounced negative effects. Taking into account the specifics of large cities (multi-apartment housing stock, centralized utilities, and infrastructure), even a small relative population decline should be thoroughly investigated and considered in policymaking and planning.

Notes

1 Cities with enterprises or other facilities connected to the military or nuclear spheres that are subjected to the regime of state secret.
2 GULag – acronym of Main Administration of Camps, the government agency in charge of the Soviet forced-labour camp system (1930-1950). The GULag has got punitive and economic functions. Prisoners were assigned distinct economic tasks, including the exploitation of natural resources and colonization of remote areas as well as realization of large-scale industrial construction projects.
3 After the collapse of the Soviet Union in 1991, the Chechen Republic of Ichkeria (Chechnya) in the North Caucasus proclaimed independence from Russia. The first Chechen military campaign (1994–1996) was an attempt by the Russian federal center to regain control over Chechnya. This was followed by the second Chechen military campaign (1999–2009), officially acknowledged as a counter-terrorist operation. Military operations led to numerous casualties, as well as massive out-migration from Chechnya (see, e.g., Le Huérou et al., 2017).

References

Batunova, E. & Gunko, M. 2018. Urban Shrinkage: An Unspoken Challenge of Spatial Planning in Russian Small and Medium-sized Cities. *European Planning Studies*, 26(8), 1580–1597.

BBSR—Bundesinstitut für Bau-, Stadt- und Raumforschung. 2018. Wachsen und Schrumpfen von Städten und Gemeinden. https://gis.uba.de/maps/resources/apps/bbsr/index.html?lang=de

Bernt, M. 2016. The Limits of Shrinkage: Conceptual Pitfalls and Alternatives in the Discussion of Urban Population Loss. *International Journal of Urban and Regional Research*, 40(2), 441–450.

Bolotova, A. 2014. *Conquering Nature and Engaging with the Environment in the Industrialized Russian North.* PhD Thesis. University of Lapland Faculty of Social Sciences. Rovaniemi, Finland: Lapland University Press.

Bruno, A. 2016. *The Nature of Soviet Power.* Cambridge: Cambridge University Press.

Cottineau, C. 2016. A Multilevel Portrait of Shrinking Urban Russia. *Espace, Populations, Sociétés,* https://journals.openedition.org/eps/6123

Haase, A., Bernt, M., Grossmann, K., Mykhnenko, V. & Rink, D. 2016. Varieties of shrinkage in European cities. *European Urban and Regional Studies*, 23(1), 86–102.

Haase, A., Rink, D., Grossmann, K., Bernt, M. & Mykhnenko, V. 2014. Conceptualizing urban shrinkage. *Environment and Planning A*, 46(7), 1519–1534.

Le Huérou, A., Merlin, A., Regamey, A. & Sieca-Kozlowski, E. (eds.) 2017. *Chechnya at War and Beyond.* New York & London: Routledge.

Mkrtchyan, N. 2011. Russia's Migration Balance by Regions: The Twenty Post-Soviet Years. *Regional Research of Russia*, 1, 195–198.

Pallagst, K., Aber, J., Audirac, I., Cunningham-Sabot, E., Fol, S., Martinez-Fernandez, C., Moraes, S., Mulligan, H., Vargas-Hernandez, J., Wiechmann, T., Wu T. & Rich, J. (eds.) 2009. *The Future of Shrinking Cities: Problems, Patterns and Strategies of Urban Transformation in A Global Context.* (No. 2009-01). https://escholarship.org/uc/item/7zz6s7bm

Pallagst, K., Mulligan, H., Cunningham-Sabot, E. & Fol, S. 2017. The Shrinking City Awakes: Perceptions and Strategies on the Way to Revitalisation? *Town Planning Review*, 88(1), 9–13.

Plane, D. & Rogerson, P. 1994. *The Geographical Analysis of Population with Applications to Planning & Business.* New York, NY: John Wiley & Sons.

Rosstat. 2019. Chislennost' naseleniya Rossijskoj Federacii po municipal'nym obrazovaniyam na 1 yanvarya 2019 goda. https://rosstat.gov.ru/compendium/document/13282

Turok, I. & Mykhnenko, V. 2007. The Trajectories of European Cities, 1960–2005. *Cities*, 24(3), 165–182.

Weaver, R., Bagchi-Sen, S., Knight, J. & Frazier, A. E. 2017. *Shrinking Cities: Understanding Urban Decline in the United States.* New York, NY: Routledge.

Wiechmann, T. 2008. Errors Expected—Aligning Urban Strategy with Demographic Uncertainty in Shrinking Cities. *International Planning Studies*, 13(4), 431–446.

Wolff, M., Fol, S., Roth, H. & Cunningham-Sabot, E. 2013. Shrinking Cities, villes en décroissance: une mesure du phénomène en France. *Cybergeo: European Journal of Geography*, http://journals.openedition.org/cybergeo/26136

11 Diverse landscape of urban and regional shrinkage in Russia

Preconditions versus preconceptions in planning and policy

Elena Batunova and Maria Gunko

Introduction

Attempts have been made to provide a comprehensive definition of "shrinkage" as a trajectory of territorial development, but until now, scholars agreed only on depopulation as its main feature (Haase et al., 2014; Bernt, 2016). Over a long period of time, the phenomenon under different labels—"shrinkage", "decline", "decay"—has been discussed in the literature within the urban (e.g., Bontje and Musterd, 2012) and neighborhood perspectives (e.g., Grigsby et al., 1987) since the negative consequences which draw most public attention are mainly manifested at the local level. Though less prominent, studies on shrinking regions in different national contexts also started to appear because global economic and demographic processes began to increasingly affect vast territorial units (European Parliament, 2008; Šimon and Mikešová, 2014). However, the interrelation of different governance levels in managing shrinkage is scarcely discussed in the literature. At the same time, scholars agree that at the urban level, shrinkage is always a result of a complex interplay between multi-scale factors which can hardly be challenged locally (Hoekveld, 2012; Martinez-Fernandez et al., 2012; Haase et al., 2014; Hospers, 2014).

In academic literature, the differences in planning strategies that a shrinking city can adopt are usually linked to awareness, degree of shrinkage's acceptance, political situation, and human agency (Bernt et al., 2014; Hospers, 2014; Pallagst, Fleschurz and Said, 2017). Thus, a shrinking city (i.e. the City Administration) is seen as a somewhat independent entity that can choose an appropriate strategy under the local actors' political willingness and awareness. Such "localist" perspective may, to some extent, hold for countries where cities enjoy relative freedom in planning and policy-making (e.g., Stone, 2015) and where the tradition of regional planning is weak and uncommon (Hollander, 2018). However, in most cases, regional and even national policies largely shape the conditions in which shrinking cities operate (Martinez-Fernandez et al., 2012; Pallagst, Fleschurz, Nothof, and Uemura, 2019). Thus, when discussing shrinkage on the local level, the scope of local authorities' decision-making capacity in various cultural and national contexts should be taken into account.

DOI: 10.4324/9780367815011-14

In European countries, the demographic situation has been characterized by low fertility and aging over the last decades. Moreover, the region is predicted to be the most depopulating geographic area in the world (United Nations, 2019). While aging and natural population decline are relevant for most European counties, the intensity and direction of migration differ, depending predominantly on the economic development (Wolff and Wiechmann, 2018). Cities in Northern and Western Europe benefit from compensatory migration flows and can implement marketing strategies to confront depopulation, Eastern and Southern Europe's cities suffer from both natural population decline and out-migration. Eastern European countries are also affected by the postsocialist transition, whereby the collapse of state socialism and new rationalities of neoliberalism "worked as a cause and as a catalyst for urban shrinkage" (Haase, Rink and Grossmann, 2016; Ringel, 2018), leading to the fact that urban shrinkage here is a rule rather than an exception (Turok, and Mykhnenko, 2007; Haase, Rink, and Grossmann, 2016). To challenge depopulation and the diminishing number of the working-age population, some European Union countries started with the implementation of policies oriented at the support of natality. Later, the focus shifted toward the support of in-migration and to the advancement of "productivity, modernization, and technological progress" (European Parliament, 2008).

Like other European countries, Russia's overall demographic situation is characterized by aging and natural population decline in most regions (Batunova and Perucca, 2020). However, the fundamental difference is in high mortality rate and the poor state of health, resulting in lesser life expectancy (Eberstadt, 2010). The positive international in-migration does not compensate for the overall depopulation; moreover, in-migration (both international and internal) is predominantly restricted to few economically prosperous regions and cities (Zaionchkovskaya, Mkrtchian and Tyuryukanova, 2014), indicating sharp spatial polarization within the country (e.g., Golubchikov, Badyna, and Makhrova, 2014).

Against the above background, the current study's objective is to explore different patterns of shrinkage and emerging responses in Russia. The aim is to foreground if and how urban shrinkage is addressed in planning and policy documents adopted by different governance levels. The research draws empirical evidence from the official regional and local statistical data on socio-economic development and national as well as selected regional and local planning and policy documents. "National", "regional", and "local" in this chapter refer to the country's administrative-territorial division and the corresponding levels of governance.

The chapter is organized as follows. Firstly, we explain the country's existing demographic situation, causes, and dynamics of national, regional, and urban population change. Here we also discuss policies approved at the national level that address demographic issues. Secondly, we present the methodological approach. Thirdly, we analyze the relationship between regional and local (urban) population decline in selected cities and regions on the one hand and policies approved at the regional and local levels on the other. The final section provides reflections on the results and concludes with a description of possible policy directions for managing urban shrinkage in Russia.

Demographic issues and Russian national policies

Demographic situation in Russia

The Russian demographic crisis, which had begun in the 1990s, a little after the collapse of state socialism, was caused by a significant drop in fertility and a simultaneous sharp increase in mortality. Against the general background of global reasons underlying depopulation, i.e., globalization, post-Fordist deindustrialization, and demographic transitions (Ringel, 2018), Russia's negative population trends were inescapable as a result of historical events in the country's past. These events transformed the population structure, causing negative population trends, which were further reinforced by the politico-economic system's change (Eberstadt, 2010).

The total population in Russia declined from 148.2 million people in 1990 to 144.4 million people in 2018 (by 3.8 million people or 2.6%).[1] The population number reached its minimum of 142.7 million in 2008 and then started slightly increasing. The pattern with an excess of the number of deaths over the number of births at the national level in Russia, the so-called "Russian cross", has been characteristic for the 1990s–mid-2000s repeating itself at the regional level with a few exceptions. In 2009–2017, the annual population change was positive, but in 2018 the country entered a new demographic decline period (Rosstat) which was further reinforced by the Covid-19 pandemic (Kobak, 2021). The growth of fertility observed between 2009 and 2017, was only a temporary phenomenon given the population structure–a relatively high number of women that entered the fertile age.

Within the imbalance between fertility and mortality in Russia, it is mortality that plays a critical role. While the birth rate in Russia is comparable with that in most European countries, the mortality rate is considerably higher (Pant, 2017). Moreover, the gap between male and female life expectancy in Russia is the largest in the world, about ten years.[2] The mortality crisis in Russia at the beginning of the 1990s had several causes that reinforced each other. Firstly, it is related to the mass stress and psychological trauma–due to the collapsed familiar socialist order and economic crises (Yurchak, 2006)–which resulted in a sharp increase in alcohol consumption and drug abuse (Shkolnikov et al., 1998). Secondly, the chaos of transition and weakened state institutions brought about increased criminalization of all spheres of life with the rising homicide and suicide rates (Lysova and Shchitov, 2015). Both above ran parallel to the decline in the quality of healthcare provision (Dunn, 2008). In the following years, the mortality rate has been decreasing but still remains high due to the restructuring of the healthcare system (for more details, see Chapter 13 in this volume), low popularity of a healthy lifestyle, poor nutrition, and environmental issues (Pant, 2017).

Even though Russia was the second-largest host country for international migrants over several years after the collapse of state socialism and remains one of the most significant migrant attractors in the world (*World Migration Report*, 2018), immigration even during the period of massive repatriation from former Soviet Republics could not compensate for the natural population decline. In 1992–2018, the country grew by 9.5 million people due to migration but lost about 13.5 million people due to natural population decline (Rosstat). This chapter does not discuss the qualitative aspects of migration, focusing on quantitative changes and the population's spatial distribution within the country. However, it is essential to note that such a phenomenon as "brain drain" is a significant aspect of the Russian demographic change.

Against the overall demographic situation at the national level, an important factor influencing population change in Russian cities and regions is the country's increasing population mobility. In the Soviet era, people were "fixed" to their places of living, the change of which was limited significantly by the state through the institute of *propiska*. In the post-Soviet period, regulations were significantly loosened; people could decide more freely where to live, which ultimately affected internal migration patterns (Andrienko and Guriev, 2005; Vakulenko, 2016). The most important current internal mobility trend is the so-called "western drift" (Zaionchkovskaya, Mkrtchian and Tyuryukanova, 2014)—migration from the Far East and Siberia toward the European part of Russia. The destination for internal migrants coincides with those for international migrants' including large cities in Central and Northern-Western Federal Districts, but above all Moscow and Saint-Petersburg (Rozanova, 2016). Only a few regions outside of the mentioned areas, such as Krasnodarsky krai and Belgorodskay oblast' with favorable conditions for agriculture or oil and gas provinces of Tyumen oblast' in Siberia, are characterized by a positive migration balance (Rosstat).

Given the general national and regional trends, about 70% of Russia's cities have lost population between 1989 (the year of the last Soviet population census) and 2018. Between 1989 and 2010, around 50% were experiencing both natural population decline and out-migration. In 10%, the positive migration balance could not compensate for the natural population decline, and in another 10%, depopulation was due to out-migration despite the natural population growth (Cottineau, 2016).

Russian national policies focused on demographic issues

Since negative population trends are a nationwide phenomenon with a majority of Russian cities experiencing depopulation, the issue should logically be addressed at the national level. Moreover, the national level of policymaking and planning is currently the most influential one in Russia. Though in the early post-Soviet period, decentralization and democratization trends set in at the local level, they saw a harsh roll-back since the beginning of the 2000s (Gelman and Ryzhenkov, 2011; Gelman, 2018). Currently, cities and regions officially exercise somewhat autonomy and bear responsibility for own development; however, the hierarchical fiscal policy and the so-called "power vertical" constrain their actual decision-making power, increasing dependence on the federal center (ibid). Moreover, some issues of planning for shrinkage, e.g., right-sizing amenities and public services due to the changing population number and demographic structure, are addressed according to standards defined at the national level.

Despite the severe demographic crisis in the 1990s, the demographic issue was not a concern for the national policies. The "Demographic Policy Concept of the Russian Federation for the Period Until 2025" was approved only in 2007. It is oriented at improving the indicators of the natural population change–birth and death rates, longevity. Even though migration is mentioned as a part of the demographic policy, little attention is paid to it within the Demographic Policy Concept. The core focus of the Demographic Policy Concept is on the increase of fertility. Proposed measures bear economic nature; among them most noticeable are various one-off payments at the birth of children—e.g., Maternal capital paid at the birth of a second child,[3] subsidies for the improvement of families' housing conditions.

The "Concept of the State Migration Policy of the Russian Federation for the Period Until 2025" was adopted in 2012. In 2019, it was replaced by the "Concept of State Migration Policy of the Russian Federation for 2019–2025". The latter became the first national policy document that emphasized that boosting international in-migration is essential to sustain a favorable economic situation given the decline in the working-age population. The document was linked to projections published by the Russian State Statistical Service. According to them, the working-age population will decline by five million people in 2015–2020 (Rosstat).

The demographic policies at the national level envisage many ministries' work and consider many aspects influencing demographic development; however, they do not take into account the existing differences in the demographic development of Russian regions. National policies address all cities and regions equally, but they cannot equally "save" them from population decline. Cities and regions compete for the same people and the same investments, discouraging partnership and collaborative planning and policymaking (Dyadik, 2016). Thus, the regional level is supposed to act as an intermediate between the national and the local levels. In this regard, regional policies should have a crucial role in matching general measures developed at the national level with the local issues, features, and interests.

State Resettlement from the North policy (hereafter–Resettlement policy). Unlike the previously described policies aimed at promoting population growth, there is also a counter-policy initiative. Since the mid-1990s, special national and regional[4] policies have been developed for northern Russian territories (the list of which is defined by the state) to resettle the population from regions with harsh natural conditions (Nuykina, 2011). The general mechanism is that the state allocates housing certificates for purchasing housing to individuals based on a list of criteria–e.g., length of work experience in the North, health condition—which allows acquiring housing in non-northern regions of Russia. Resettlement policy's overall contribution to out-migration from the North is estimated as minor, less than 6% of the total out-flow in 2003–2017 (Denisov, 2017). However, it presents an important national guideline against which the regional and local policies should be formulated.

Methodological approach

Understanding the causes of depopulation in Russian cities is limited by the scarcity and unreliability of official data (Vishnevsky and Zakharov, 2010; Batunova and Perucca, 2020). Though the available data in general terms allows identifying which component–natural population decline or out-migration—contributes more to regional and urban depopulation. This provided us with an opportunity to classify regions and cities before evaluating their planning and policy documents. The study uses data on population collected by the Russian State Statistical Service for 83 regions for the period 1990–2017 and 1,095 cities for the period 1989–2017.

Prior to the selection of case studies, we made a regional classification using available data. Firstly, based on the data about regional population change in 1990–2017, regions were divided into two groups: (1) shrunk, in which the total population number in 2017 was lower than in 1990; and (2) grown, where total population number increased during the same period. Components of population change for the period 1995–2017[5] for every region were analyzed. Based on this analysis, we defined six regional population change patterns: three for shrunk regions and three for grown regions. These patterns depend on the combination of contributors to regional population change during the whole period: the balance of fertility and mortality and in- and out-migration. The classification of regions is as follows (see Figure 11.1):

1 "Shrinking regions"—mortality exceeded fertility, out-migration exceeded in-migration;
2 "Depopulating regions"—mortality exceeded fertility, in-migration exceeded out-migration (in-migration could not compensate for natural population decline);
3 "Donating regions"—fertility exceeded mortality, out-migration exceeded in-migration (natural population growth could not compensate for out-migration);

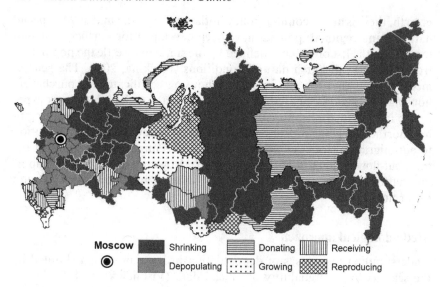

Figure 11.1 Classification of Russian regions according to the components of population change in 1990–2017.

4 "Growing regions"—fertility exceeded mortality, in-migration exceeded out-migration;
5 "Receiving regions"—mortality exceeded fertility, in-migration exceeded out-migration (migration compensated for natural population decline);
6 "Reproducing regions"—fertility exceeded mortality, out-migration exceeded in-migration (natural population growth compensated for out-migration).

Secondly, we divided Russian cities into two groups: grown and shrunk in 1989–2017. This approach has limitations as it does not identify all cities in which depopulation is currently an ongoing process since many cities started to depopulate in the 21st century, and their population number is still higher than that in 1989. Furthermore, in some cases, the available data does not consider administrative changes when cities expand their boundaries to increase or at least maintain the population number. Although a more precise analysis will probably identify more shrunk cities, it will not change the general picture.

Upon reviewing the results of the regional classification and the location of shrunk and grown cities–we used a combination of "extreme" and "most similar" case selection strategies (Seawright and Gerring, 2008). "Extreme" case study selection means that the case studies exemplify the most striking samples of urban shrinkage, while the "most similar" case study selection means that they also exhibit some resemblance in terms of structural characteristics. The outline of possible options was based on three formal criteria: (1) cities having a status of *gorodskoy okrug* meaning

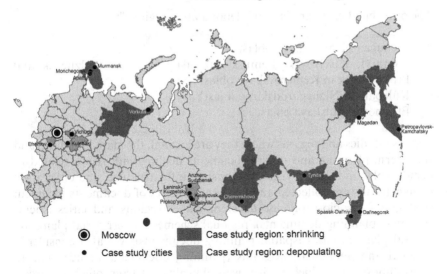

Figure 11.2 Case studies.

direct subordination to the regional government and more autonomy in policymaking (in comparison to *gorodskoe poselenie*); (2) population number 30,000 inhabitants and over meaning that the city has a more or less developed infrastructure and housing; and (3) cities where population dynamics is less than the value of the first quartile of population dynamics distribution corresponding to depopulation at an average pace of 1% and more annually. The latter margin seems too significant to be ignored by both local and regional governments. Out of 20 cities that corresponded to the chosen criteria, 2 cities were excluded that the necessary data and documents for further analysis were absent.

Selected 18 cases are located in 11 regions from different parts of the country (see Figure 11.2), allowing comparisons of regional and urban policies in diverse geographical conditions. Ten of these cities are located in seven "shrinking regions". Among these "shrinking regions" in three regions—Magadanskaya oblast', Murmanskaya oblast', and Kamchatsky krai—there were no growing cities, even the regional capitals depopulated in 1989–2017. The list of cases is as follows:

- Tynda in Amurskaya oblast';
- Cheremkhovo in Irkutşkaya oblast';
- Petropavlovsk-Kamchatsky (regional capital) in Kamchatsky krai;
- Magadan (regional capital) in Magadanskaya oblast';
- Murmansk (regional capital), Apatity and Monchegorsk in Murmanskaya oblast';
- Dalnegorsk and Spassk-Dalniy in Primorsky krai;
- Vorkuta in Komi Republic.

The other eight cases are in four "depopulating regions":

- Vichuga in Ivanovskaya oblast';
- Osinniki, Kiselyovsk, Leninsk-Kuznetsky, Anzhero-Sudzhensk and Prokopyevsk in Kemerovskaya oblast';
- Kulebaki in Nizhegorodskaya oblast';
- Efremov in Tulskaya oblast'.

For case studies and regions where they are located, the main long-term and short-term planning and policy documents dealing with issues related to demography and the provision of housing, infrastructure, and cityscapes' management were collected and analyzed. The set of documents included: strategies for socio-economic development of regions and cities (socio-economic planning documents), regional schemes for territorial planning, general plans of cities (spatial planning documents). We also considered regional and municipal programs (planning documents of mixt nature depending on the subject). They were downloaded from official sources: State Automatic Information System "Governance" (http://gasu.gov.ru/), Federal Governmental Information System on Territorial Planning FGIS TP (https://fgistp.economy.gov.ru/); Catalogues of Russian national and regional normative documents "Consultant Plus" (http://www.consultant.ru/) and "Garant" (http://www.garant.ru/), as well as the official websites of regions and cities. Moreover, we analyzed structures of regional governments responsible for "localizing" national policies and programs.

The evaluation of regional policies and government structures was based on the priorities set (at least formally) by the national government concerning the demographic issue—development of human capital, economic development, and quality of the living environment. The focus on these three directions is expressed through numerous national strategies, projects, and programs (e.g. "State National Policy Strategy", "National Security Strategy", "National Goals and Strategic Objectives of the Development", "Bases of the State Policy of Regional Development", "Strategy of Spatial Development", "Demographic Policy Concept", Priority project "Formation of a Comfortable Urban Environment", etc.) and the recently introduced "National Projects of Russia 2019–2024" formed in three areas: "Human Capital", "Comfortable Living Environment", and "Economic Growth".

Results

Regional and urban patterns of population change

In 1990–2017, 58 out of 83 (or 70%) Russian regions lost population. The number of shrunk cities was 774 among 1,095 analyzed (the same 70%); 664 shrunk cities (or 86%) were in shrunk regions (see Table 11.1). The biggest number of shrunk cities was in "depopulating regions" (328 cities or 42% of

Table 11.1 Distribution of shrunk cities among different categories
of regions

Region's type	Number of regions of this type	Number of shrunk cities in the regions of this type
Shrunk Regions	**57**	**664**
Shrinking	33	318
Depopulating	21	328
Donating	3	18
Grown Regions	**25**	**110**
Growing	7	7
Receiving	14	95
Reproducing	4	8

shrunk cities), 318 shrunk cities (or 41%) were located in "shrinking regions".
In these two types of regions, shrunk cities made up the majority. A big
conglomerate of shrunk cities was in "receiving regions" (95 cities or 12%
of shrinking cities). This clearly demonstrates that a positive migration balance at the regional level does not guarantee population growth in all cities.
Moreover, 73 of 95 shrunk cities (77%) in "receiving regions" were located in
the Moscow metropolitan area–the region with the highest concentration of
economic activity and population and the densest settlement and transport
networks. In most shrunk regions, both shrunk and grown cities are present; generally, grown cities in shrunk regions are regional capitals.

Planning and policymaking for shrinkage

Structure of regional governments. The analysis of regional governments'
structures demonstrated that the focus was on economic development in
most regions. Usually, regional governments include several ministries,
departments, or agencies responsible for economic development, investments, labor, innovations, or entrepreneurship. The "living environment"
issue also seems significant. There were ministries or agencies responsible
for spatial development, architecture, construction, and land use in most
regions. Firstly, this is explained by the importance of land distribution
and regulation of land use imposed by the state through the introduction
of the Land Code and Urban Planning Code. Secondly, there is a need for
regulation over construction, which is considered one of the most important economic activities (Batunova and Gunko, 2018). Thirdly, the issue of
a "comfortable urban environment" over the recent years has evolved into
the new leitmotif for Russian urban planning and development (Zupan and
Gunko, 2019).

Generally, Russian national and regional governments provide sectoral policies addressing different issues mentioned in the national demographic policy through a specific ministry/department/agency programs.
For example, measures aimed at raising fertility defined at the national

level can be a shared responsibility between regional agencies such as the Department for Social Welfare, the Department for Education and Science, the Department for Labor and Employment, the Department for Public Health, the Construction Department and the Main Financial Department (Kemerovoskaya oblast'). Thus, the direction of "human capital" in the regional governments' structure is addressed by sectoral ministries responsible for diverse social issues such as education, health care, sport, or culture. However, the demographic issue *per se* is poorly reflected in the structures of the regional governments. There are ministries or departments in some regions, which are responsible indirectly for the demographic development focusing primarily on labor force provision (e.g., Employment and Migration Policy Agency of the Kamchatsky krai or the Committee of the Ivanovskaya oblast' on Labor, Promotion of Employment and Labor Migration). The only example when the demographic issue is mentioned in an agency's title is the Ministry for Health and Demographic Policy of Magadanskaya oblast'.

In some regions, specific government structures coordinating activities of different agencies in the field of demographic policy are established: e.g., Governor Commissions on Demographic Policy in Ivanovskaya oblast'; Family, Children, and Demographic Policy Commission in Tulskaya oblast'; Permanent Working Group on the Implementation of Demographic Policy in the Komi Republic. However, the analysis of such structures' activity indicates that their actual influence on regional policy development and implementation is negligible. They do not produce demographic projections, analyze spatial aspects of depopulation and intraregional differences. Their activities are restricted to monitoring the demographic situation and submitting regular reports, which in most cases date to 2015–2016 with no further information about their recent work.

Regional planning and policies. Policy and planning decisions at the regional level are reflected in the long-term strategic and territorial planning documents, as well as in regional programs. The primary two long-term documents are the strategy for socio-economic development and the Scheme for regional territorial planning. All regions included in the current study have these documents.

In the analyzed documents, little attention is paid to the causes and consequences of depopulation. Though in some regions shrinkage (and related terms) is mentioned as a threat to future development, the prognosis for future regional development on average includes an over-optimistic estimation of the demographic situation with improvement due to the efforts and measures of the national government. As an exception, the Strategy for socio-economic development and the Scheme for territorial planning of Magadanskaya oblast' can be mentioned. Both long-term documents emphasize demographic issues. These documents propose measures reacting to depopulation, such as the transformation of settlement system with the closure of some settlements (complete resettlements of residents and closure of all facilities), optimization of public services and amenities, development of online education, telemedicine, and mobile medical groups. However, the Strategy points that such measures are impossible to implement at the regional level without an appropriate national policy.

Less than half of researched regions have developed a concept or a program for demographic policy despite the negative demographic situation (Magadanskaya oblast', Irkutskaya oblast', Nizhegorodskaya oblast', Tulskaya oblast', and Kamchatsky krai). Few existing regional Concepts of demographic policy highlight the complexity of the demographic issue. The Concept for Demographic Policy of Kamchatsky krai points out that the Russian Federation's Demographic Policy Concept does not consider regional specifics of migration or propose a differentiated approach to managing it. Furthermore, it also hopelessly notes that attracting migrants cannot help solve the regional demographic problem due to the finitude of the country's demographic resources. Other regions claim to have a "demographic policy", which refers to the measures stated in different sectoral programs on education, health care, housing, etc. However, regional programs have a short planning horizon and are aimed at solving current problems, which does not allow taking into account the long-term perspective of regional demographic situation change and the transformation of the regional settlement system. Moreover, similar to national policies, which do not differentiate regions, regional policies ignore spatial differences at the local level.

Overall, it is obvious that in most researched regions, the negative demographic situation can hardly be ignored longer. However, the dependence of regions on the national government and the lack of a legal mechanism that could be adapted for managing decline nullify attempts to find solutions to the current situation at the regional level.

Local planning and policies. As was mentioned above, local planning and policymaking have little room for maneuvering, even less than the regional level has against the national one. Undoubtedly, local interests, responsibilities, and human agency advocate changes (Gunko et al., 2021). However, the content and wording of local planning and policy documents are predominantly framed by the requirements of higher tiers of governance since the local level is financially dependent on subsidies from the regional or federal budgets.

At the local level, the focus of policies is mostly on the direction of "living environment" due to their responsibilities in land use management, as well as the implementation of the "comfortable urban environment" projects (Zupan and Gunko, 2019). The second important direction is economic development which is seen by most municipalities as the primary goal. Despite the presence of long-term strategic and planning documents, most municipalities do not attribute the demographic situation to risks for future development. The insufficiency and low quality of demographic projections taken together with the lack of understanding of certain demographic trends' underlying reasons lead to false hopes of returning to growth. In some cities, local planning and policy documents do not present their own demographic projection basing proposed aims, goals, and measures on average regional figures.

Foremost, causes for demographic change on the local level are simplified. For instance, the municipal program "Providing Affordable and Comfortable Housing to Residents of Petropavlovsk-Kamchatkiy" claims that there are two causes of low fertility—lack of prospects for obtaining (acquiring) housing and low income. This statement is further developed into the notion about a necessity to provide new land plots for housing construction. Similar logic underlies most general plans that propose new residential areas and are oriented toward increasing housing construction. Moreover, most documents assume that the improvement of the overall economic situation, increase of employment and creation of new jobs, economic incentives for the development of entrepreneurship will lead to population growth or at least slowdown depopulation.

There is almost no correlation between regional and local levels of policymaking and planning regarding the demographic issue. In most cases, documents of local and regional levels are contradictory, with depopulation being addressed differently. For example, in Magadanskaya oblast' at the regional level, the demographic issue is considered a serious obstacle for future development and a problem requiring actions. But in the strategy for socio-economic development of Magadan, city demographic issue is completely ignored, even though it lost 38.9% of its population since 1989. This may be partly explained by the strict division of responsibilities between different levels of governance and different sectoral policies with the absence of methodologies for an integrated approach.

Among the researched cases, the only city that realistically perceived the situation and had a specific policy to mitigate the negative consequences of depopulation (so-called "controlled shrinkage") is Vorkuta in the Komi Republic. Spontaneous out-migration and the state Resettlement policy, which is actively supported on the local level (Shiklomanov et al., 2019), contributed here to a large amount of non-privatized, municipal housing stock—ca 60% in comparison to the national average of 10% (Gunko, Eremenko, and Batunova, 2020). Much of this housing stock is abandoned or partially vacant. The need for cutting costs to maintain the local budget and a strong local political leadership led to the official acceptance of shrinkage with supporting measures for the cityscape. These included densifying the center parallel to disconnection from utilities and demolition of abandoned housing on the peripheries (ibid). In other researched cases, we saw no variations in planning and policymaking approaches despite the differences in the cases' size, status, geographic location, and presence of the Resettlement policy. Thus, Vorkuta's exceptional approach is a rare deviation from the pro-growth agenda. It is fostered by the need to manage vacant municipal property and a strong human agency of local policymakers who creatively translated the existing approaches and effectively utilized available funds (Gunko, Eremenko, and Batunova, 2020; Shiklomanov et al., 2019).

Discussion and conclusions

Currently, most Russian regions and cities are shrinking and will pursue the same path in the future due to the complex combination of factors causing population change in the country. The excess of deaths over births is the main driver of depopulation both at the national level and in most cities and regions. This precondition, being a significant challenge for policymaking, needs to be understood and addressed properly.

With the growing re-centralization of Russian urban planning and policymaking mirroring the patterns of the Soviet system, it becomes increasingly hard for a city or a region to act independently, as well as find its own resources to cope with emerging issues (Kinossian and Morgan, 2014, Zupan and Gunko, 2019). It is thereby important to understand at which stage of depopulation acceptance the national level is. While in many European countries, the move toward acceptance of depopulation as an inevitable process through formulating policies not for countering it but for adapting is obvious (Hospers, 2014), in Russia, the birth support policy with less attention to in-migration is still seen as a priority (Demographic Policy Concept, 2007). This is despite the actual projections and evaluations confirming the impossibility to reach demographic growth based on existing national demographic resources. However, we consider this approach the earliest step of shrinkage's acceptance and assume that countering depopulation is the main strategy for all governance levels.

When national policies are place-unspecific, the regional level should compensate for this shortcoming making national strategies, plans, and programs more "territorial". In the Russian case, though, regions seem to be "weak links" in the "power vertical". Based on the results obtained through the analysis of cases, it is fair to say that the significant differences in demographic development demonstrated at the regional level are almost not reflected in regional policies. Rare exceptions point to the need for better management of demographic processes at the national level but without mentioning the need for a wider autonomy of the regions in policymaking. Moreover, there seems to be no synergy between regional and local planning and policies. The understanding of the demographic changes and subsequent challenges by regional and local authorities may differ significantly.

The analyzed documents indicate that depopulation is foremost ignored or there is hope for improving the situation based on external input. Instead of thorough research, the development and implementation of regional and local policies and planning measures are based on preconceptions and fears of depopulation that do not allow a realistic evaluation of the situation at hand and a proper projection of future development. Generally, the lack of relevant data, financial resources, and qualified specialists becomes obstacles to understanding population change's nature at the regional and local levels (Batunova and Gunko, 2018). It is believed that cities lose population mostly due to job-related out-migration and other economic reasons. This preconception prioritizes the search for new investors and opportunities to boost the local labor markets as the main strategy against depopulation. Searching for ways to improve the economic situation and increasing liveability through housing construction and the creation of a "comfortable urban environment" are the two approaches for managing depopulation that are actively supported by the national government.

The presence of state Resettlement policy seems to have little influence on agenda-setting in the researched cases, five of which were in the North. Even though the state Resettlement policy partly fosters depopulation in the Russian North, the state does not provide legal mechanisms, organizational, and financial support for municipalities to cope with depopulation's negative consequences in affected cities. Moreover, it seems that there is a deep contradiction between the Resettlement policy and the overall aspirations of the state in terms of population growth, "forcing" regions and cities of the North to develop pro-growth planning and policy documents regardless of out-migration. The only exception, Vorkuta in the Komi Republic, is a rare example where the local strategy runs contra state's agenda.

Overall, countering shrinkage can hardly be successful in most cases in Russia. Instead, shrinking cities and regions should acknowledge the causes and consequences of their depopulation and focus on adaptation to new conditions. Coping with depopulation and challenges caused by it requires effective integration of all three governance levels with the shift from short-term measures (addressing the current problems) to a long-term strategy considering how the future changes would impact the society, economic development, and liveability. Thus, if the state continues the concentration of power at the national level, it should take responsibility and put more effort into developing differentiated approaches for managing urban shrinkage corresponding to the variety of regional and urban conditions.

Notes

1 The Republic of Crimea being a disputed territory is not included in the present study.
2 According to Rosstat, in 2017, male life expectancy was 67.5 years and female life expectancy was 77.6 years.
3 In 2019, it was ca. 7,000 dollars.
4 Ymalo-Nenets avtonomny okrug, Khanty-Mansi avtonomny okrug, Krasnoyarsky krai, Magadan oblast', Tuymenskay oblast', Sakha Republic.
5 The period for which the data on the components of population change was available.

References

Andrienko, Y. & Guriev, S. 2005. *Understanding Migration in Russia*. Washington, D.C.: World Bank Policy Note.

Batunova, E. & Gunko, M. 2018. Urban Shrinkage: An Unspoken Challenge of Spatial Planning in Russian Small and Medium-sized Cities. *European Planning Studies*, 26, 1580–1597.

Batunova, E. & Perucca, G. 2020. Population Shrinkage and Economic Growth in Russian Regions 1998–2012. *Regional Science Policy and Practice*, 12(4), 595–609. https://doi.org/10.1111/rsp3.12262

Bernt, M. 2016. The Limits of Shrinkage: Conceptual Pitfalls and Alternatives in the Discussion of Urban Population Loss. *International Journal of Urban and Regional Research*, 40(2), 441–450.

Bernt, M., Haase, A., Großmann, K., Cocks, M., Couch, Ch., Cortese, C. & Krzysztofik, R. 2014. How does(n't) Urban Shrinkage get onto the Agenda? Experiences from Leipzig, Liverpool, Genoa and Bytom: Urban Shrinkage on European Policy Agendas. *International Journal of Urban and Regional Research*, 38(5), 1749–1766.

Bontje, M. & Musterd, S. 2012. Understanding Shrinkage in European Regions. *Built Environment*, 38(2), 153–161.

Concept of the state migration policy of the Russian Federation for the period until 2025. Approved by the President of the Russian Federation of June 13, 2012. Available at: <http://www.consultant.ru/> [Accessed 10 September 2019].

Cottineau, C. 2016. A Multilevel Portrait of Shrinking Urban Russia. *Espace Populations Sociétés* 2015/3–2016/1. http://eps.revues.org/6123

Demographic policy concept of the Russian Federation for the period until 2025. Approved by the Decree of the President of the Russian Federation of October 9, 2007 No. 1351. Available at: <http://www.consultant.ru/> [Accessed 10 September 2019].

Denisov, E. 2017. Migratsionnie protsessi v gorodakh Rossiyskogo severa v 1990–2010ᵉ [Migration Processes in Cities of the Russian North in the 1990s–2010s]. *Regionalnye issledovaniya*, 2, 44–55.

Dunn, E. C. 2008. Postsocialist Spores: Disease, Bodies, and the State in the Republic of Georgia. American Ethnologist, 35(2), 243–258.

Dyadik, V. 2016. Strategic planning at the municipal level: Russian challenges and Nordic practices. *Barents Studies: Peoples, Economies and Politics*, 1(2), 75–95.

Eberstadt, N. 2010. Russia's Peacetime Demographic Crisis: Dimensions, Causes, Implications. National Bureau of Asian Research, Project MUSE database. Available at: https://www.nbr.org/publication/russias-peacetime-demographic-crisis-dimensions-causes-implications/ [Accessed 23 February 2019].

European Parliament. 2008. *Shrinking Regions: A Paradigm Shift in Demography and Territorial Development*. Brussels: Directorate-General for Internal Policies of the Union.

Gelman, V. 2018. Bringing Actors back in: Political Choices and Sources of Post-Soviet Regime Dynamics. *Post-Soviet Affairs*, 34, 282–296.

Gelman, V. & Ryzhenkov, S. 2011. Local Regimes, Sub-national Governance and the "Power Vertical" in Contemporary Russia. *Europe-Asia Studies*, 63, 449–465.

Golubchikov O., Badyina A. & Makhrova A. 2014. The Hybrid Spatialities of Transition: Capitalism, Legacy and Uneven Urban Economic Restructuring. *Urban Studies*, 51(4), 617–633.

Grigsby, W., Baratz, M., Galster, G. & Maclennan, D. 1987. The Dynamics of Neighborhood Change and Decline. *Progress in Planning*, 28(1), 1–76.

Gunko, M., Eremenko, Y. & Batunova, E. 2020. Strategii planirovaniya v usloviyakh gorodskogo szhatiya v Rossii: issledovanie malikh I srednikh gorodov [Planning Strategies in the Context of Shrinkage in Russia: Evidence form Small and Medium-sized Cities]. *Mir Rossii*, 29(3), 121–141.

Gunko, M., Kinossian, N., Pivovar, G., Averkieva, K. & Batunova, E. 2021. Exploring Agency of Change in Small Industrial Towns Through Urban Renewal Initiatives. *Geografiska Annaler, Series B: Human Geography.* https://doi.org/10.1080/0435368 4.2020.1868947.

Haase, A., Rink, D. & Grossmann, K. 2016. Shrinking Cities in Post-socialist Europe: What can We Learn from their Analysis for Theory Building Today? *Geografiska Annaler: Series B, Human Geography*, 98(4), 305–319.

Haase, A., Rink, D., Grossmann, K., Bernt, M. & Mykhnenko, V. 2014. Conceptualizing Urban Shrinkage. *Environment and Planning A*, 46(7), 1519–1534.

Hoekveld, J. J. 2012. Time-space Relations and the Differences Between Shrinking Regions. *Built Environment*, 38(2), 179–195.

Hollander, J. 2018. *A Research Agenda for Shrinking Cities.* Cheltenham/Northampton, MA: Edward Elgar.

Hospers, J.-G. 2014. Policy Responses to Urban Shrinkage: From Growth Thinking to Civic Engagement. *European Planning Studies*, 22(7), 1507–1523.

Kinossian, N. & Morgan, K. J. 2014. Development by Decree: The Limits of "Authoritarian Modernization" in the Russian Federation. *International Journal of Urban and Regional Research,* 38, 1678–1696.

Kobak, D. 2021. Excess Mortality Reveals Covid's True Toll in Russia. *Significance*, 18(1), 16–19. https://doi.org/10.1111/1740-9713.01486.

Lysova, A. & Shchitov, N. 2015. What is Russia's Real Homicide Rate? Statistical Reconstruction and the 'Decivilizing Process'. Theoretical Criminology, 19(2), 257–277.

Martinez-Fernandez, C., Kubo, N., Noya, A. & Wiechmann, T. 2012. *Demographic Change and Local Development: Shrinkage, Regeneration and Social Dynamics.* Paris: OECD Publishing. http://dx.doi.org/10.1787/9789264180468-en.

Nuykina, E. 2011. *Resettlement from the Russian North: An Analysis of State-induced Relocation Policy.* Rovaniemi: Arctic Centre, University of Lapland.

Pallagst, K., Fleschurz, R., Nothof, S. & Uemura, T. 2019. Shrinking Cities: Implications for Planning Cultures? *Urban Studies.* Epub ahead of print 17 December 2019. https://doi.org/10.1177/0042098019885549

Pallagst, K., Fleschurz, R. & Said, S. 2017. What Drives Planning in a Shrinking City? Tales from two German and two American Cases. *Town Planning Review*, 88(1), 15–28.

Pant, H. 2017. *Russia's Demographic Trajectory: Dimensions and Implications.* ORF Occasional Paper.

Ringel, F. 2018. *Back to the Postindustrial Future: An Ethnography of Germany's Fastest Shrinking City.* New York and Oxford: Berghahn Books.

Rosstat. Russian Federal State Statistics Service website. Available at: <http://www.gks.ru/> [Accessed 10 September 2019].

Rozanova, M. S. (eds.) 2016. *Labor Migration and Migrant Integration Policy in Germany and Russia.* Saint Petersburg: Saint Petersburg State University.

Seawright, J. & Gerring, J. 2008. Case Selection Techniques in Case Study Research: A Menu of Qualitative and Quantitative Options. *Political Research Quarterly*, 61, 294–308.

Šimon, M. & Mikešová, R. 2014. *Population Development and Policy in Shrinking Regions: The Case of Central Europe.* Prague: Institute of Sociology, Academy of Sciences of the Czech Republic.

Shiklomanov, N., Streletskiy, D., Suter, L., Orttung, R. & Zamyatina, N. 2019. Dealing with the Bust in Vorkuta, Russia. *Land Use Policy.* https://doi.org/10.1016/j.landusepol.2019.03.021

Shkolnikov, V. M., Cornia, G. A., Leon, D. A. & Meslé, F. 1998. Causes of the Russian Mortality Crisis: Evidence and Interpretations. *World Development*, 26(11), 1995–2011.

Stone, C. 2015. Reflections of Regime Politics: From Governing Coalitions to Urban Political Order. *Urban Affairs Review*, 51(1), 101–137.

Turok, I. & Mykhnenko, V. 2007. The Trajectories of European Cities, 1960–2005. *Cities*, 24(3), 165–182.

United Nations, Department of Economic and Social Affairs, Population Division. 2019. *World Population Prospects 2019*, Volume II: Demographic Profiles (ST/ESA/ SER.A/427).

Vakulenko, E. S. 2016. Econometric Analysis of Factors of Internal Migration in Russia. *Regional Research of Russia*, 6, 344–356.

Vishnevsky, A. G. & Zakharov, S. V. 2010. Chto znaet i chego ne znaet rossiyskaya demograficheskaya statistika [What Russian Demographic Statistics does and does not Know]. *Voprosi statistiki*, 2, 7–17.

Wolff, M. & Wiechmann, T. 2018. Urban Growth and Decline: Europe's Shrinking Cities in a Comparative Perspective 1990–2010. *European Urban and Regional Studies*, 25(2), 122–139.

World Migration Report. 2018. International Organization for Migration, Geneva.

Yurchak, A. 2006. Everything Was Forever, Until It Was No More: The Last Soviet Generation. Princeton: Princeton University Press.

Zaionchkovskaya, Zh., Mkrtchian, N. & Tyuryukanova, E. 2014. Russia's immigration challenges. In *Russia and East Asia: Informal and gradual integration* (pp. 200– 243). New York: Routledge.

Zupan, D. & Gunko, M. 2019. The Comfortable City Model: Researching Russian Urban Planning and Design Through Policy Mobilities. *Gorodskie issledovaniya I praktiki*, 4(3), 7–22.

12 The transformation of local labor markets in shrinking cities in Russia from 2010 to 2017

Evgenii Antonov

One of the key questions that arise when studying the phenomenon of shrinking cities is the relationship between economic and demographic characteristics, as well as their mutual influence. Previous studies indicate that shrinkage is a complex process (Haase et al. 2014), whereby demographic and economic shrinkage often reinforce each other (Hoekveld 2012), but the underlying cause is not obvious (Großmann et al. 2013). Although shrinkage is usually associated with deterioration of key socio-economic characteristics and features, in some cases, exceptions may be observed. For example, sometimes a decrease in the population is accompanied by an increase in economic activity (Hartt 2018) that may be due both to individual reasons in each particular city and the overall complexity of spatial ties between cities, especially those that are a part of larger urban agglomerations.

In the current chapter, based on empirical evidence from Russia, the focus is on the (inter)relationships between urban population dynamics and the state of the local labor market (hereinafter, LLM). The latter is assessed based on, firstly, the physical availability of jobs and, secondly, average wages. The research question is: Are cities facing demographic shrinkage characterized by a worse situation on the LLM than that in growing cities? There are practically no studies of this kind in relation to shrinking cities.

Most studies conducted on shrinking cities implicitly state that urban population dynamics is related to the state of the LLM since the crisis in the economy, accompanied by a reduction in the number of jobs, leads to higher unemployment rates and lower-income levels resulting in out-migration of economically active population groups (Wolff & Wiechmann 2018). On average, the narrowed natural reproduction of the population and the LLM crisis, with consequent out-migration, predetermine steadily negative demographic trends in many Russian cities, turning them into shrinking ones (Mkrtchan 2017). At the same time, there are also counterexamples: the adverse state of the LLM may not necessarily lead to urban shrinkage and vice versa (Antonov 2019b).

The task of finding a (statistically significant) relationship between the state of the LLM and urban population dynamics is not particularly difficult for countries with a developed statistical data accounting system. But in Russia, finding and collecting relevant indicators at the urban (municipal) level for a

DOI: 10.4324/9780367815011-15

large sample of cities is quite challenging (Vishnevsky & Zakharov 2010). As a result, in the 1990s and 2000s, most of the research on LLM in Russia was limited to the analysis of the situation in separate cities (or their groups based on territorial or sectoral characteristics) and was based on periodic data from sample surveys. Most prominent and fruitful were studies of the LLM in single-industry towns (Vlasova et al. 1999; Lyubovny et al. 2001; Kuznetsova 2003; Nekrasova 2012; Mikryukov 2016). Recently, studies of labor mobility associated with commuting and fly-in fly-out (FIFO) type of employment (in Russian tradition "*otkhodnichestvo*") are becoming numerous (Veliky 2010; Nefedova 2015; Plyusnin et al. 2013; Mkrtchyan & Florinskaya 2016), especially for large urban agglomerations (Makhrova & Bochkarev 2018; Averkieva et al. 2015; Makhrova & Kirillov 2015). At the same time, there have been very few comprehensive studies linking labor migration and the LLM (for example, Antonov 2016) or studies that attempt to assess the state and dynamics of the LLM by drawing on a large sample. These topics are usually bypassed or are addressed superficially in general overviews of socio-economic development in cities (Treyvish & Nefedova 2010; Nefedova et al. 2016).

Research of the LLM in Russian cities for the period between 1990 and 2000 noted divergent trends in their transformation under the influence of several key factors. The first one of these indicators is city size. The labor markets of large cities, especially capitals including regional capitals, was and remains on average better in terms of most characteristics (for example, employment rate, wages) than that in small cities.[1] There has been a more intensive decrease in the number of employees in small and medium-sized cities during the post-Soviet period and a noticeably higher unemployment rate. A second indicator is economic (industrial) specialization. In cities with "successful" industries, such as the oil and gas industry, the state of the LLM has been on average better than that in other cities, especially those specializing in forestry and the timber industry, machinery, and construction materials production. In the latter, the most adverse situation was observed during domestic (transitional) and world economic crises in the 1990s and from 2008 to 2009. Third, the position of a city in the settlement system, in addition to its indirect impact on the economy of urban enterprises as a marketing factor, has an impact on the sustainability of the LLM from the viewpoint of opportunities for external labor supply such as commuting and FIFO workers. Finally, specific institutional conditions determine the balance between small business and the shadow economy in the LLM, both in terms of quantitative indicators of employment and wages, as well as qualitative parameters of available jobs.

Having outlined the main trends, it is important to note that, in the works that addressed the issues related to the development of the LLM in Russian cities, population dynamics were not viewed as a separate factor. Perhaps this was not considered as particularly necessary since in the 1990s and 2000s, the vast majority of Russian cities were depopulating with varying degrees of intensity; for the period 1989–2010, over 68% of Russian cities lost population, between 2010 and 2017—more than 70% (Rosstat) (for a more detailed overview see

Chapter 10 by Efremova & Averkieva in this book). However, given the currently high level of socio-spatial polarization in Russia (for example, Nefedova 2009), the trends on the LLM of growing and shrinking cities may be very divergent. Thus, the current contribution aims to shed more light on this particular issue. The generalizations drawn from the results may also be, to some extent, applicable to cities of Central and Eastern Europe (CEE) that face similar trends of deepening socio-spatial inequalities (Lang et al. 2015).

Methodology and data

The current study regarding the processes of formation and transformation of the LLM in Russian cities and in shrinking ones covers the post-Soviet period with a focus on the 2010s. In order to understand the situation at hand, I have used data for all urban municipalities in Russia, highlighting the situation of the LLM of Russian shrinking cities. The core and most agreed-upon feature of urban shrinkage is depopulation (Bernt 2016); thus, the term "shrinking cities" hereafter refers to all Russian cities that lost population during the period 2010–2017 regardless of the pace of depopulating.

The analysis is based on several sources of information. First, the longest statistical series for the period 1992–2013 are drawn from the database *Economy of Russian cities* (Economy of 2021) with data on the number of employees at large and medium-sized enterprises and organizations,[2] their wages, as well as the number of overall unemployed persons. This enables tracking key trends in urban employment and unemployment. Second, information on the sectoral structure of employment, available only for the late 2000s—early 2010s, is drawn from the *Municipal Units Database* (hereinafter referred to as MUDB, Municipal Units Database 2021). Third, information is presented in reports of the Federal Tax Service's regional divisions web sites (Federal Tax Service 2011,data from Form 5—Personal income tax) which enables a detailed analysis of the LLM for 2017, based on such indicators as the number of persons employed in the full range of organizations within the municipality and average wages. The above are the only available open sources of information characterizing the situation of the LLM of a large sample of Russian urban municipalities.

Main stages of post-Soviet local labor markets' transformation in Russia (1990s–2000s)

The initial situation of the LLM of most cities at the end of the Soviet period was roughly the same but slightly differentiated by territory: almost full employment was achieved through employment at state enterprises and public organizations; unemployment was extremely low. The employment at private small and medium-sized enterprises was minimal. After the collapse of state socialism, there was an ongoing process of labor resources reallocation between the main sectors of the economy: (1) large and medium-sized enterprises; (2) small and medium-sized businesses; (3) informal employment[3]; (4) temporary employment sector.

Even though the post-Soviet period is characterized by a decline in manufacturing, this trend should not be regarded as a fully fledged deindustrialization. Despite the dramatic shift of workers from industry into the service sector, the Russian economy to date remains largely industrial, combining more and less successful sectors of industry (Crowley 2016).

The transitional crises (beginning of the 1990s) associated with the change of the economic system from command toward the market led to an almost universal reduction of employment at the large and medium-sized enterprises and organizations. Decline in the number of jobs due to liquidation of inefficient enterprises was evident in cities of all sizes, geographical locations, and economic (industrial) specializations. Virtually every city had one or a group of main enterprises that were not able to adapt to the new conditions. They were partly replaced, including absorption of the forcibly released employees, by small business together with non-institutionalized businesses in the informal sector of the economy. However, these processes did not compensate for job losses in all cities. Therefore, the unemployment rate increased sharply. It was directly determined by the size of the city: in small cities, the situation of the LLM was on average worse than in their larger counterparts. This was due to the size of the urban economy—in large cities, consumer demand created conditions for small business development, while in small cities, demand was not high enough.

The 1998 crisis and subsequent recovery of the economy in the 2000s significantly differentiated cities depending on their economic specialization and size. The largest cities of the country (especially Moscow, St. Petersburg, and other million-plus cities) and their agglomerations received a significant additional resource of development in the form of increased rental income, accumulating in them and generating additional demand, particularly in the service industries. This was accompanied by a corresponding increase in the employment rate. Enterprises in cities with a "successful" economic specialization showed a stable state or a moderate decrease (as part of ongoing business optimization processes) in employment, and in some industries, there may have also been moderate growth (for example. oil and gas, retail and services). Nevertheless, even during this relatively favorable period, employment in the industrial sector continued to decline in most cities, being greatest in small ones. The second most important trend in the transformation of employment was the intensification of "optimization" (that is, reduction) of employment in the public sector—education, healthcare, public administration. The decline in employment in these sectors was also more intense in small cities, while large cities, including regional centers, seemed to have had more opportunities to maintain the level of employment.

By the *end of the 2000s*, the overall reduction in employment at large and medium-sized enterprises in cities, excluding those with a population over one million, averaged 20%. By that time, about 1/5 of labor resources of the country moved toward the informal sector, and at least 20% were employed in small business compensating for the loss of employment at the large and medium-sized enterprises typical for the Soviet period. It is at the end of the 2000s, as a response to reduced employment in large enterprises and in the public sector, that employment became concentrated in the small and medium-sized business and the informal sector. This has become the main adaptation mechanism of the LLM.

During this period, there was also a large-scale redistribution of jobs between cities of different sizes (and over the territory of the country in general). It resulted in an increase in geographical disparities in the availability of jobs and the intensification of internal labor migration. Taking into account the growth of the LLM in Moscow and St. Petersburg, the total increase in the number of jobs in million-plus cities in the period between 1998 and 2007 was 7%, while in all other cities of the country there was a 12% decrease.

The *economic crisis of 2008–2009* did not have a cardinal influence on the direction of the LLM transformations. The crisis mostly affected single-industry towns and other industrial centers, where the enterprises were forced to cut down the number of employees. The effect was also noticeable for major cities and regional centers, where employment was also reduced because of some decline of the "overblown" service sector.

The general trend of changes in the sectoral structure of employment in the 1990s and 2000s in cities was a decrease in employment in the industrial sector, especially in manufacturing industries. In its place came diverse employment in services, which was often informal.

Main trends and structural changes in urban employment in the beginnings of 2010s

What happened to the LLM of Russian cities both growing and shrinking in the first half of the 2010s? It is rather difficult to answer this question due to the extremely small amount and poor quality of available statistical data. Termination of the *Multistat* database renewal, which provided data on the city level, made it difficult to assess trends of urban development including that in regard to the LLM after 2013. One of the very few sources left that allows analysis of the LLM is the MUDB; however, it narrows the range of studied urban municipalities (towns/cities) to those that have a status of urban *okrug*. Thus, the data for the majority of cities in this database are absent. As a result, key trends of the LLM evolution were analyzed by mainly using the example of the largest cities, while small ones that generally do not have a status of urban *okrug* were not considered. The sample narrows to 366 units[4] (about a third of all cities); however, the bulk of the Russian urban population live in them.[5]

Employment data for large and medium-sized enterprises (excluding small businesses) indicate the ongoing process of reduction of jobs in the period 2010–2016 in most cities (Figure 12.1). Similar trends are noted in the small business sector on the national scale. In 2010–2013 reduction processes slowed but accelerated in 2014–2016 due to the recession in the national economy. Among the shrinking cities (about 2/3 of those surveyed), the overall decrease in the number of jobs was slightly larger than among the cities that experienced

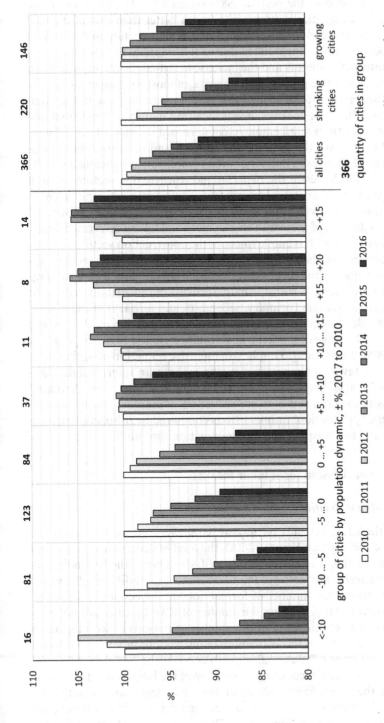

Figure 12.1 Dynamics of employment in large and medium-sized enterprises in urban okrugs by grouped according to population dynamic in 2010–2016 (2010 = 100%).

Source: Own calculations based on MUDB.

Note: Data for 366 urban okrugs of Russia.

population increase over the period—88% versus 93%. The situation of the LLM of fast-growing cities is, on average, much better than that of the rest, despite the negative trends in 2014–2016. The overall reduction in the number of official jobs in cities is most likely due to the overflow of labor resources toward the informal economy and the increase in self-employment, which can be regarded as a negative process characteristic of most Russian cities.

Overall, the tertiary sector of Russian cities saw an increase in the number of jobs in 2010–2016; however, it was localized mainly in large cities and resulted from growth in retail employment (a record increase of 35%); financial activity (+11% over the period); and in hotel and restaurant businesses (+4%). Structural changes in industrial urban employment indicated an unfavorable trend—job losses in the material and construction sectors, in both absolute terms and, to an even greater extent, relative rates of employment in trade. Low-quality growth in the tertiary sector of the economy proceeded along with a reduction of industrial employment and falling investment activity (the indicator of which is employment in the construction sector).

The compression of the LLM in shrinking cities, resulting from the stagnation of the economy in recent years, has been accompanied by structural changes in employment. A transformative process of redistribution of labor resources between sectors (primary, secondary and tertiary) in these cities in the last decade has occurred close to the American-Asian model, but not as intense as at the stage of market transformations in the 1990s (*Typology of regions…* 2008). This scenario implies an intensive reduction of agricultural employment in favor of the services sector without an intermediate stage of growing industrial employment (European model). Moreover, shrinking cities of Russia experienced a severe reduction of employment in the sectors of material production[6] (minus 10% as compared to 2010), construction (minus 7%), and in transport and communications (minus 8%).

Current state of Russian local labor markets

Availability of jobs in cities

The size of the LLM depends on several parameters. Among them are the population of the city (see Figure 12.2A), the level of economic activity, and geographical position (in relation to other LLMs). One of the key indicators that can characterize the status of the LLM is the availability of jobs, which refers to the ratio of the number of formal[7] jobs to the economically active population (EAP).[8] Since data on the number of EAP by cities are not published, an alternative estimation of the number of persons of working age is used. The average value of the availability of official jobs in the formal sector in Russia in 2017 was 0.79, and taking into account the additional estimate for shadow employment (according to a survey of the labor force data)—0.9.

In general, the geography of jobs availability has clear "north–south" and "urban–rural" gradients. The first one is due to both the already mentioned FIFO employment and the high cost of living in the northern territories,

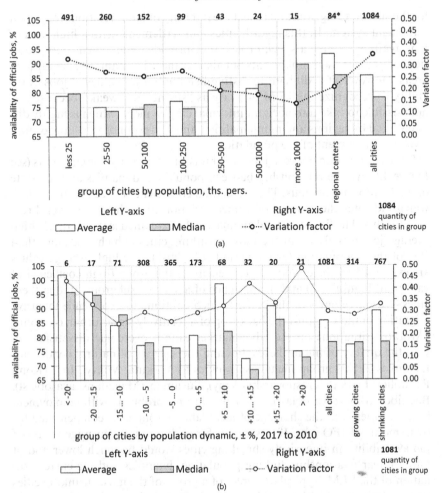

Figure 12.2 Availability of official jobs in cities of different sizes (A) and in cities with different population dynamics (2010–2017) in 2017 (B).

Source: Own calculations based on Federal Tax Service data and MUDB.

Note:

* Except Anadyr (data lack).

where opportunities for living without formal employment and a stable source of income are limited, thus encouraging an outflow of the unemployed. In most regions, not only large (usually regional capitals) but also medium-sized and small cities are cores of employment, concentrating the vast majority of jobs. While the role of medium-sized and small cities in the overall national labor market is not very significant, the same cannot be said about their role for the surrounding rural area.

The high concentration of jobs and, consequently, the high level of availability of jobs can be an indicator of the "organizing power" of settlements for the surrounding area and reflect the intensity of labor relations for a local settlement system. In this regard, cities and settlements with a high level of jobs availability play a special role—they provide employment opportunities not only for their inhabitants but also for the inhabitants of surrounding areas. However, it is difficult to distinguish the border values of jobs availability to recognize the LLM of a city as "strong" or "weak" because it is not possible to take into account informal employment, which is highly differentiated by regions and municipalities, and correspondingly—the overall level of employment of the population.

If the relationship between job availability and the size of the city exists (see Figure 12.2A), the relationship between population dynamics and the state of the LLM is not obvious. The correlation coefficient between them is –0.05, which indicates the absence of a linear relationship. There are several reasons for this. The first statistical anomaly to be explained is the relatively high average job availability in the most shrinking cities (which lost more than 10% of their population for the period 2010–2017). The high average values are explained by the extreme heterogeneity of the group (high coefficient of variation, Figure 12.2) and the presence of a specific subgroup among these cities—resource cities of the North, which continue to experience intensive migration outflow against the backdrop of a relatively stable urban economy. Favorable situations on the labor market cannot overcome dissatisfaction with the quality of life in these cities and the severity of natural conditions. Examples of this group include such cities as Vorkuta, Vuktyl, Inta (Komi Republic), Igarka (Krasnoyarskyi kray), Susuman (Magadanskaya oblast), Bodaibo (Irkutskaya oblast), etc. There are no problems with employment in these cities, and the shortage of local labor resources is compensated for by numerous FIFO jobs. If it were not for such cities, the average values of job availability in intensively shrinking cities would be much lower and fit into a linear relationship, linking population dynamics and the current situation of the LLM. Typical examples of a group of the most shrinking cities with undeveloped LLMs are old-industrial, single-industry shrinking cities in structural crises (Table 12.1). The crisis in main industries has led to a sharp reduction in the number of jobs while alternative jobs have not appeared. In these cities, the unfavorable situation of the LLM is one of the factors fostering out-migration that determines further shrinkage.

The second statistical anomaly, distorting the linear dependence, is characteristic not of shrinking but growing cities. Their LLMs are very uneven (high value of the coefficient of variation, Figure 12.2). Among these cities, there are both cities with a very high provision of jobs due to the presence of large enterprises and government institutions and cities with an extremely low number of jobs (Table 12.1). Population growth in the cities is not associated with internal economic growth and, thus, by the creation of new jobs, but rather by the orientation of the population toward the external, neighboring LLMs. This primarily applies to "cities" that are part of large urban agglomerations, thus being suburbs of cities with a population over one million. A

Table 12.1 Characteristics of the fastest growing and shrinking cities and availability of jobs

Town	Main branch	Population, ths. pers. 2020	Population dinamics, 2020 to 2010, +%	Availability of official jobs, % 2019
Shrinking cities with developped local labor market				
Igarka	Transport	4,3	−30,1	207,7
Susuman	Gold mining	6,7	−25,6	169,6
Bodaibo	Gold mining	12,0	−21,9	151,7
Shrinking cities with undeveloped local labor market				
Efremov	Chemical industry	54,9	−18,7	71,6
Scopin	Engineering industry	26,8	−15,2	68,9
Novotroitsk	Metallurgy	90,2	−14,8	75,4
Growing cities with developped local labor market				
Krasnogorsk	Services and public management	175,6	50,2	162,8
Magas	Services and public management	12,2	386,4	112,3
Domodedovo	Diversed	184,2	36,0	126,0
Growing cities with undeveloped local labor market				
Mikhailovsk (Stavropol krai)	Diversed	95,7	34,9	32,8
Goryachy Klyuch	Recreation	69,7	21,7	44,8
Ivanteevka	Diversed	81,3	38,6	53,9

separate group is made up of resort cities in the south of the European part of Russia (Goryachy Klyuch, Gelendzhik, Anapa, etc.), which actively attract migrants and significant informal employment in the recreational sphere.

Salary in cities

The second key parameter characterizing the state of the LLM is the level of wages. Together with the availability of jobs, these two parameters determine the possibility of employment with adequate remuneration for work. In a market economy, these two parameters should determine the mobility of labor resources (Andrienko & Guriev 2004; Radu & Straubhaar 2012)— both migrations with the change of permanent residence and the intensity of recurrent labor migrations. Theoretically, the higher the relative gap in the level of remuneration in the city and the surrounding area, the more intense the migration should be.

This study examined the ratio of the average wage in the city in relation to the regional average (regional average = 100%) depending on population size (Figure 12.3A) and its dynamics (Figure 12.3B). In this case, the average

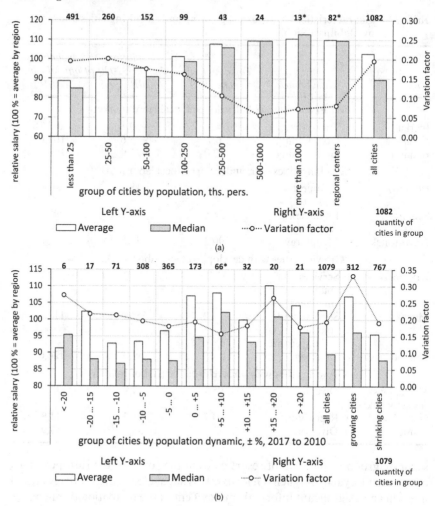

Figure 12.3 Relative salary (100% = average salary of a city in a region) in cities of different sizes and regional centers (A) and in cities with different population dynamics (2010–2017) in 2017 (B).

Source: Own calculations based on Federal Tax Service data and MUDB.

Note:
* Except Moscow and Saint-Petersburg, which are regions themselves.

wage in cities increases linearly with city size (Figure 12.1A) peaking in large cities, including regional centers. Within groups of various sizes, the situation is quite differentiated, especially among small and medium-sized cities (a relatively high variation factor for cities with a population of up to 250,000 people is observed). The dependence between the dynamics of the population and the level of wages (Figure 12.3B), as in the case of job availability, is distorted due to the small group of the most shrinking cities. The average high level of wages there is due to the few high values in the resource cities of the North. However, excluding outliers, in general, the level of wages in shrinking cities is significantly lower than that in cities experiencing population growth (median value 87% versus 96%, respectively).

Conclusions: understanding the specific of the local labor markets in shrinking cities

The analysis of the dynamics and accessibility of jobs in Russian cities, as well as the level of wages in growing and shrinking cities, allows for the following conclusions. The average state of the LLM (in terms of access to jobs) in most shrinking cities is worse than that in growing ones. This leads to the formation of a downward spiral (Stryjakiewicz et al. 2012), where problems of the LLM may lead to the formation of a job-related out-migration. This, in turn, leads to a reduction in the internal size of the economy, making cities less attractive for business due to reduced demand, ultimately leading to further compression of the economy and employment. However, for a rather large group of resource cities in the North and suburban territories, there is no direct correlation between the population dynamics and the state of the LLM.

Statistics on official employment cannot fully reflect the situation with the availability of jobs in cities, since in the vast majority of them, there seems to be a significant sector of informal employment. Compression of employment in the formal sector, which continues in shrinking cities, leads to the overflow of labor resources into the informal sector of employment and shuttles that portion of the labor force out of the labor market. It is estimated that about a third of the workforce in Russia is self-employed or employed informally (Antonov 2019a). The bulk of informal employment is concentrated in the wholesale and retail trade (31.3% of all informal workers); in agriculture (16.4%); construction (10.7%); manufacturing (10.7%); and transport (10.4%).

Adaptation to the crisis of official employment in shrinking cities of Russia occurs through the intensification of recurrent labor migration. In cities outside large urban agglomerations, this form of adaptation (FIFO) temporarily minimizes and delays the migration outflow of the population. In the 2010s, there was an increase in labor mobility of the population, provoked by problems of the LLM, which was especially evident in the intensification of labor migrations directed to Moscow, St. Petersburg, and the Siberian regions (Averkieva et al. 2016) where oil and gas are produced (Tyumen region with Autonomous Okrugs).[9]

The quality of jobs and the structure of the labor market in most Russian cities, with the exception of the largest and regional centers, deteriorated in the 2010s. Amid declining employment in industries requiring a high level of qualification (processing industry, skilled services), only branches of wholesale and retail trade grew rapidly. In small cities that are intensively shrinking, the structure of employment in the official sector is turning into two components. The first group consists of employment in public sector institutions—education, healthcare, public administration, etc. The stability of employment in this group is mainly due to state funding. The second group of official employment comprises trade and the provision of low-skilled services. Its growth was due to an average increase in consumer spending by the population, which stopped after 2013. In the face of a decline in real incomes of the population, it is obvious that this sector of employment in Russian cities cannot be sustained and the decline will inevitably continue.

The low level of salary in small cities, most of which are shrinking, is one of the key factors for the migration outflow of the working age population, especially young people (Vakulenko & Mkrtchyan 2019). A large gap in wages with regional centers, together with other reasons, leads to an almost total non-return of youth to their hometowns after education in professional institutions (Mkrtchyan & Vakulenko 2019). Middle-aged staff who have gained work experience and are highly qualified also seek to relocate to large growing cities, where there are more employment opportunities (Karachurina & Mkrtchyan 2018).

As a result, we can say that between large and small Russian cities—between growing and shrinking cities—a steady socio-economic gradient continues to be maintained. It is reflected in labor markets too. Throughout post-Soviet history, shrinking cities (with some exceptions) are characterized by less attractive employment conditions, a variety of jobs, and lower wages. This situation is an objective reflection of the ongoing spatial polarization of Russia (Nefedova 2009; Zubarevich 2019).

Systemic state (regional or local) policy regarding the support of the LLM in structurally weak cities of Russia is absent. In acute periods of crisis, such as in 2008–2009, the state took short-term measures to stabilize the labor markets, primarily in single-industry towns, by creating temporary jobs and stimulating the diversification of the economy (Ovcharova & Zubarevich 2010; Efimova 2011) in order to avoid mass lay-offs and associated political unrest (Crowley 2016). This may have alleviated short-term problems but did not change the general trajectory of divergence between cities. The vast majority of Russian cities do not have the resources to cope with their own economic problems, including stabilizing the LLM due to the profound re-centralization of governance witnessed since the early 2000s (Gelman 2018). There is now less local autonomy through the re-establishment of political control over local governance and a reform of the taxation system in favor of centralizing financial resources (Gelman & Ryzhenkov 2011; Kinossian 2017).

Acknowledgment

The study was prepared based on research within the framework of the state ordered research theme of the Institute of Geography RAS, no. 0148-2019-0008 (AAAA-A19-119022190170-1) Problems and prospects of Russia's territorial development in terms of its unevenness and global instability.

Notes

1 According to the established Russian classification by G. M. Lappo, small cities are those with populations of 50 thousand people and less; medium-sized cities—from 50 to 100 thousand people. See full classification in Table 10.1 of Chapter 10.
2 Russian statistical accounts highlight a special category of large and medium-sized enterprises and organizations for which the main annual statistics are available. They, together with small enterprises (delimited according to two indicators—revenue up to 800 million rubles (2020) and employment up to 100 people) and individual enterprises form the so-called "full range of enterprises and organizations".
3 Informal employment includes any labor relations that are not formalized in accordance with the current Russian labor legislation.
4 The time period available for analysis is limited to 2016 due to a change in the national classifier of types of economic activity.
5 57.6% of Russian town's population (2017).
6 Agriculture, forestry, mining, and manufacturing industry.
7 Formal jobs are formal jobs for which the employer pays taxes.
8 The indicator of job availability can be higher than 1 if the number of jobs exceeds the working-age population. Thus, the higher the value of the indicator, the more favorable is the situation on the LLM.
9 Although there are no official statistics on the scale of FIFO employment for cities, available estimates based on regional data indicate an increase in this phenomenon in the 2010s. The share of FIFO employment in some cities exceeds 50% and more (Table 12.1).

References

Andrienko Y. & Guriev S. 2004. Determinants of Interregional Mobility in Russia. *Economics of Transition*, 12(1), 1–27.

Antonov E. V. 2016. Labor Mobility of the Population of Russia (According to the 2010 All-Russian Census Data). *Moscow University Bulletin: Series 5: Geography*, 2, 54–63 (in Russian).

Antonov E. V. 2019a. The Dynamics of Employment and Regional Labour Markets Situation of Russia in 2010–2017. Vestnik of Saint Petersburg University. *Earth Sciences*, 64(4), (in Russian). https://doi.org/10.21638/spbu07.2019.404

Antonov E. V. 2019b. Development and Current State of Urban Labour Markets in Russia. *Population and Economics*, 3(1), 75–90. https://doi.org/10.3897/popecon.3.e34768

Averkieva K. V., Antonov E. V., Denisov E. A. & Faddeev A. M. 2015. Territorial Structure of the Urban System in the Northern Sverdlovsk Oblast. *Regional Research of Russia*, 5(4), 349–361. https://doi.org/10.1134/S2079970515040036

Averkieva K. V., Kirillov P. L., Makhrova A. G., Medvedev A. A., Neretin A. S., Nefedova T. G. & Treivish A. I. 2016. Between the home and… home. *The Return Spatial Mobility of the Population in Russia*. Moscow, Novyi khronograf Publ. (in Russian). http://ekonom.igras.ru/wp-content/uploads/2022/01/BHAH2016.pdf

Bernt, M. 2016. The Limits of Shrinkage: Conceptual Pitfalls and Alternatives in the Discussion of Urban Population Loss. *International Journal of Urban and Regional Research*, 40(2), 441–450.

Crowley, S. 2016. Monotowns and the Political Economy of Industrial Restructuring in Russia. *Post-Soviet Affairs*, 32(5), 397–422.

Data forms of statistical tax reporting by Form 5—personal income tax in the context of municipalities (in Russian). (2021). https://www.nalog.ru

Economy of the cities of Russia. "Multistat" database (in Russian). (2021). http://www.multistat.ru/?menu_id=9310014

Efimova, E. A. 2011. Labor Market in Single-Industry Towns: Current Situation and Prospects. *Prostranstvennaya Ekonomika*, 1, 119–135 (in Russian). https://doi.org/10.14530/se.2011.1.119-135

Federal Tax Service of the Russian Federation (FTS of the Russian Federation). *Statistics and analytics*. (2021). https://www.nalog.gov.ru/rn77/related_activities/statistics_and_analytics/forms/

Gelman, V. 2018. Bringing Actors back in: Political Choices and Sources of Post-Soviet Regime Dynamics. *Post-Soviet Affairs*, 34(5), 282–296.

Gelman, V. & Ryzhenkov, S. 2011. Local Regimes, Sub-national Governance and the "Power Vertical" in Contemporary Russia. *Europe-Asia Studies*, 63(3), 449–465.

Großmann, K., Bontje, M., Haase, A. & Mykhnenko, V. 2013. Shrinking Cities: Notes for the Further Research Agenda. *Cities*, 35, 221–225.

Haase, A., Rink, D., Grossmann, K., Bernt, M. & Mykhnenko, V. 2014. Conceptualizing Urban Shrinkage. *Environment and Planning A*, 46, 1519–1534. https://doi.org/10.1068/a46269

Hartt, M. D. 2016. How Cities Shrink: Complex Pathways to Population Decline. *Cities*, 75, 38–49. https://doi.org/10.1016/j.cities.2016.12.005

Hoekveld J. J. 2012. Time-space Relations and the Differences Between Shrinking Regions. *Building and Environment*, 38(2), 179–195. https://doi.org/10.21638/11701/spbu07.2018.104

Karachurina, L. & Mkrtchyan, N. 2018. Age-specific Net Migration Patterns in the Municipal Formations of Russia. *GeoJournal*, 83, 119–136. https://doi.org/10.1007/s10708-016-9757-4

Kinossian, N. 2017. State-led Metropolisation in Russia. *Urban Research & Practice*, 10(4), 466–476.

Kuznetsova, G. Y. 2003. Geographical Study of Single-industry Settlements in Russia: Thesis.... Cand. of Geogr. Sciences. Moscow: Moscow Pedagogical University (in Russian).

Lang, T., Henn, S., Sgibnev, W. & Ehrich, K. (eds.) 2015. *Understanding Geographies of Polarization and Peripheralization: Perspectives from Central and Eastern Europe and Beyond*. London & NY: Palgrave Macmillan. https://doi.org/10.1057/9781137415080

Lyubovny, V. Y., Kuznetsova, G. Y. & Lycheva, T. M. et al. 2001. Methodological Recommendations on the Formation of Territorial Special Programs to Promote Employment Development in Single-industry Cities. *Library of Local Governance*, 41, Moscow: MNF (in Russian).

Makhrova, A. G. & Bochkarev A. N. 2018. Analyzing Local Labor Markets Through Commuting (A Study of Moscow Municipalities). *Vestnik of Saint Petersburg University. Earth Sciences*, 63(1), 56–68.

Makhrova, A. G. & Kirillov P. L. 2015. Seasonal Fluctuations in Population Distribution Within Moscow Metropolitan Area Under Travelling to Second Homes and Labor Commuting: Approaches and Estimations. *Regional Studies*, 47(1), 117–125 (in Russian).

Mikryukov, N. Y. 2016. Factors, problems and models of development of monocities in Russia: Thesis....Cand. of Geogr. Sciences. Moscow: Lomonosov Moscow State University, Faculty of Geography (in Russian).

Mkrtchan, N. V. 2017. The Youth Migration from Small Towns in Russia. *Monitoring of Public Opinion: Economic and Social Changes*, 1, 225–242 (in Russian).

Mkrtchyan, N. V. & Florinskaya Y. F. 2016. Socio-economic Effects of Labor Migration from Small Towns of Russia. *Economic Issues*, 4, 103–123 (in Russian).

Mkrtchyan, N. & Vakulenko, E. 2019. Interregional Migration in Russia at Different Stages of the Life Cycle. *GeoJournal*, 84, 1549–1565. https://doi.org/10.1007/s10708-018-9937-5

Municipal Units Database of Rosstat (MU DB) (in Russian). (2021). http://www.gks.ru/dbscripts/munst/munst.htm

Nefedova, T. G. 2009. Polarization of Russian Space: Areas of Growth and "Black Holes": Economical. *Jekonomicheskaja nauka sovremennoj Rossii*, 1(44), 62–77 (in Russian). https://elibrary.ru/item.asp?id=12109800

Nefedova, T. G. 2015. Employment of Population and a Phenomenon of Seasonal Work in the Stavropol Krai. *Moscow University Bulletin. Series 5: Geography*, 2, 93–100 (in Russian).

Nefedova, T. G., Averkieva, K. V. & Makhrova, A. G. (eds.) 2016. *Between the Home and… Home: Return Spatial Mobility of the Population of Russia*. Moscow: A New Chronograph (in Russian). https://doi.org/10.15356/BHAH2016

Nekrasova, E. V. 2012. Optimization of Internal Migration as a Mechanism for Solving the Problems of Monotowns in Sverdlovsk Region. *Economy of the Region*, 2, 315–320 (in Russian).

Ovcharova, L. N. & Zubarevich, N. V. 2010. Anti-crisis Activities of Social Support at the Regional Level. *Uroven' zhizni naselenija regionov Rossii*, 9(151), 52–59. https://elibrary.ru/item.asp?id=15855039

Plyusnin, Y. M., Zausaeva, Y. D., Zhydkevich, N. N. & Pozanenko, A. A. 2013. *Seasonal Workers*. Moscow: A New Chronograph (in Russian).

Radu, D. & Straubhaar, T. 2012. Beyond "Push-Pull": The Economic Approach to Modelling Migration. In Martiniello, M. & Rath, J. (eds). *An Introduction to International Migration Studies: European Perspectives*. Amsterdam University Press. https://doi.org/10.2307/j.ctt6wp6qz.5; www.jstor.org/stable/j.ctt6wp6qz.5

Stryjakiewicz, T., Ciesiółka, P. & Jaroszewska, E. 2012. Urban Shrinkage and the Post-socialist Transformation: The Case of Poland. *Built Environment*, 38(2), 197–213.

Treyvish, A. I. & Nefedova, T. G. 2010. Cities and Countryside: State and Ratio in the Space of Russia. *Regional Studies*, 2, 42–57 (in Russian).

Typology of regions by structure of employment in the economy (1960, 1975, 2002). 2008. In *National Atlas of Russia*, 3: 462–463 (in Russian). https://nationalatlas.ru/tom3/462-463.html

Vakulenko, E. & Mkrtchyan, N. 2019. Factors of Interregional Migration in Russia Disaggregated by Age. *Applied Spatial Analysis and Policy*. https://doi.org/10.1007/s12061-019-09320-8

Veliky, P. P. 2010. Neootkhodnichestvo, or Excessive People of the Contemporary Village. *Sociological Research*, 9, 44–49 (in Russian).

Vishnevsky, A. G. & Zakharov, S. V. 2010. What Russian Demographic Statistics does and does not Know. *Voprosi Statistiki*, 2, 7–17 (in Russian).

Vlasova, N. Y., Kuznetsova, G. Y., Lycheva, T. M. & Lyubovny, V. Y. 1999. Issues of Employment in Monotowns of Russia. In Proceedings of the International Scientific-Practical Conference *Current State and Prospects of Employment and the Labor Market of Russia* (in Russian). Moscow.

Wolff, M. & Wiechmann, T. 2018. Urban Growth and Decline: Europe's Shrinking Cities in a Comparative Perspective 1990–2010. *European Urban and Regional Studies*, 25(2), 122–139. https://doi.org/10.1177/0969776417694680

Zubarevich, N. V. 2019. Inequality of Regions and Large Cities of Russia: What was Changed in the 2010s? *Obshhestvennye nauki i sovremennost'*, 4, 57–70. https://doi.org/10.31857/S086904990005814-7

13 Giving birth in dying towns

Healthcare shrinkage in a depopulating Russian region

Anastasia Novkunskaya

Introduction

Long-term depopulation is considered to be a general feature of urban shrinkage (Bernt, 2016). Along with demographic decline caused by low fertility, aging populations, and intensive out-migration, loss of urban characteristics and functions is viewed as typical for shrinking cities. This is manifested in economic decline and reduction in public services such as healthcare, education, and social security. In particular, research shows that in shrinking cities public services decline (Bierbaum, 2020) owing to decreasing demand and increasing costs against a backdrop of lower tax revenues (Winthrop & Herr, 2009). The objective of the current chapter is to explore the changes in healthcare functioning in the context of shrinking small towns in a depopulating Russian region (Tverskaya Oblast'), where the changes to services provision is influenced both by shrinking-related causes and general restructuring of the healthcare system taking place worldwide (Kuhlmann & Annandale, 2012).

In post-Soviet Russia, healthcare organization and policy reflect contradictions in social and public services restructuring over recent decades. On the one hand, healthcare is still considered to be an important part of the postsocialist political agenda: the Russian government explicitly claims to be responsible for citizens' well-being and invests considerable resources in modernizing public services (President of Russia, 2012; 2014). On the other hand, the concrete policies implemented do not appear to be attuned to multiple localities—the spatial and social specificities of different regions and cities across the country. The above contradiction becomes even more acute in the case of maternity care, given that demographic issues are recognized as one of the priorities of Russian state policy (Rivkin-Fish, 2010) and the rates of maternal and infant mortality are one of the key indicators of the regional authorities' performance.

Against the above background, the aim of this chapter is, firstly, to analyze inconsistencies in public policies—discrepancies between the state demographic (pronatalist) and healthcare policies in shrinking towns. Secondly, the study aims to highlight the individual experiences of people

DOI: 10.4324/9780367815011-16

who are planning for and giving birth and healthcare practitioners who provide maternity services. Addressing the above aims helps to better understand the everyday life in shrinking towns drawing on empirical evidence from Tverskaya Oblast', which emphasizes both regional and national features of healthcare functioning.

The first section of the chapter provides a review of policies aimed at managing public services in shrinking cities and towns. The second section discusses the regional background to the study and focuses on socioeconomic aspects of the area, characterized by Russian experts as being economically semi-depressed (Zubarevich, 2016:92). Section three addresses key stages of reforms to maternity care in Russia in general and outlines some salient features of the organization and provision of healthcare services. The field of maternity and infant care in Russia is quite volatile in terms of policies and relates directly to demographic issues, which are now recognized as a priority in state policy (Rivkin-Fish, 2010). Section four provides research methodology, which employs a qualitative case study involving semi-structured, in-depth interviews with residents and with professionals working in medical facilities in the region. Section five presents the results of the study focusing on five dimensions of healthcare services shrinkage in shrinking towns, elaborated through analysis of the empirical data. The conclusions seek to provide an overview of the issues at hand.

Analytical framework of "right-sizing" and shrinking healthcare services

Changes to public services provision in shrinking cities are partly a result of policy decisions, for which there are two feasible models of implementation. The first is underpinned by the rationality of "right-sizing", an approach generally adopted in planning debates relating to the built environment and infrastructure of shrinking cities (Hummel, 2014). It focuses on "managing low population densities and increasing per-capita costs in the areas of service and infrastructure provision" (Coppola, 2019) by reducing the physical footprint, infrastructure, and services in shrinking cities to match the new population number (LaFrombois, Park & Yurcaba, 2019). The second model is general withdrawal and changes to the provision of public services under budget austerity, neoliberal rationalities, and in the case of healthcare, technical specifics of the system (Kuhlmann & Annandale, 2012; Evans, 2020). Thus, the second type of policies does not necessarily correlate healthcare rearrangement with the actual demographic situation.

Healthcare optimization in terms of reduction of beds and personnel or centralization of services is a common process for many countries (McCourt et al., 2016), and in general, such policies make assumptions that the availability of remaining facilities, quality of services, and affordable transportation will compensate for the withdrawal (Baillot & Evain, 2013). Healthcare is considered part of the state's demographic agenda, and measures taken to rearrange it are often related to the size of the population, its dynamics, and structure.

A growing body of literature has examined various urban development strategies to deal with depopulating areas and shrinking cities worldwide (Hummel, 2014; Camarda et al., 2015; Beal et al., 2017). In particular, these studies describe various planning principles employed to realize "right-sizing" strategies (Beal et al., 2017:193), which refer to the policies that aim "to maintain a high quality of life in shrinking cities so that current residents will not leave" (Hummel, 2015:2), by matching available resources with residents' demands (ibid.:3). Healthcare in this framework is analyzed as both a resource that must be matched with the changing population and as a probable reason for the outflow of residents if its quality does not meet their expectations and needs (Camarda et al., 2015:129). In practice, healthcare restructuring sometimes causes inequality of access (Charreire et al., 2011:505), so health planning must balance safety and quality on the one hand and spatial accessibility and proximity on the other (ibid.:504).

Despite the severe demographic crisis in the 1990s, the demographic issue had not become the concern of national policies before the mid-2000s (Kainu et al., 2017) when the *Demographic Policy Concept of the Russian Federation for the period until 2025* was approved. Its core aims are tied to improving indicators of natural population change, including those related to the scope of healthcare provision, birth and death rates, and longevity. However, at the same time, depopulation *per se* has seldom been properly addressed in healthcare policies as these have been poorly linked to demographic policies. It is fair to say that demographic policies are not explicitly correlated with healthcare restructuring.

As previous research has demonstrated, restructuring policies relating to urban shrinkage are more successful if they are both attuned to the demographic situation, local specificities, and spatial implications. It is important to note that such policies should not only consider the issue of depopulation but also take residents' perspectives into account (Camarda et al., 2015; Hummel, 2015). Otherwise, changes to healthcare that contrast with the implications of demographic policy result in multiple contradictions and challenges for public services providers and recipients. However, norms that define the level of public services and the distance to their location discussed in the urban planning regulations only apply to new construction (Code of rules, 2016). There is no particular policy or right-sizing strategy to match available public services (including healthcare) with the demands of decreasing and aging populations (Hummel, 2015:3).

Regional background

Tverskaya Oblast', in the Central Federal District of Russia, can be considered a "left behind" region since it suffers from both economic decline and a decreasing population. The latter is determined not only by an imbalance between fertility and mortality rates but also by the region's geographical location between Moscow and St Petersburg (Tver' is located 180 km from Moscow and 520 km from Saint-Petersburg), the two poles of in-migration (Rozanova, 2016). Depopulation of Tverskaya Oblast' over the recent decades has been significantly greater than across the country in general: population losses for the period 1989–2019 amounted to 24.47% (1.68 million down to 1.26 million), while in the Russian Federation, in general, the population decreased by 0.4% (from 147.4 million to 146.7 million) during the same period (Rosstat, 2019). However, depopulation has been highly uneven. Tver' itself, as the regional center, has suffered less from depopulation than Tverskaya Oblast' in general, with a reduction from 0.45 million people in 1989 to 0.42 million in 2019 (ibid.). At the same time, small and medium-sized cities in the region have shrunk significantly, some at an average annual pace of 1% or higher since 1989 (Batunova & Gunko, 2018). Overall, in 2019 the region was among those with the highest natural population decline across the country, at 8.5% (Khasanova et al., 2019: 63).

The tendency toward depopulation is aggravated by Tverskaya Oblast's unstable economic position, as it is highly dependent on federal subsidies. Therefore, many areas of public services and social infrastructure exert an appreciable budgetary burden (Fomkina, 2015:59), sometimes resulting in funding shortages. Combined with the loss of population and budget austerity, the lack of material and social resources may aggravate the shrinkage of healthcare services, resulting in total closure of medical facilities, their decreased autonomy, or reductions in the breadth of services provided (ibid.:62).

The top-down nature of healthcare regulation, mimicking Russia's wider governance (Gel'man, 2015) with limited autonomy for health institutions and regional authorities, offers limited opportunities to attune the process of healthcare restructuring to local specificities. Consequently, there is no mechanism in the process of healthcare restructuring to account for patients' actual needs; health services appear to be unequally accessible, differing in both quantity and quality from place to place (Panova, 2019). Figure 13.1 illustrates these quantitative changes in Tverskay Oblast' with the example of hospital beds. The number of beds is shown by the solid line and calculated per 10,000 residents (dotted line), although there is no national standard for the number of beds per person.

There is appreciable interregional diversity across the country, with wide demographic, cultural, and economic variation. Although Tverskaya Oblast' is not representative of the whole country in terms of demographic characteristics, its health policy applies nationally, thus some of the outcomes of the current study may be generalized, especially given that the region's allocated number of obstetric beds (9.83 per 10,000 women in 2017) correlates with the national average (9.73 per 10,000 women in 2017).

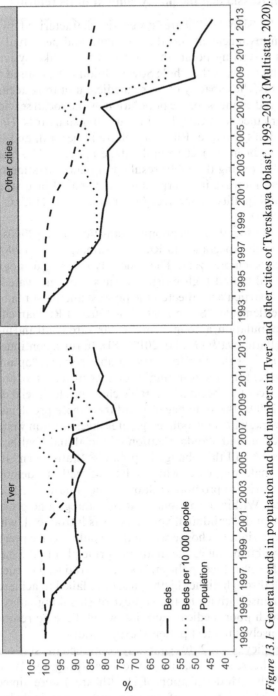

Figure 13.1 General trends in population and bed numbers in Tver' and other cities of Tverskaya Oblast', 1993–2013 (Multistat, 2020).

Healthcare in Russia and the policy context of its reform

The current Russian healthcare system is characterized by centralized and top-down arrangements on the one hand and perpetual volatility on the other. As a starting point, the so-called Semashko system of healthcare implemented across the whole Soviet Union presupposed a plan-based economy; that is, the capacity of every medical unit took accurate account of the size of, and changes to, the population, while social services were provided in accordance with calculated norms and standards. However, since the Soviet Union's collapse, Russian welfare in general, and healthcare in particular, have been subject to multi-stage reforms. The healthcare policy implemented during the 1990s resulted in regionalization of healthcare services and an increase in interregional and spatial inequalities in terms of access to and quality of services (Kochkina et al., 2015; Cook, 2017:13; Kainu et al., 2017).

In general, the social and economic transition of the 1990s contributed to a decline in life expectancy in Russia (Walberg et al., 1998), which particularly affected remote areas. The generally negative demographic trend toward low fertility and high mortality rates was aggravated by internal migration that particularly affected the poorest and most remote areas and the smallest settlements: "Since 1989, about 70% of Russian cities have lost some of their population, among them over 90% are small and medium-sized (SMS) ones" (Batunova & Gunko, 2018:1581). In turn, continued depopulation was used as a justification for further public services "optimization". By the turn of the century, the resulting inequality in access to, and quality of, healthcare services had become a notable challenge for social development.

In the mid-2000s, the state began to address these problems with a new policy framework, which introduced greater investment in welfare and centralized regulation and standardization of social and healthcare services. Political scientists call this change in policy a "statist turn" (Cook, 2014). In particular, maternity care, which relates directly to demography (and the national priority to promote childbirth), became a special object of the state's attention. Within the national "Health" foreground project launched in 2006, a system of "childbirth vouchers" was introduced, which guaranteed patients the right to choose where they could receive pregnancy monitoring and give birth, under the financing principle of "the money follows the patient". However, these policy measures, introduced to encourage population growth through higher birth rates, have failed to achieve their goals owing to their insensitivity to the context of shrinking cities (with aging populations) with poor medical facilities, which have no resources to staff and equip themselves to meet patients' expectations.

The health policy of the 2010s generally continued the statist approach to regulating healthcare services, and the new state programs launched at this time (particularly "Modernization of Healthcare") were aimed at healthcare improvement. The new set of state policy measures included additional investments in the material provision and renovation of medical facilities, construction of new, technologically developed hospitals and centers, and implementation of a routing scheme (framed as regionalization of healthcare). In particular, the organization and provision of maternity care were considerably shaped by Decree №572 issued by the Ministry of Health (2012), setting the order for "routing" (*marshrutizatsia*) of pregnant women. Decree № 572n laid down rules on pregnant women's hospitalization, depending on their risk of complications and pathologies associated with pregnancy or childbirth, and defined in the process of "pregnancy monitoring". This required risky cases to be routed to a facility equipped to assist with specific pathologies, illnesses, or complications.

As a result, since 2012, a three-level system of medical facilities has been adopted for maternity care to provide different services, different equipment and receive different financing in accordance with their assigned status (Ministry of Health, 2012):

First-level facilities comprise basic, low-capacity obstetric wards (fewer than 30 beds, accommodating fewer than 500 births per year) and are equipped to assist with low-risk childbirths. These are usually maternity wards (employing 2–4 obstetrician-gynecologists and 4–6 midwives) in central district hospitals. In Tverskaya Oblast', most maternity facilities are assigned first-level status, and hence must refuse to hospitalize many pregnant women and route them to second- or third-level maternity units. Therefore, they inevitably lose a number of medical cases and consequently receive less revenue.

Second-level facilities provide maternity care to patients at moderate risk of complications during pregnancy, labor, and the postpartum period. These are usually maternity wards in central inter-district hospitals in medium-sized and large cities or independent maternity homes accommodating more than 500 births per year and equipped with intensive care and resuscitation units. There are only four independent maternity hospitals and three other second-level maternity wards in the Tver' region, including two located in the regional center itself.

Third-level maternity hospitals and perinatal centers are medical organizations that ensure life-saving interventions for mothers and newborns. Women with high-risk pregnancies are admitted to such facilities, which are equipped with advanced instruments and apparatus and highly skilled personnel. These are usually independent maternity homes, research institutes, or perinatal centers offering high-tech services, which provide the only such care in each regional center. There is only one perinatal center in the region, constructed in 2010, where the number of assisted childbirths is increasing annually, indicating the growing centralization of maternity care.

Such optimization of maternity services and routing systems are found in other countries. A risk-oriented model is quite reasonable for both financing and prevention of medical complications. However, such reorganization should be designed to account for spatial specificities and will be effective and positively evaluated by patients only if adequate support services are provided. For example, French researchers demonstrate how a similar model of services optimization can be successful if it considers the time and quality of transportation, as well as introduce patient-oriented care (Baillot & Evain, 2013:7). In the case of remote Russian areas, these policy measures have not specifically addressed the processes of depopulation and urban shrinkage of small and medium-sized settlements. Consequently, the amount of material resources distributed across the regions has not been perfectly attuned to the actual needs of their residents. The leveling of maternity services has placed greater burdens on expectant mothers seeking to access preferred medical facilities, giving rise to challenges such as increased costs, traveling time, organizational inconveniences, and associated risks, which will be addressed in the results section.

In conclusion, the routing model implemented in Tverskaya Oblast' has unintentionally led not only to the centralization of maternity care but also to impoverishment of the healthcare infrastructure and reductions in both medical beds and personnel. As a result, optimization has affected the poorest (and usually remotest) areas, which have been unable to meet the demand for self-repayment owing to the lack of population, reduced rates of remuneration from the mandatory health insurance system, and extremely high transportation costs. Recent policies have also resulted in the spatial polarization of health services between regional centers and their peripheries since only residents of the former have both better accessibility and more choice options to acquire public services. Hence, health policies do not explicitly correlate with the decreasing population and local specificities but considerably affect the daily routines of patients and health practitioners, offering scope for further analysis.

Data and methodology

The empirical evidence presented in this chapter is derived from a case study of Tverskaya Oblast' conducted during 2018–2019. The study focuses on the perspectives of families using maternity services, health professionals working in maternity facilities or providing related health services, state officials, and other specialists identified as experts on the issue.

In order to address the individual perspectives of both families with small children (the youngest being no more than three years old) living there and healthcare practitioners working in the region, a qualitative methodology was adopted as being most sensitive to personal perspectives and informal practices (Guba & Lincoln, 1998). The empirical data collected consists of expert interviews with a representative of the investigation committee (1); a ministry of health officer (1); healthcare managers (4); and representatives of patients and professional associations (2). Another set of data consists of semi-structured interviews with healthcare practitioners (obstetrician-gynecologists, neonatologists, midwives, nurses, and feldshers) working in

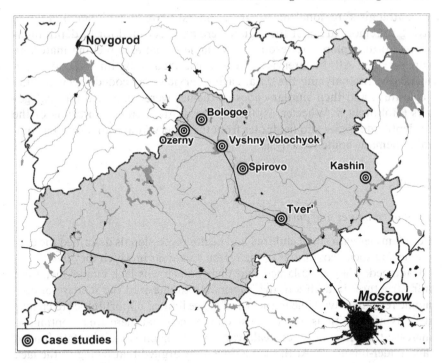

Figure 13.2 Map of Tverskaya Oblast' showing cities studied as cases and the regional capital Tver'.

maternity units, antenatal clinics, and ambulance services in Tverskaya Oblast' (12). The interviews were conducted in the region's six cities and towns, including its capital. Figure 13.2 shows the cities studied as cases and the regional capital Tver' on the map of Tverskaya Oblast', Table 13.1 summarizes the population dynamics in the case locations.

In addition, the empirical data includes analysis of policy, legal, and regulatory documents, and observations at medical conferences on perinatology, obstetrics, and midwifery, and obstetrician-gynecologists' and

Table 13.1 Population dynamics across the cases under investigation during the last decades (Rosstat, 2020)[1]

Settlement/year	1990 (1989)	2000 (2001)	2010	2020
Tver'	449,000	454,900	403,606	425,072
Vyshny Volochyok	64,789	59,700	52,370	45,481
Bologoe	35,926	33,200	23,494	20,498
Kashin	21,186	19,900	16,171	13,757
Spirovo	6,896	6,859	6,267	5,787
Closed territorial unit "Ozerny"	(Vypolzovo)[2]	10,689	10,882	10,661

neonatologists' regional meetings (29 hours). Parents with small children living in remote areas of the region were also recruited for in-depth interviews (6) to explore the resources available to patients seeking maternity care services. All the interviews were transcribed verbatim and coded thematically. In analyzing the data, each interview was coded (open coding) separately, and then similar codes or themes were combined into categories (Holloway & Wheeler, 2010). All the written data (transcripts of the in-depth interviews and fieldnotes from the observations) underwent a common thematic content analysis.

Results

Lack of resources and poverty

Both families with small children and health professionals described the economic and social conditions of the region and areas in which they live as disadvantaged. They complained that their hometowns lack employment and infrastructure. In addition, the informants emphasized the scarcity of economic resources available to them and hence limited opportunities to change their living conditions. Doctors also complained that there are no options to increase their income since healthcare in the region is mainly state-funded (and under-financed). At the same time, they argued that owing to the lack of resources, the state does not provide medical practitioners (who are state employees) with housing, as it used to do during the Soviet period. As a result, doctors must combine work at different jobs to earn sufficient income:

> And so, in addition to the maternity ward, I have two other positions. I work in our prison, in the [women's] colony as an obstetrician-gynecologist, and in Bologoe in a private clinic. I go there, work there, in order to somehow make ends meet in this life. Because the salary is worthless here.
> (Female obstetrician-gynecologist, born 1969)

Low wages in the state-funded healthcare system have frequently been cited as a key challenge for healthcare practitioners and thus for young doctors' recruitment. As a consequence of this (self-estimated) underpayment, many inhabitants of small towns opt to move to a regional center or even to other regions, resulting in an outflow of young people and promising professionals who can find better working conditions in nearby regions.

Outflow of population and professionals

As stated above, the lack of well-paid employment was reported as the key reason for the population outflow from remote areas of the region, especially among young people and promising professionals. This tendency, in turn, affects the working conditions of the remaining healthcare practitioners.

In particular, the problem of "ageing" among doctors emerged as crucial since practitioners do not have younger colleagues to replace them. Limited social resources are available for health services, leading to growing workloads for doctors, nurses, and midwives. In addition, lack of personnel makes it difficult to maintain facilities and provide services. Thus, with a lack of doctors, even hospitals and wards designed to provide more complex medical services are forced to reduce the range of services provided, and in some cases, even close down:

> *If basically, there used to be some specialists who could work there,* [but now] *the old personnel have gone, and the young doctors… there is a lack of them … Well, look, this is what concerns obstetrics. If you take the 572 order* [of the Ministry of Health], *it says that only facilities with a round-the-clock duty of an obstetrician-gynecologist constitute the second level* [of maternity services]. *And if there are only two or one doctor? 24/7 work.*
>
> (Head of a remote advisory center with offsite anesthesiology and resuscitation obstetric teams; senior non-staff specialist in emergency obstetric care for the region, born 1950, Tver')

There are other consequences of the outflow of professionals. It is mostly women who agree to work under conditions of underpayment. The medical profession in Russia is generally quite feminized owing to comparatively low wages and the non-autonomous position of medical professionals (Schecter, 2000; Riska, 2008), most of whom are state employees:

> *There are more women working in ambulances. Men didn't come because of the low wages – they mostly went to Moscow. They all went there, about 8–10 years ago, eight, probably, a lot of people left eight years ago, somewhere around 20 people, 30. I can't say for sure.*
>
> (Ambulance paramedic in Vyshny Volochyok, born 1978)

However, some medical specialties require physical labor, and a lack of male workers may affect the speed and quality of services provided, which is critical for ambulance workers.

Poor infrastructure

In remote areas of the region, poor infrastructure is another important feature of healthcare services. Lack of funds affects not only the salaries of medical practitioners, as previously described, but also the material conditions, technical equipment, and even medicines available to patients. Patients reported that even basic utilities may be absent from the units, which is especially critical during the postpartum period when a mother is in a quite vulnerable position, especially if she or the newborn has any complications or procedures:

Do you know what conditions were in the maternity ward? ... there were husbands, who had to carry their wives if they had a Caesarean... a woman [without a partner] *who had a Caesarean had to find a man – two pairs of male hands, actually, to reach the second floor. And the fact that there are no bathtubs. ... but after giving birth* [on the second floor] *there is no shower there, although I really wanted to wash us. There was a bidet with a broken tap. So once again you can't sit down... nothing works. And you also need to get up close, adapt – we used to go there with a plastic bottle.*

(Mother of two-year-old child, lives in Vyshny
Volochyok, born 1991 in Spirovo)

Lack of material resources can affect not only patients' comfort, but also their safety. Healthcare shrinkage may result in deterioration of the facilities' medical and technical equipment; hence, such organizations may refuse to accept patients who are expected to have complications. Lack of necessary medical instruments and equipment in small and remote maternity facilities becomes even more dramatic in emergency cases of sudden complications or pathology, which are always possible during childbirth. This was the case described by the head of a remote advisory center, whose professional responsibility is to transport pregnant women with sudden and severe complications or operate on them on-site if there are appropriate conditions:

A district hospital that doesn't have an operating room or an anesthesiologist – well, they have a gynecologist there. Well, there's nothing else. And imagine, the ambulance came there with a [patient with] *hemorrhage.* [A patient is] *diagnosed with placental abruption – what's next? Well, they call the sanitary aviation.* [But] *sanitary aviation is not a teleportation ... so in order to fly to a remote area by helicopter it takes time to prepare ... It will take at least two hours. During this time, you can die two or three times.*

(Head of a remote advisory center with offsite anesthesiology
and resuscitation obstetric teams, born 1950, Tver')

As the empirical data reveals, poor infrastructure, as a "side effect" of healthcare services' shrinkage, affects not only the maternity care system but also all other medical services. The head of the remote advisory center with offsite anesthesiology and resuscitation obstetric teams explained how the healthcare infrastructure has become so poor across the region. He emphasized that the whole system of healthcare was designed during the Soviet period, based on the principle of leveling of care combined with a wide coverage of basic services provided in feldsher-midwifery points (facilities providing only paramedic services). The system as it used to work provided access to care, which has since been reduced due to the closure of both small paramedical facilities and district hospitals, without accounting for the distances between remaining facilities.

Necessity and costs of travel

Poor healthcare infrastructure in remote areas of the Tver' region profoundly shapes the strategies adopted by its inhabitants to acquire necessary medical services. As a result of health services' shrinkage, patients, including pregnant women and families with small children, must travel to other districts or even to a regional center, which is the only place to access many medical services. It is worth mentioning again that pregnancy and childbirth in Russia are quite medicalized, so pregnant women are expected to receive regular checkups and be monitored by an obstetrician-gynecologist once or twice a month, depending on the stage of gestation. At the same time, as my informants stressed, they must cover all the travel costs themselves, as well as managing visits to other facilities. This sometimes results in a strategy to avoid additional travel during pregnancy, particularly if there are other small children in the family, which considerably influences patients' choices. They may opt for less qualified and less desirable care to reduce the frequency and distance of travel. However, childbirth itself is considered to be a more critical moment that deserves additional investment in terms of both money (paying for maternity services) and readiness to travel. A doctor described pregnant women's strategies to obtain the desired maternity care:

> They, [patients from] *Bologoe, generally go somewhere ... they either go to Valdai* [city in the Novgorodskaya Oblast'], *or go to us, or give birth in their hometown, or in Borovichi.*
>
> (Obstetrician-gynecologist, born 1969, works in maternity unit in Vyshny Volochyok)

Similar strategies are quite widespread among families with small children, since, as the informants argued, there is no suitable medical care for infants in their hometowns: "We have a therapist. As for all other doctors, they, respectively, are only in Tver'" (Mother of three children, born in 1990, lives in Spirovo). Traveling long distances (up to 200 km) with a small child, especially by public transport, is quite challenging for young families, and hence they sometimes opt to pay for services, if available. However, private healthcare is not very developed in these areas, due mainly to the lack of paying capacity among residents, but also to the absence of health professionals who wish to work in remote areas. Hence, in some cases, there is no choice at all in obtaining the necessary services.

Risks of travel

Other problems with the inevitable travel to obtain necessary services are the poor quality roads and unpredictable childbirth process. In order to reduce the risks of traveling while having contractions, pregnant women often try to arrive at maternity facilities in advance since they appreciate that the routes to maternity wards are uncomfortable and unpredictable. From the health practitioners' point of view, the system itself is not well attuned to transporting women with contractions to the facilities they need to attend. First, they emphasize that there are too few vehicles equipped to transport women in labor since they use the regular ambulance service for transport which may be busy when contractions start. Second, childbirth is described as a very unpredictable process in terms of accurately estimating and predicting its course:

> *Well, it* [the routing model] *was invented correctly; it is a necessary thing for people with pathology, with premature childbirth,* [which gives the opportunity] *to give birth in specialized institutions, where they have all the equipment, which we do not have … But there are nuances with transportation. Because there are no special brigades to carry people there* [to the perinatal center]. *Our regular* [district] *ambulance carries them in between their regular calls … But you cannot know what time exactly a woman will give birth, because it is impossible to predict.*
>
> (Obstetrician-gynecologist, born 1969, works in a maternity unit in Vyshny Volochyok)

As a result, patients often fail to get to the desired facility if they have chosen another district hospital. Hence, the decision to travel is described as quite risky, with no guarantee of obtaining the desired service.

Conclusion

The aim of this chapter was to analyze the inconsistencies between the state demographic (pronatalist) and healthcare policies in Russia by showing the failure of healthcare restructuring to take into account the realities of spatially distributed localities and ongoing depopulation. By addressing the neglected issue of inhabitants' everyday lives in shrinking towns, the study has uncovered some unintended consequences of inconsistent healthcare restructuring— crises in governance and lack of policies attuned to local specificities, along with budget austerity. The closure of healthcare facilities is the most extreme feature of public services restructuring in a Russian shrinking town.

By design, the cuts in healthcare personnel and beds, as well as intraregional inequality of access to medical services, are supposed to be partially compensated for by the routing scheme introduced in 2012. However, owing to the lack of medical transport, equipment, and personnel, new challenges for both patients and health practitioners have emerged. Budget austerity causes poor working conditions for employees of medical facilities and medical practitioners. This leads to a population outflow, as people migrate to secure better paid jobs. Depopulation considerably affects healthcare services provision, mainly due to the growing lack of medical personnel. The lack of social resources, in turn, aggravates the working conditions of those who remain: their workloads increase, and they cannot always manage their tasks optimally.

Healthcare shrinkage in the investigated context is evident in the poor infrastructure of healthcare services in general and maternity services in particular. Patients reported on the impoverished material conditions of maternity facilities, while doctors complained that some facilities often lack necessary equipment and medicines. Thus, new gaps in the healthcare system are emerging because some facilities cannot provide basic services, including urgent care. For many patients who need maternity or infant care, the necessity to travel becomes inevitable. However, in some cases, especially associated with sudden or severe complications, the current healthcare arrangements create new risks arising from the necessity to transport patients. The smallest maternity facilities, which are the only ones available in the small towns of the Tverskaya Oblast', often lack the equipment and personnel necessary to provide emergency care. This results in delays, which sometimes become an issue of life or death.

This case study highlights the failure of the current federal and regional healthcare restructuring to address population dynamics and more broadly jeopardizes achieving the goal of desired natural population increase. There is no clear and articulated policy of adapting healthcare services to the actual needs of the people who seek medical help. The case of maternity services is particularly topical because Russian state policy is generally pronatalist, yet the healthcare system, which is among the most important elements in achieving higher fertility and lower mortality, remains insufficient, poor, and insensitive to the needs of small local and regional towns. "Right-sizing" and qualitative change in healthcare provision in shrinking towns to match the new population needs is absent. Policies, which take no account of the reality of shrinking towns, fail to serve the population's interests and impose additional costs on them (Hummel, 2014; 2015). This, in turn, negatively affects the daily routines of families living in shrinking towns, as well as the working conditions of healthcare practitioners, thus fostering out-migration and further depopulation (Camarda et al., 2015).

Acknowledgments

I express my deepest gratitude to Maria Gunko for her inspiration and invaluable assistance with geographical aspects of this study, as well as for her comments and edits, which greatly improved the manuscript. I also want to thank Alexander Sheludkov for the processing of statistical data and prompt technical assistance in preparation of the publication.

Notes

1 For 1990 and 2000 years the data from the population is based on the national population census of 1989 and 2001, respectively.
2 There is no separate statistics for "Ozerny" before 1992, when it was accounted as a part of Vypolzovo.

References

Baillot, A. & Evain, F. 2013. Les maternités: un temps d'accès stable malgré les ferme-tures. *Journal de gestion et d'économie médicales*, 31(6), 333–347.

Batunova, E. & Gunko, M. 2018. Urban Shrinkage: An Unspoken Challenge of Spatial Planning in Russian Small and Medium-sized Cities. *European Planning Studies*, 26(8), 1580–1597.

Béal, V., Fol, S., Miot, Y. & Rousseau, M. 2017. Varieties of Right-sizing Strategies: Comparing Degrowth Coalitions in French Shrinking Cities. *Urban Geography*, 40(2), 192–214.

Bernt, M. 2016. The Limits of shrinkage: conceptual pitfalls and alternatives in the discussion of urban population loss. *International Journal of Urban and Regional Research*, 40(2), 441–450.

Bierbaum, A. H. 2020. Managing Shrinkage by "Right-Sizing" Schools: The Case of School Closures in Philadelphia. *Journal of Urban Affairs*. https://doi.org/10.1080/07352166.2020.1712150

Camarda, D., Rotondo, F. & Selicato, F. 2015. Strategies for Dealing with Urban Shrinkage: Issues and Scenarios in Taranto. *European Planning Studies*, 23(1), 126–146.

Charreire, H., Combier, E., Michaut, F., Ferdynus, C., Blondel, B., Drewniak, N., ... & Zeitlin, J. 2011. Une géographie de l'offre de soins en restructuration: les territoires des maternités en Bourgogne. *Cahiers de géographie du Québec*, 55(156), 491–509.

Code of rules 42.13330.201. *Urban and rural planning and development*. Actual edition of the Building codes and rules 2.07.01-89, Issued in 2016: http://docs.cntd.ru/document/456054209?fbclid=IwAR3BY8MPAaZyNwvwg_q8EXxwdp6wkPedIEaFePDDbxdPy5sORZDzr0VeJAA)

Cook, L. J. 2014. "Spontaneous privatization" and its political consequences in Russia's postcommunist health sector. In Cammett, M. & MacLean, L. M. (eds.) *The Politics of Non-State Social Welfare* (pp. 217–236). Ithaca, NY: Cornell University Press.

Cook, L. J. 2017. Constraints on universal health care in the Russian federation: Inequality, informality and the failures of mandatory health insurance reforms. In Ilcheong, Y. (eds.) *Towards Universal Health Care in Emerging Economies* (pp. 269–296). New York: Palgrave Macmillan.

Coppola, A. 2019. Projects of Becoming in a Right-sizing Shrinking City. *Urban Geography*, 40(2), 237–256.

Evans, D. 2020. What Price Public Health? Funding the Local Public Health System in England Post-2013. *Critical Public Health*. https://doi.org/10.1080/09581596.2020.1713302

Fomkina, A. A. 2015. Mezhrayonnyye tsentry sotsial'noy infrastruktury: novyy pod-khod k ikh vydeleniyu (na primere Tverskoy oblasti). Vestnik Moskovskogo univer-siteta. Geografiya, [Inter-district Centers of Social Infrastructure: A New Approach to Their Allocation (on the example of the Tver region)]. *Bulletin of Moscow University*, 5 (6), 57–64.

Gel'man, V. 2015. *Authoritarian Russia: Analyzing Post-Soviet Regime Changes*. University of Pittsburgh Press.

Guba, E. & Lincoln, Y. 1998. Competing Paradigms in Qualitative Research. In Denzin, N. & Lincoln, Y. (eds.) *The Landscape of Qualitative Research* (pp. 198–221). London: Sage.

Holloway, I. & Wheeler, S. 2010. *Qualitative Research in Nursing and Healthcare*. John Wiley & Sons.

Hummel, D. 2014. Right-sizing Cities in the United States: Defining its Strategies. *Journal of Urban Affairs*, 37(4), 397–409.

Hummel, D. 2015. Right-sizing Cities: A Look at Five Cities. *Public Budgeting & Finance*, 35(2), 1–18.

Kainu, M., Kulmala, M., Nikula, J. & Kivinen, M. 2017. The Russian Welfare State System. In Aspalter C. (ed.) *The Routledge International Handbook of Welfare State Systems* (pp. 291–316). Oxon: Routledge.

Khasanova, R. R., Florinskaya, Yu. F., Zubarevich, N. V. & Burdyak, A. Ya. 2019. Demografiya I sotsial'noe razvitie regionov v pervom kvartale 2019 g. (po rezul'tatam regulyarnogo Monitoringa INSAP RANKhiGS). *Ekonomicheskoye razvitiye [Demography and Social Development of Regions in the First Quarter of 2019 (Based on the Results of Regular Monitoring of the INSAP RANEPA). Economic Development of Russia]*, 26(6), 62–80.

Kochkina, N. N., Krasil'nikova, M. D. & Shishkin, S. V. 2015. Dostupnost' i kachestvo meditsinskoy pomoshchi v otsenkakh naseleniya [Accessibility and quality of medical care in population estimates, Higher School of Economics]. Vysshaya shkola ekonomiki. Seriya WP8 "Gosudarstvennoye i munitsipal'noye upravleniye" [State and Municipal Administration].

Kuhlmann, E. & Annandale, E. 2012. Researching transformations in healthcare services and policy in international perspective: an introduction. *Current Sociology*, 60(4), 401–414.

LaFrombois, M., Park, Y. & Yurcaba, D. 2019. How U.S. Shrinking Cities Plan for Change: Comparing Population Projections and Planning Strategies in Depopulating U.S. Cities. *Journal of Planning Education and Research*, https://doi.org/10.1177/0739456X19854121

McCourt, C., Rayment, J., Rance, S. & Sandall, J. 2016. Place of Birth and Concepts of Wellbeing. *Anthropology in Action*, 23(3), 17–29.

Ministry of Health of the Russian Federation. 2012. The Order "On Approval of the Procedure for Providing Medical Care in the Profile" Obstetrics and Gynecology (except for the use of assisted reproductive technologies) N 572n of 1.11.2012.

Multistat. Multifunctional statistical portal. 2020. Jekonomika gorodov Rossii [The economy of Russian cities]. Available at: http://www.multistat.ru/?menu_id=9310004 (Accessed 03 June 20).

Panova, L. 2019. Access to Healthcare: Russia in the European Context. *The Journal of Social Policy Studies*, 17(2), 177–190.

President of Russia. 2012. Ukaz «O sovershenstvovanii gosudarstvennoy politiki v sfere zdravookhraneniya» ot 7 maya 2012 g. № 598 [Decree "On Improving the State Policy in Health Care" dated May 7, 2012 No. 598]. Available at: http://kremlin.ru/catalog/keywords/125/events/15234 (Accessed 03 June 20).

President of Russia. 2014. *Annual Address to the Federal Assembly*. Available at: http://kremlin.ru/events/president/news/47173 (Accessed 03 June 20).

Riska, E. 2008. The Feminization Thesis: Discourses on Gender and Medicine. *NORA—Nordic Journal of Feminist and Gender Research*, 16(1), 3–18.

Rivkin-Fish, M. 2010. Pronatalism, Gender Politics, and the Renewal of Family Support in Russia: Toward a Feminist Anthropology of "Maternity Capital". *Slavic Review*, 69(3), 701–724.

Rosstat. 2019. *Socio-economic characteristics of the Tverskaya Oblast*. Available at: https://gks.ru/region/docl1128/Main.htm (Accessed 03 June 20).

Rosstat. 2020. *Population of the Russian Federation by municipality*. Available at: https://www.gks.ru/folder/12781?print=1 (Accessed 03 June 20).

Rozanova, M. S. (eds.) 2016. *Labor Migration and Migrant Integration Policy in Germany and Russia*. Saint Petersburg State University.

Schecter, K. 2000. The politics of health care in Russia: the feminization of medicine and other obstacles to professionalism. In *Russia's Torn Safety Nets* (pp. 83–99). New York: Palgrave Macmillan.

Walberg, P., McKee, M., Shkolnikov, V., Chenet, L. & Leon, D. A. 1998. Economic Change, Crime, and Mortality Crisis in Russia: Regional Analysis. *BMJ*, 317(7154), 312–318.

Winthrop, B. & Herr, R. 2009. Determining the Cost of Vacancies in Baltimore. *Government Finance Review*, 25(3), 38–42.

Zubarevich, N. 2016. Trendy v razvitii krizisa v regionakh. Ekonomicheskoye razvitiye Rossii [Trends in the Development of the Crisis in the Regions]. *Economic Development of Russia*, 23(3), 89–92.

14 Review of spatial planning instruments in a Russian shrinking city

The case of Kirovsk in the Murmansk region

Daria Chigareva

Introduction

According to the United Nations, the total urban population and the share of urbanized territories in the world are increasing yearly (UN 2019), but this growth is uneven. While some cities exhibit a steady population growth, others are experiencing longstanding depopulation, and the number of the latter is steadily increasing (Martinez-Fernandez et al., 2016). Over the past 50 years, more than 370 cities with a population over 100,000 people lost at least 10% of their population (Oswalt, 2005). In academic debates, several concepts were used to define such cities—"shrinking", "declining", "decaying", "de-urbanizing"– the choice of the term depended only on the preferences of the researcher (Haase et al., 2014). In the mid-2000s, cities that experienced longstanding depopulation were labeled "shrinking" following the use of this term by scholars including Philip Oswalt and members of the Shrinking Cities International Network. Nevertheless, scholars did not come to an agreement on the conceptualization of shrinkage in quantitative and qualitative terms. The scale and pace of depopulation, its time frame, and accompanying changes in the city's socioeconomic and planning aspects were determined differently.

Shrinkage does not only mean depopulation *per se*. It is usually accompanied by changes in the functioning of the city often causing structural crisis (Fol, 2012). The lack of efficient and timely measures may lead a shrinking city into a "vicious circle" (Martinez-Fernandez et al., 2012). However, despite the relevance and prevalence of the phenomenon, not only is a generally accepted concept and theory of urban shrinkage lacking, but so is a universal and effective recipe for managing it. Shrinking cities are located in various environmental conditions and cultural contexts, have different planning structures, and are subject to divergent policy and planning approaches.

Urban shrinkage is predominantly linked to major economic reasons (Hollander, 2018) such as deindustrialization and the new territorializing of capital. Likewise, economic restructuring is one of the causes of urban shrinkage in Russia. However, natural population decline is an equally influential factor, making Russian shrinking cities, along with those in Japan and South

DOI: 10.4324/9780367815011-17

Korea, "atypical" in the "Westcentric" (more specifically "USAcentric") perspective (see e.g. Hollander, 2018). Overall, strong negative demographic tendencies since the collapse of state socialism—natural decline and aging of the population (with a short break in the middle of 2010) (RBC, 2019)—taken together with centripetal migration flows from smaller cities to larger ones (Kachurina, 2012) define the fact that about 70% of Russian cities have suffered from depopulation since 1989 (Batunova and Gunko, 2018).

Despite being a prevailing trend of Russian urban development, shrinkage is largely tabooed by planners and policymakers. Urban growth is declared in official planning and policy documents as being one of the ways for "positive" reporting of municipal authorities to higher levels of governance (Batunova and Gunko, 2018; Gunko, Eremenko, and Batunova, 2020). Statistical manipulations, lack of awareness and capacity for independent decision-making, as well as developers' lobbying, turn orientation toward growth into a dominant paradigm. Notwithstanding this, there are a few cases of more circumspect approaches toward planning in shrinking cities despite the lack of comprehensive methodological and legislative frameworks (Gunko and Batunova, 2019).

This chapter focuses on spatial planning instruments for managing the physical consequences of urban shrinkage. Its aim is to research the trajectory of shrinkage and summarize spatial planning instruments utilized in the Russian context along with corresponding limitations. The empirical evidence is drawn from the city of Kirovsk located in Murmanskaya oblast (hereafter—Murmansk region) above the Arctic circle. Despite being located in the Arctic, Kirovsk is an "average" Russian shrinking city in terms of the patterns, causes, and consequences of shrinkage. The choice of Kirovsk as a case study is based on its exceptional (under Russian conditions) response to urban shrinkage. The City Administration not only acknowledged shrinkage but also adopted a specific set of spatial planning instruments to manage its outcomes.

Spatial planning instruments for managing the physical consequences of shrinkage

In the public discourse, urban shrinkage is often viewed as a negative phenomenon, a step toward "death" of a city. However, as any other phenomenon, it has both negative and positive aspects. Urban shrinkage can and should be perceived as a challenge but a challenge that may lead to renewal. Population outflow and associated de-densification is a chance for introducing improvements to the housing market, decreasing the pressure on the environment, developing green zones, and catalyzing new local initiatives. As previous research indicates, residents of some shrinking cities are often no less satisfied with the quality of life than those in growing cities (Delken 2008). However, space in shrinking cities needs to be managed differently than that in growing ones.

Currently, some cities focus predominantly on the supply side assuming that an increase in the volume of housing stock will inevitably lead to an increase in demand (Right-sizing America's Shrinking Cities, 2007). This is precisely the approach proposed in most general plans of Russian cities. In reality, this approach only works for the benefit of developers, except in situations when most of the housing stock is in poor condition and new construction is needed to resettle residents to provide them with higher-quality residential areas. With intensive depopulation and closure of businesses, the share of vacant areas often increases leading to abandonment and spatial fragmentation (Mah, 2012; Hollander, 2018). This becomes an additional burden for the municipal budget as additional funds are required for maintenance, as well as, if necessary, for demolition and reclamation (Semnasem. ru, 2019). Abandoned and stagnating territories also create a negative image of the city and reduce overall security, as the number of "street eyes" that regulate the interaction of people in the territory decreases (Jacobs and Rawlins, 2011). Thus, the lack of adequate instruments for managing the cityscape of a shrinking city contributes to uneven living conditions and increases the gap between the most affluent and the poorest groups of the population (Martinez-Fernandez et al., 2012).

Spatial planning instruments in shrinking cities include optimization of infrastructure (right-sizing); demolition and redevelopment of vacant premises; greening (for example, urban gardening and farming) of vacant land lots; and temporary use of vacant lots and buildings. Reconsidering the use of space, they allow the city to become more livable and better balance budget issues. Thereby shrinkage can be turned from a problem into an opportunity for further development and achievement of a balance between the size and quality of the built-up space.

For comparative analysis, I have summarized spatial planning instruments used in selected shrinking cities where information about spatial reorganization and implementation of specific instruments was publicly available. For this purpose, available urban planning documents and existing research were reviewed for 28 cities located in ten countries. Part of this information was provided by representatives of UN Habitat:

- USA—Detroit, Cleveland, St. Louis, Buffalo, Youngstown, Cincinnati, Newark, Canton (Oswalt, 2005; Re-Imagining a More Sustainable Cleveland, 2008; Schilling and Logan,2008; Lyons, 2009; Mallach and Brachman, 2013);
- Germany: Schwedt, Halle, Dresden, Leipzig, Aschersleben, Chemnitz (Haase, Rink, & Grossmann, 2016; Delken, 2008)
- South Korea—Daegu (Joo and Seo, 2018);
- UK—Manchester, Liverpool, Sheffield (UK) (Cunningham-Sabot and Fol, 2009);
- Australia—Mount Isa, Broken Hill (Australia) (Martinez-Fernandez et al., 2012);

- Spain—Bilbao (Mallach and Brachman, 2013);
- Japan—Hakodate (Oswalt, 2005);
- Brazil—Nilópolis, São Caetano do Sul (Moraes, 2013);
- Ukraine—Makeevka (Ukraine) (Oswalt, 2005);
- Cherepovets, Vorkuta, Novotroitzk (Russia) (Semnasem.ru, 2019).

Spatial planning instruments are summarized and grouped according to their purpose in Table 14.1. The sources of information are listed above.

Table 14.1 Systematization of spatial instruments for managing shrinking cities

Instrument	Sub-instrument	Example of implementation
Residents resettlement	Resettlement within the city borders for the purpose of densification and compact development Resettlement into other cities	Vorkuta (Russia)
Demolition	Demolition of housing, industrial, and public buildings (e.g., schools, kindergartens, hospitals) which become vacant due to depopulation and deindustrialization	Makeyevka (Ukraine); Schwedt, Stendal, Halle, Dresden, Chemnitz, Leipzig (Germany); Youngstown, Buffalo (USA); Daegu (South Korea)
Redevelopment of existing building	Reconstruction of housing Redevelopment of existing building with introduction of new functions (cultural, commercial, tourist)	Manchester, Liverpool (UK), Detroit, Canton, St Louis, Cleveland (USA), Aschersleben, Leipzig, Dresden (Germany), Mount Isa (Australia), Bilbao (Spain)
Development of the historical center	Working with cultural heritage sites and attracting residents to the historical center	Hakodate (Japan); Cincinnati, Buffalo, St. Louis, Cleveland (USA); Broken-Hill (Australia); Daegu (South Korea); Dresden (Germany)
Reclamation of the deconstructed areas	Greening and development of recreational zones Complete deurbanization with a possible return to rural land use	Daegu (South Korea), St. Louis, Cleveland (USA), Sheffield (UK);
Optimization of the transport infrastructure	Development of external communication Improving the transport connectivity of certain parts of the city Optimization of public transport routes	Nilopolis, Sao Caetano do Sul (Brazil), Newark Cleveland (USA), Hakodate (Japan), Cherepovets (Russia)
Optimization of utility networks	Optimization of utility networks	Vorkuta (Russia), Makeevka (Ukraine), Halle (Germany)

It must be noted that each case where specific spatial planning instruments were employed, as well as their combination, was unique in terms of the existing legislation, the capacity of local authorities for independent policymaking, and the required investment.

Methodological approach

Case study

Kirovsk is located in the Arctic zone of the Russian Federation in the Murmansk region, 205 km south of the regional capital, Murmansk. It has a strong connection through a common industrial heritage, infrastructure, and daily working commutes with the nearby city of Apatity[1], located 20 km to the southwest.

In 2019, the population of Kirovsk was 26,200. Depopulation, due to natural population decline and out-migration, is the main trend of its demographic development, which has been evident since the 1990s. Out-migration is partially institutionalized within the framework of resettlement from the North program—Federal law No. 125-FZ of 25.10.2002 *On housing subsidies to citizens leaving the far North and equivalent localities* (Federal Law, 2002). However, this program has not been very successful in the Murmansk region due to a lack of support from the regional government (Nuykina, 2011).

Kirovsk is a company town (*monogrod*) with the "city-forming" enterprise JSC Apatite established in the 1930s. In 2013, JSC Apatite was incorporated into PhosAgro Chemical Holding, one of the world's leading producers of phosphorus fertilizers. The location of the enterprise is associated with rich deposits of Apatitee-nepheline ores in the surrounds of Kirovsk. Despite its industrial profile, Kirovsk is also one of the country's three well-known mountain-ski resorts—BigWood (owned and managed by PhosAgro), which attracts tourist inflows and subsequent development of small businesses.

Methods

To review and summarize spatial planning instruments employed in Kirovsk, both desk and field research were carried out. Field research, which took place in January 2020, included observations and semi-structured interviews. Interviews were carried out with nine respondents who possessed relevant knowledge about urban planning not available in the public domain as it was (partially) confidential. Respondents included representatives of the local administration (Head of the Commission for Municipal Property, Chief Architect, and several other administrative staff); the Deputy Director for Information Policy and Public Relations for the Kirovsk branch of JSC Apatite; and representatives of the Local History Museum. The aims of the interviews were threefold. Firstly, to understand the discourse about shrinkage, including that of the City Administration, and to compare the latter with official urban planning documents. Secondly, to obtain data not available in the public domain. Thirdly, to determine the existing restrictions for implementing specific spatial planning instruments aimed at managing the effects of urban shrinkage and cityscape reorganization.

Desk research comprised an analysis of various documents and statistical data, including:

1 Statistical data provided by the Federal statistics service, the Federal statistics service for the Murmansk region, as well as semi-censored statistical data provided by the administration of Kirovsk
2 Long- and short-term planning documents that defined the spatial development of the city
3 News reports in the regional and local media covering events related to the spatial development of the city.

Shrinkage and spatial planning in Kirovsk

Population development in Kirovsk

On average, Russian Arctic cities are much larger than their international counterparts since they were established and developed under the Soviet centralized planning system (Laruelle, 2019), which means that they were initially largely dependent on state support (Gaddy and Ickes, 2013). The collapse of state socialism and subsequent transition to a market economy, implementation of austerity politics, and liberalization of migration led to significant demographic changes in Russia in general and in the Russian Arctic in particular. As an exception to the booming oil and gas provinces, the Russian Arctic is rapidly depopulating (Heleniak, 2009; Heleniak, 2017). The population of the region has declined by 20% due to the large-scale migration outflow, as well as negative natural population decline, resulting in many small settlements across the Russian Arctic being abandoned (Heleniak, Turunen, and Wang, 2020). The processes that affect population dynamics in Kirovsk are identical to those that define population change in other shrinking cities of the Russian Far North. Since 1989, the population of Kirovsk has decreased by about 40% (Territorial body of the Federal state statistics service for the Murmansk region, 2020) (see Figure 14.1).

Kirovsk experienced a decrease in birth rates due to a decrease in the number of women in the most reproductive age group (from 20 to 29 years), as well as a decrease in the population of working age men and women (Kirovsk. ru, 2019). Natural population decline, in general, is due to a high mortality rate which is common in Russia on average (Makarentseva and Khasanova, 2018). The causes of high mortality in Kirovsk are linked to health issues such as cardiovascular illnesses (55.5%) and malignant neoplasms (13%). In turn, the above have been linked to an adverse environmental situation in the city. Kirovsk is exposed to emissions from the "city-forming" enterprise

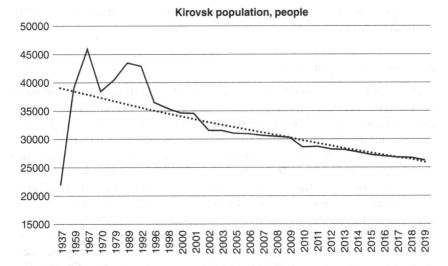

Figure 14.1 Kirovsk's population 1937–2019.

Source: Murmansk region data, Federal Statistical Service.

JSC Apatite, as a result of which the prevalence of cardiovascular illnesses and various chronic diseases of the respiratory system is almost 2.5 times higher than in the Murmansk region on average (Revich et al., 2014). The company has largely been investing in modernization of its fuel and energy complex (TASS, 2017); however, to date the environmental issues remain.

Migration outflow from Kirovsk is related to the natural environment and economic situation. Firstly, Murmansk region experience harsh natural conditions–long cold winters with snow storms and polar night lasting in some parts of the region up to around one month. Surveys indicate that currently Northerners do not see clear advantages in living above the Arctic Circle and would like to migrate to warmer and more comfortable parts of the country (Alishevskaya, 2019). Secondly, due to the economic restructuring following the transition from state socialism to the market in the 1990s, younger citizens began to leave the city. This led not only to the decline of the number of inhabitants *pe se* but also to decreases in birth rates and aging. Though the intensity of outflow from Kirovsk varied throughout the postsocialist period, by the end of the 1990s, when the economic situation in the country in general and at JSC Apatite, in particular, began to stabilize, out-migration slowed. In the early 2000s, when JSC Apatite was incorporated into PhosAgro, the migration balance was around zero. However, due to the reduction in net profit, the company started optimizing production with a reduction in employees, causing intensification of depopulation. Currently, the unemployment rate in Kirovsk is one of the highest in the region (Committee for Labor and Employment of Murmansk Region, 2019). The composition of the unemployed population is mainly represented by socially vulnerable and marginalized groups of the population: young people without work experience, disabled people, and women (Barasheva, 2010).

Development of the cityscape

In 2008, the Federal company RosNIPI Urbanistiki (St. Petersburg) developed the current general plan for Kirovsk until 2023, which was to guide the further development of the city's planning structure, taking into account the planned socioeconomic development, as well as the development of transport and engineering infrastructure. According to the general plan, the city's population should grow, reaching 32,000 by 2023 (in 2008 the population was 30,000). Despite the long-term trend of population decline, the authors of the general plan assumed that the city should be developed within the framework of a growth paradigm. However, in Kirovsk, the situation with spatial development is far from obvious. What on paper may be seen as "expansion" or sprawl, is, in reality, a remedy for previous omissions. The general plan officially included some of the urban areas built up in the 1980s in the city borders. Overall, the expansion of the urban borders was planned at the expense of the forest land owned by Kirovsk forestry—69.63 ha. This land was to be allocated to the following functional categories: industrial area (20 ha), residential area (32.1 ha), and lands of the reserve fund (3.67 ha). The latter includes all lands in state property that have not been provided for any type of private ownership and use.

During interviews in January 2020, representatives of the administration proclaimed that they perceive the general plan as nothing more than a "paper created by non-residents", which in 2023 will lose its significance. This is a typical discrepancy between official documents and actual policymaking in the realm of urban development in Russia due to the need of following the general national discourse, as well as reduced possibilities to develop own urban planning documentation within the current legislation (Gunko, Eremenko, and Batunova, 2020).

In 2019, the existing housing stock in Kirovsk amounted to 890,100 sq. m. according to the data provided by the City Administration, of which 779,100 sq. m. was private property and 116,700 thousand sq. m. was municipal property. In addition, 20,633.17 sq. m. of the municipal housing stock (444 apartments) was recognized as vacant, of which 146 apartments were located in the older part of the city—Kukisvumchorr district (see Figure 14.2).

Most of the city's housing stock, represented by buildings dating to the 1970s and older, is considered to have an "operational period" which is coming to an end. This implies significant deterioration of load-bearing structures. In 2015, according to the former Head of Administration, Vladimir Dyadik, of 276 apartment buildings, 134 had a wear rate of up to 60% (Severpost.ru 2015). At the end of 2019, the area of "housing in emergency condition"[2] included 17114.6 sq. m. or 1.9% of the total housing.

Since 2011, the city has implemented municipal *blagoustroystvo* programs. As part of these programs, new playgrounds and sports fields were

Figure 14.2 Abandoned housing in Kukisvumchorr district of Kirovsk.
Source: Photo made by the author.

established in the city, along with recreational areas and public spaces. Since 2017, the city has launched a municipal program *Formation of a contemporary urban environment* (in line with the national modernization project *Formation of a comfortable urban environment*), aimed at improving the cityscape. The program includes building courtyards for apartment buildings and public spaces with high-quality small architectural firms, play and sports grounds, lighting, and landscaping. The "city-forming" enterprise also contributed to improving the aesthetic state of the city, including the funding of murals on residential buildings. At the same time, the *Formation of a contemporary urban environment* program does not deal with housing and infrastructure, which require extensive maintenance in Kirovsk.

Spatial planning instruments

The spatial planning instruments employed in Kirovsk are reviewed in summary provided in the theoretical section of the chapter.

Resettlement

Residents living in apartment buildings in "emergency condition" located in Kukisvumchorr district were resettled and the buildings were demolished or are being prepared for demolition. There is a further possibility of resettling residents not only from buildings in "emergency condition" but also from partially vacant buildings in order to provide them with better utilities and higher comfort. This approach will limit the urban sprawl of Kirovsk thus bringing it closer to a "compact" city model with the concentration of development closer to the current city center.

Generally, resettlement is a tool relevant for many shrinking cities, but it requires regulatory preparation not only at the municipal level but also at higher levels of governance. In Russia, there is a lack of legal framework for resettling residents from partially empty residential buildings if these are not acknowledged as being "housing in emergency condition" by the special state commission.

Demolition and reclamation

Three residential buildings were demolished primarily in Kukisvumchorr district through programs funded by the City Administration. Moreover, massive work took place in 2013, with the demolition of the first Apatite-nepheline processing plant that ceased operation in the 1990s. The ruins of the latter were located on the picturesque shore of the mountain lake in the central part of the city. The demolition was funded and took place under the control of PhosAgro. The demolished site, however, was not subjected to reclamation since, according to the Deputy Director for Information Policy and Public Relations of the Kirovsk branch of JSC Apatite, there was no investor who would be willing to carry out any sort of activity on this site. PhosAgro funded the demolition and reclamation of the site under an abandoned building of the Mining College through a social partnership contract (variation of a private-public partnership) signed with the City Administration.

For the municipality to carry out any kind of demolition, be that of residential or non-residential premises, a transfer of private property to municipal ownership is required. Thus, each case requires close communication and negotiation between the municipality and the owner. However, the most complicated case is when the owner is unknown, a phenomenon common in Russia due to the chaotic privatization following the collapse of state socialism. Here two variations are possible according to the legislation—ownerless (*beskhozyaynaya*) and escheated[3] (*vymorochennay*) property which ultimately affects the procedure taken by the administration, though both types require filing a costly and time-consuming court case.

Conclusion

Comprehensive management of the negative effects of urban shrinkage in Russia is an exception rather than a common practice. Architects, urban planners, and representatives of city administrations alike have little experience in this direction. Thus, the existing crisis in most cases worsens over time. Kirovsk is one of the few Russian cities which acknowledges shrinkage and conducts purposeful work to reduce the physical negative consequences of depopulation and structural crises. Dealing with shrinkage in Kirovsk is organized in line with the principle of public-private partnerships. The "city-forming" enterprise provides the City Administration with additional financial and managerial resources. The reason for such a partnership is the desire of the city-forming enterprise to diversify the economy of the city through tourism; thus, creating a more comfortable and attractive cityscape is one of the mutual goals on the agenda.

Drawing on the analysis of international approaches and spatial instruments, the current study highlights the Russian pathway in managing the cityscape under conditions of shrinkage. As the current research demonstrated, the adopted instruments are limited to a demolition-first approach (Béal, Fol, and Rousseau, 2016; Weaver and Knight, 2018). The demolition-first approach accompanied in some cases with resettlement may be seen as "universal" since it is found in shrinking cities located in various countries. It requires a minimum level of adaptation to the local specificities and does not need specific expertise. On the contrary, such instruments and approaches as redevelopment, reclamation of deconstructed areas, right-sizing of infrastructure and utility networks, and valorization of cultural heritage sites are less common due to their high capital intensity and need for specific expertise in implementation. The case of Kirovsk highlights the fact that even if a city is willing to tackle urban shrinkage, it is often limited to bluntly dealing with its physical aspect. In Kirovsk, this includes demolition of housing and improving public spaces. At the same time, the city is not able to deal directly with environmental degradation due to its ongoing industrial activities and having no resources to remedy the remnants.

Furthermore, it is important to highlight that implementation of specific spatial planning instruments and approaches in Russia, in general, is largely limited by the current legislation and complex property issues (Gunko, Eremenko, and Batunova, 2020). It is not solely funding and expertise, but more fundamental issues such as local policymaking capacity and the legal framework that are obstacles for more effective urban planning in Russian shrinking cities. Nevertheless, the experience of such cities as Kirovsk allows us to contribute to the assemblage of instruments, competencies, and legal procedures needed in managing urban shrinkage in a significant number of Russian cities.

Overall, even though challenging the effects of urban shrinkage in Kirovsk is an ongoing process, researchers and urban actors can already see and evaluate the positive results of cityscape reorganization. In 2020, Kirovsk was listed in the top 20% of company towns that are recognized by the Russian Government as most comfortable for residents (Monogoroda. ru 2020). At the same time, it is vital to note that reorganization of the cityscape is only one aspect of enhancing liveability in a shrinking city. Changes to spatial planning have to be accompanied by socioeconomic measures aimed at improving the quality of life of the population. These include providing high-quality infrastructure and housing, securing jobs, and providing opportunities for a variety of leisure activities.

Notes

1 The Kirovsk-Apatity is a coupled urban area, forming a "second regional center" of Murmansk region. Both cities are currently shrinking at a similar pace. However, unlike Kirovsk, urban shrinkage is ignored within planning and policymaking in Apatity (Gunko et al. 2020).
2 Housing in "emergency condition" is an official category in the Russian legislation applicable to buildings that are characterized by significant damage and deformations, deterioration of load-bearing structures, and in danger of collapse.
3 Escheated property—term used in civil law which denotes property left by a deceased person that no one claims or can claim by will or inheritance.

References

Alishevskaya, L. 2019. Iz Murmanskoy oblasti ludi chatsche pereezzhaut v Sankt-Peterburg, Moskvu... i *Krasnodar*. *Kp.ru*, 28 December. Available at: https://www.kp.ru/daily/27074.7/4144066/

Barasheva, T. 2010. Rynok truda i zanyatost v Murmanskoy oblasti [Labor market and employment in the Murmansk region]. *Voprosy Strukturizatsii Ekonomiki*, 2, 125–129.

Batunova, E. & Gunko, M. 2018. Urban shrinkage: an unspoken challenge of spatial planning in Russian small and medium-sized cities. *European Planning Studies*, 26(8), 1580–1597.

Béal, V., Fol, S. & Rousseau, M. 2016. De quoi le "smart shrinkage" est-il le nom? Les ambiguïtés des politiques de décroissance planifiée dans les villes américaines [What is being called "smart shrinkage"? The ambiguities of planned shrinkage policies in American cities]. *Géographie, Économie, Société*, 18(2), 211–234.

Committee for Labor and Employment of Murmansk Region. 2019. Labor and labor market. Available at: http://murman-zan.ru/

Cunningham-Sabot, E. & Fol, S. 2009. Shrinking cities in France and Great Britain: A silent process? In Pallagst, K. et al. (eds.) *The Future of Shrinking Cities* (pp. 17–27). Available at: https://escholarship.org/uc/item/7zz6s7bm

Delken, E. 2008. Happiness in Shrinking Cities: A Research Note. *Journal of Happiness Studies*, 9(2), 213–218.

Federal Law 2002. *Federal Law of the Russian Federation "On housing subsidies to those leaving the Far North and areas equated to it"* 25.10.2002 no. 125-FZ. https://base.garant.ru/12128598/

Fol, S. 2012. Urban Shrinkage and Socio-spatial Disparities: Are the Remedies Worse than the Disease?. *Built Environment*, 38(2), 259–275.

Gaddy, C. G. & Ickes, B. 2013. *Bear's Traps of Russia's Road to Modernization*. Routledge.

Gunko, M. & Batunova, E. 2019. Kak otvetit na depopulyaciyu Rossijskih gorodov [How to respond to depopulation of Russian cities]. *Vedomosti.ru*, 22 December. Available at: https://www.vedomosti.ru/opinion/articles/2019/12/22/819389-depopulyatsiyu-gorodov?fbclid=IwAR3vG1fDSgJ8mQ6Wo1EST5YGNa2qY4bRKChgtdPuPZjHejJ8xOPtBB3i8HE

Gunko, M., Eremenko, U. & Batunova, E. 2020. Strategii planirovaniya v uslovi-yah gorodskogo szhatiya v Rossii: issledovanie malyh i srednih gorodov [Planning strategies in the context of urban shrinkage in Russia: research of small and medium-sized cities]. *Mir Rossii: Sociologiya, Etnologiya*, 29(3), 121–141.

Haase, A., Grossmann, K., Rink, D. & Bernt, M. 2014. Conceptualizing urban shrink-age. *Environment and Planning A*, 46(7), 1519–1534.

Haase, A., Rink, D. & Grossmann, K. 2016. Shrinking Cities in Post-socialist Europe: What can we Learn from Their Analysis for Theory Building Today? Geografiska Annaler: Series B, *Human Geography*, 98(4), 305–319.

Heleniak, T. 2009. Growth Poles and Ghost Towns in the Russian Far North. In Rowe, E. W. (ed.) *Russia and the North* (pp. 129–163). Ottawa: University of Ottawa Press.

Heleniak, T. 2017. Boom and Bust: Population Change in Russia's Arctic Cities. In Orttung R. (ed.) *Sustaining Russia's Arctic Cities: Resource Politics, Migration, and Climate Change* (pp. 67–87). Berghahn Books.

Heleniak, T., Turunen, E. & Wang, S. 2020. Demographic changes in the Arctic. In Coates, K. & Holroyd, C. (eds.) *The Palgrave Handbook of Arctic Policy and Politics* (pp. 41–59). Cham: Palgrave Macmillan.

Hollander, J. 2018. *A Research Agenda for Shrinking Cities*. Cheltenham: Edward Elgar Publishing Ltd.

Jacobs, J. & Rawlins, D. 2011. *The Death and Life of Great American Cities*. Westminster: Books on Tape.

Joo, Y. & Seo, B. 2018. Dual Policy to Fight Urban Shrinkage: Daegu, South Korea. *Cities* 73, 128–137.

Kachurina, L. 2012. Urbanizatsis po-rossiyski [Urbanization the Russian way]. *Otechestvennie zapiski* 3, 10–24.

Kirovsk.ru. 2019. Prognoz sotsialno-ekonomicheskogo razvitiya munitsipalnogo obrazovaniya gorod Kirovsk s podvedomstvennimi territoriyami na 2019 god i planovy period 2020–2021 [Forecast for socio-economic development of the Kirovsk Municipality for 2019 and planned period 2020–2021]. Available at: https://kirovsk.ru/files/npa/adm/2018/1466/pril_post_1466.pdf

Laruelle, M. 2019. The Three Waves of Arctic Urbanisation: Drivers, Evolutions, Prospects. *Polar Record*, 55(1), 1–12.

Lyons, S. 2009. *Buffalo's Demolition Strategy*. Cornell University ILR School. Available at: https://digitalcommons.ilr.cornell.edu/cgi/viewcontent.cgi?referer=https://www.google.com/&httpsredir=1&article=1054&context=buffalocommons

Mah, A. 2012. *Industrial Ruination. Community, and Place*. Toronto: University of Toronto Press.

Makarentseva, A. & Khasanova, R. 2018. *Smetrnost i prodolzhitelnost zhizni naseleniya Rossii [Mortality and life expectancy of the population of Russia]*. SSRN. http://dx.doi.org/10.2139/ssrn.3169745

Mallach, A. & Brachman, L. 2013. *Regenerating America's Legacy Cities*. Cambridge, MA: Lincoln Institute of Land Policy.

Martinez-Fernandez, C., Weyman, T., Fol, S., Audirac, Y., Sabot-Cunningham, E., Wiechmann, T. & Yahagi, H. 2016. *Shrinking Cities in Australia, Japan, Europe and the USA*. Amsterdam: Elsevier.

Martinez-Fernandez, C., Wu, C-T., Schatz, L., Taira, N. & Vargas-Hernandez, J. 2012. The Shrinking Mining City: Urban Dynamics and Contested Territory. *International Journal of Urban and Regional Research*, 36(2), 245–260.

Monogoroda.rf. 2020. Poryadka 20% monogorodov priznany territoriyami s blago-priyatnoy gorodskoy sredoy [Around 20% of company towns have been recognized as having a comfortable urban environment]. Available at: https://realty.ria.ru/20200113/1563370770.html

Moraes, S. 2013. Inequality and Urban Shrinkage: A Close Relationship in Brazil. In *Shrinking Cities. International Perspectives and Policy Implications.* London: Routledge.

Nuykina, E. 2011. *Resettlement from the Russian North: An Analysis of State-Induced Relocation Policy.* Lapland University Press.

Oswalt, P. 2005. *Shrinking Cities.* Available at: http://shrinkingcities.com/fileadmin/shrink/downloads/pdfs/SC_Band_1_eng.pdf

RBC. 2019. *Estestvennaya ubyl rossiyan stanet rekordnoj za 11 Let. Pochemu Rossiya pro-dolzhaet teryat naselenie* [*The Natural Decline of Population in Russia Will be Record-High in 11 years: Why Russia Continues to Lose Population*]. RBC, 13 December. Available at: https://www.rbc.ru/economics/13/12/2019/5df240d49a79475d8876bdc3

Re-imagining a More Sustainable Cleveland. 2008. Re-Imagining a More Sustainable Cleveland. Citywide Strategies for Reuse of Vacant Land. Available at: https://community-wealth.org/sites/clone.community-wealth.org

Revich, B., Kharkova, T., Kvasha, E., Bogoyavlensky, D., Korovkin, A. & Korolev, I. 2014. Sotsialno-demographiecheskie ogranicheniya ustoychivigo razvitiya Murmanskoy oblasti [Socio-demographic limitations for sustainable development of Murmansk region]. *Problemy prognozirovaniya,* 2, 127–135.

Right-Sizing America's Shrinking Cities. 2007. Results of the Policy Charrette and Model Action Plan. Available at: http://cudcserver2.cudc.kent.edu/projects_research/research/rsc_final_report.pdf

Schilling, J. & Logan, J. 2008. Greening the Rust Belt: A Green Infrastructure Model for Right Sizing America's Shrinking Cities. *Journal of the American Planning Association,* 74(4), 451–466.

Semnasem.ru. 2019. Upravluaemoe szhatie. Kak Rossiiskie goroda prisposablivautsa k novoi zhizni posle togo, kak ottuda massovo uezhaut ludi. Obyasnyaem na primere Vorkuti [Controlled shrinkage. How Russian cities adapt to a new life after being massively abandoned by people. Explanation based on Vorkuta case]. *Semnasem.ru,* 20 February. Available at: https://semnasem.ru/vorkuta/

Severpost.ru. 2018. Segodnya mintrans reshit sudbu dorogi Apatitey-Kirovsk [Today, the Ministry of Transport will decide on the fate of the Apatitey-Kirovsk road]. *Severpost.ru,* 10 April. Available at: https://severpost.ru/read/64634/

TASS. 2017. *V ekoproekty v zapolyarnom monogorode Kirovske vlozhili svyshe 3,3 mlrd rublej* [*Over 3.3 billion rubles have been invested in eco projects in the polar single industry city of Kirovsk*]. TASS, 17 February. Available at: https://tass.ru/obschestvo/4032032

United Nations, Department of Economic and Social Affairs, Population Division. 2019. *World Population Prospects 2019, Volume II: Demographic Profiles* (ST/ESA/SER.A/427).

Weaver, R. & Knight, J. 2018. Can Shrinking Cities Demolish Vacancy?: An Empirical Evaluation of a Demolition-first Approach to Vacancy Management in Buffalo, NY, USA. *Urban Science,* 2(3), 69. https://doi.org/10.3390/urbansci2030069

Part IV

Postsocialist Europe

15 Introduction to the Postsocialist Europe section

Tadeusz Stryjakiewicz

Unlike the previous sections on China and Russia, where the process of urban shrinkage was analyzed within one country, this section attempts to offer insight into this process for the group of states collectively known as "post-socialist Europe".[1] The group, which largely consists of eastern European countries, demonstrates a "mosaic" of different pathways of urban development, including the process of shrinkage (Tsenkova and Nedović-Budić 2006; Sykora 2009; Haase et al. 2013). The distinguishing feature of all these countries in the context of the issue discussed in this book is not so much their geographical or cultural proximity as their common path of development after World War II. This path was marked by the prolonged operation of a centrally planned economy, with strong interference of the central government in the process of urbanization and spatial planning. The transition from a centralized command system to a democratic market-oriented one, however, occurred in different countries in different ways and at different speeds (e.g., gradualism vs. "shock therapy"; cf. Bontje 2004; Stryjakiewicz 2013). Moreover, the institutional framework of this process has varied between countries (e.g., membership in the European Union). Some of the states, created after the break-up of Yugoslavia, were still at war from 1992 to 1995. The divergent development paths are reflected in the breadth and depth of research on the process of urban shrinkage in addition to the quality and availability of statistical data and other reference resources (in the case of some countries, online reports are the predominant source).

From the analytical point of view, postsocialist Europe can be divided into two groups of countries:

1 The countries of Central-Eastern Europe. They are the focus of this section. This group includes the countries known as the Visegrad Group: the Czech Republic (or Czechia), Hungary, Poland, and Slovakia, which began the process of postsocialist transformation the earliest (in 1989) and became member states of the European Union as early as 2004. Additionally, this group includes the post-Soviet Baltic states: Estonia, Latvia, and Lithuania, which regained independence after the break-up of the Soviet Union (from 1990 to 1991) and joined the European Union in 2004.

DOI: 10.4324/9780367815011-19

2 The countries of South-Eastern Europe. This is a very diversified group, including Albania, Bulgaria, Romania, and the states created after the break-up of former Yugoslavia: Bosnia and Herzegovina, Croatia, Montenegro, North Macedonia, Serbia, and Slovenia. The political status of these countries is very different. While some have acceded to the EU, including Slovenia in 2004, Bulgaria and Romania in 2007, and, most recently, Croatia in 2013, the remainder are not EU member states. The status of Kosovo remains in dispute.

The process of postsocialist transformation is also exhibited by the former German Democratic Republic (or East Germany) following its reunification with West Germany in 1990. However, we leave that case aside as there are many empirical and practically oriented studies on urban shrinkage in this region.

Urban shrinkage affects all postsocialist European countries, albeit to a different extent. They demonstrate a higher intensity of urban shrinkage than the European average (cf. Martinez-Fernandez et al. 2016; Wolff and Wiechmann 2018). Bulgaria, Romania, Bosnia, and Herzegovina, as well as the Baltic states, are the "European leaders of urban shrinkage". In some countries, the scale of the process is so extensive that the majority of cities and towns can be called "shrinking" in all dimensions of this term; still, the demographical aspect seems crucial. According to the estimations of the UN, since beginning to transform their political, social, and economic systems in 1990, the countries of postsocialist Europe have collectively lost 18 million (or 6 per cent) of their population, mainly urban (UN World Cities Report 2018; UN World Population Prospects 2019; Krastew 2020).

This large scale of urban shrinkage is a great challenge for both research and policy. For many years, the issue of shrinkage was neglected in scientific and practically oriented debates in most countries of postsocialist Europe (excluding the former German Democratic Republic). Only in recent years has this gap begun to be gradually bridged. Of particular importance in this respect have been the activities of the Shrinking Cities International Research Network (SCIRN, www.shrinkingcities.org) and the implementation of such international research projects of the European Union as:

a Shrink Smart—The Governance of Shrinkage within a European Context, which was implemented under the EU 7th Framework Programme (for details, see Bernt et al. 2012; Couch et al. 2012; Haase et al. 2013); and

b CIRES—Cities Regrowing Smaller—Fostering Knowledge on Regeneration Strategies in Shrinking Cities across Europe, implemented under EU COST Action[2] (Wiechmann 2013; Wolff and Wiechmann 2018).

A few other projects are underway, including 3S RECIPE (Smart Shrinkage Solutions—Fostering Resilient Cities in Inner Peripheries of Europe; www.jpi-urbaneurope.eu) and Re-City (Reviving Shrinking Cities—Innovative

Paths and Perspectives toward Livability for Shrinking Cities in Europe; www.uni-kl.de/re-city), implemented under the European Union's Horizon 2020 research and innovation program. Some research results stemming from the above-mentioned projects are presented in this section.

The contents of this section are part of an ongoing discussion on postsocialist shrinking cities in Europe.[3] Its essence are the case studies on industrial shrinking cities located in four countries of Central-Eastern Europe (Poland, Slovakia, Czechia, and Lithuania). However, in order to offer a wider perspective of the process of urban shrinkage (including its history and main determinants) for the whole region, the case studies are preceded by a general, partly comparative overview of shrinking cities in selected postsocialist countries of both Central-Eastern and South-Eastern Europe.

Our focus on Central-Eastern Europe is justified for at least two reasons:

a the studies on shrinking cities in this part of postsocialist Europe are most advanced, which facilitates comparisons with shrinking cities in Western Europe as well as China and Russia (discussed in the previous sections of this book); and

b the experiences of Central-Eastern European countries in coping with shrinkage may serve as good (or bad) practices for the rest of postsocialist Europe.

That said, the countries of South-Eastern Europe seem worthy of at least a short mention in order to indicate a few "untouched" research areas on urban shrinkage, such as the role of the so-called urbicide or illegal suburbanization. A more detailed analysis of these countries would demand further in-depth studies and probably warrant a separate book that accounts for, among other things, the enormous differences between particular states in this region.

The section is structured as follows. After a general overview and comparison of urban shrinkage in Central-Eastern and South-Eastern Europe (Chapter 16), subsequent chapters present the "model cases" of old industrial shrinking cities in Poland (Wałbrzych—Chapter 17), Slovakia (Banská Štiavnica—Chapter 18), and Czechia (Ostrava—Chapter 19). Within wider national contexts, these chapters present the cities' disparate development pathways, determinants of shrinkage, and policies dealing with shrinkage. The authors of Chapter 18 compare approaches for coping with shrinkage in the old mining city of Banská Štiavnica with a relatively young city, Prievidza, which developed rapidly during the period of intensive socialist industrialization after World War II. Chapter 20 continues the discussion on the present-day legacy of Soviet-type industrialization by way of a case study of the Lithuanian city of Šiauliai. Here not only the collapse of industry but also the altered geopolitical position of the city affects the process of shrinkage. The location of all cities under discussion is shown in Figure 15.1.

To conclude, this section offers insight into the main drivers, pathways of change, consequences, and policy responses to urban shrinkage in

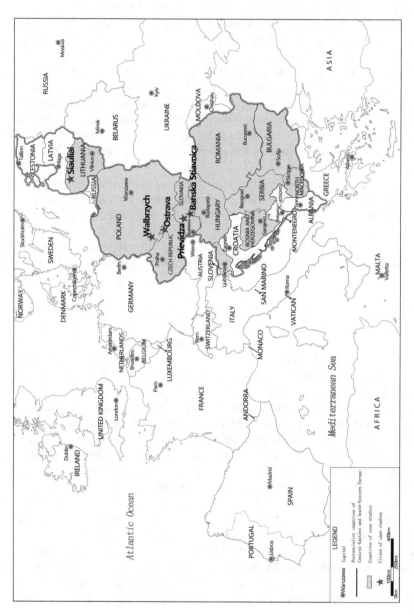

Figure 15.1 Location of case studies in postsocialist Europe.

postsocialist Europe. Whereas scientific evidence and governance of this phenomenon in the countries of Central-Eastern Europe now seem relatively "mature", the challenges of urban shrinkage in South-Eastern Europe demand further investigation (especially at the local level) and greater interest of policymakers.

Notes

1 Disregarding semantic discussions, in this section, the term "postsocialist" is used interchangeably with "postcommunist".
2 COST—European Program of Cooperation in Science and Technology.
3 The distinctive features of postsocialist cities are discussed in Sailer-Fliege (1999), Stanilov (2007), Sykora (2009), and Stryjakiewicz, Ciesiółka and Jaroszewska (2012).

References

Bernt, M., Cocks, M., Couch, C., Grossmann, K., Haase A. & Rink D. 2012. *Policy response, governance and Future Directions* (Shrink Smart Research Brief No. 2). Leipzig: Helmholtz Centre for Environmental Research—UFZ.

Bontje, M. 2004. Facing the Challenge of Shrinking Cities in East Germany: The Case of Leipzig. *GeoJournal*, 61(1), 13–21.

Couch, C., Cocks, M., Bernt, M., Grossmann, K., Haase, A. & Rink D. 2012. Shrinking Cities in Europe. *Town & Country Planning*, 81(6), 264–270.

Haase, A., Bernt, M., Grossmann, K., Mykhnenko, V. & Rink, D. 2013. Varieties of Shrinkage in European Cities. *European Urban and Regional Studies*, 12, 1–17.

Krastew, I. 2020. Czym grozi wyludnianie się Europy Wschodniej? (What is the risk of depopulation in Europe?), www.wyborcza.pl (Accessed 08 February 2020).

Martinez-Fernandez, C., Weyman, T., Fol, S., Audirac, I., Cunningham-Sabot, E. & Wiechmann, T., et al. 2016. Shrinking Cities in Australia, Japan, Europe and the USA: From a Global Process to Local Policy Responses. *Progress in Planning*, 105, 1–48.

Sailer-Fliege, U. 1999. Characteristics of Postsocialist Urban Transformation in East Central Europe. *GeoJournal*, 49(1), 7–16.

Stanilov, K. (ed.) 2007. *The Postsocialist City: Urban form and Space Transformations in Central and Eastern Europe after Socialism*. Dordrecht: Springer.

Stryjakiewicz, T. 2013. The Process of Urban Shrinkage and its Consequences. *Romanian Journal of Regional Science*, Special Issue on New Urban World, 7, 29–40.

Stryjakiewicz, T., Ciesiółka, P. & Jaroszewska, E. 2012. Urban Shrinkage and the Postsocialist Transformation: The Case of Poland. *Built Environment*, 38(2), 197–213.

Sykora, L. 2009. Postsocialist Cities. In Kitchin, R. & Thrift, N. (eds.) *International Encyclopedia of Human Geography* 8, 387–395.

Tsenkova, S. & Nedović-Budić, Z. (eds.) 2006. *The Urban Mosaic of Postsocialist Europe: Space, Institutions and Policy*. Springer Science & Business Media.

UN World Cities Report. 2018. The World's Cities in 2018. Data Booklet. htpp://un.org./en/the_world_cities_in_2018_data_booklet.pdf (Accessed 31 October 2020).

UN World Population Prospects. 2019. http://population.un.org/wpp (Accessed 31 October 2020).

Wiechmann, T. 2013. *Shrinking cities in Europe: Evidence from the COST Action Cities Regrowing Smaller (CIRES)*. Final Conference of the EU COST Action Cities Regrowing Smaller (CIRES), 12–13 September 2013. Essen. http://www.shrinking cities.eu.

Wolff, M. & Wiechmann, T. 2018. Urban Growth and Decline: Europe's Shrinking Cities in a Comparative Perspective 1990–2010. *European Urban and Regional Studies*, 25(2), 1–18.

16 Shrinking cities in postsocialist countries of Central-Eastern and South-Eastern Europe

A general and comparative overview

Tadeusz Stryjakiewicz

Introduction

The results of a comprehensive study on the trajectories of European cities by Turok and Mykhnenko (2007) led its authors to the conclusion that national distinctions seem to matter in explaining the patterns of urban shrinkage.[1] This is particularly true in relation to the postsocialist cities of Central-Eastern and, even more so, South-Eastern Europe. Postsocialist Europe is a "mosaic" of different pathways of urban development, including the process of shrinkage. This process is particularly visible since 1990, i.e., from the beginning of transformation of the political, social, and economic systems in this part of Europe. In some countries (e.g., Bulgaria, Romania, Bosnia and Hercegovina, the Baltic states), the scale of the process is so extensive that the majority of cities and towns can be called "shrinking" (in all dimensions of this term, especially demographic). This large scale of urban shrinkage is a great challenge for both research and policies. For many years, the issue of shrinkage was neglected in scientific and practically oriented debates in most countries of postsocialist Europe (excluding the former German Democratic Republic). Although this gap has been gradually bridged in recent years, further comprehensive studies on the topic are very much in order.

This chapter offers a brief general overview of the history and conditions of urban shrinkage as well as of the situation of postsocialist shrinking cities in selected countries of Central-Eastern Europe (Czechia, Hungary, Lithuania, Poland, Slovakia) and South-Eastern Europe (Bosnia and Herzegovina, Bulgaria, North Macedonia, Romania, Serbia, Slovenia).[2] It highlights both common and country-specific features of urban shrinkage. Finally, a comparison between the two parts of postsocialist Europe with regard to the scale and predominant determinants of urban shrinkage is presented. Beyond serving as a starting point for the in-depth case studies to follow in subsequent chapters, this chapter adds to the wider discussion of urban shrinkage, which, it is argued, is worth continuing and deepening in further investigations.

DOI: 10.4324/9780367815011-20

This chapter also underpins two points highlighted in our book, which can be summarized as follows:

a country-specific context matters in explaining the processes of urban shrinkage; and
b shrinking cities in postsocialist countries should not be fully compared to those of western countries in theory building (cf. Chapter 2).

At this point, it should be noted that the overview of the situation of shrinking cities presented later in this chapter is rather descriptive than explanatory, especially with regard to many postsocialist countries of South-Eastern Europe. This is due not only to a scarcity of in-depth studies on the subject but also to the low quality of data, in particular on demographic change (cf. Yudah 2019, 2020; in the present chapter, the latter issue is discussed using the case of North Macedonia). These limitations partly explain why a separate chapter on this region is absent from this book. Nevertheless, we wish to make a note of at least a few issues concerning urban shrinkage in South-Eastern Europe, which are underexplored and still worth "unpacking", such as so-called urbicide (in the case of Bosnia and Herzegovina) and shrinking illegal suburbia (in the case of Serbia).

A brief history and the conditions of urban shrinkage in postsocialist countries

A general overview of the process of urban shrinkage in Central-Eastern and South-Eastern Europe is not an easy task because the region encompasses a set of very different countries. These differences are related to the scale of particular countries, including their population (in particular, urban population), institutional arrangements (including their status as EU or non-EU members), availability of data, the extent of shrinkage, policies applied, etc. For instance, the object of our analysis ranges from a country like Poland, an EU member state with 38 million inhabitants and 829 cities and towns, to a country like North Macedonia, with just 2 million inhabitants and 40 cities and towns, and which only gained national independence at the end of 1991. As these differences make comparative analyses very difficult,[3] a mixture of detailing and generalizing approaches seems desirable.

The starting point of this study is an attempt at identification of common features that define urban processes in postsocialist Europe. Under the control of communist regimes after the Second World War, the 1950s and 1960s were a period of industry-based urban growth in most of these countries. The service sector started to gain prominence in the late 1960s (e.g., in Hungary and Serbia) or the early 1970s (e.g., in Poland), fueling further urban growth and the development of housing, mainly of a prefabricated multifamily type. This process followed similar developments taking place in Western Europe, though with a considerable time lag and qualitative differences. Almost immediately after the collapse of the communist system in 1990, radical economic restructuring took place. Once prioritized industries suddenly proved inefficient, declined, and gradually closed. The growth of services (e.g., communication, commerce, finance, law, administration, tourism, and recreation) partly made up for the employment gap. However, the emergence of locally based competitive, specialized, small-scale industries lagged behind,[4] as did the development of R&D sector. Consequently, deindustrialization became one of the most important determinants of shrinkage, and old industrial centers of mining, heavy and textile industries were particularly affected. Some examples of such shrinking cities are discussed in the following chapters.

Next to deindustrialization, suburbanization acted as another most important driver of urban shrinkage. In both Central-Eastern and South-Eastern Europe, suburbanization began at a larger scale as early as the 1980s (i.e., before the systemic transformation), especially around capital cities and some supra-regional centers (such as Brno in the Czech Republic, Győr in Hungary, and Krakow or Poznań in Poland), before intensifying during the postsocialist period. With time, suburbanization became a major contributor to the shrinkage of core cities. Its main causes include the pursuit of a higher quality of life, changing residential preferences, and rapidly growing differences in land and rental prices between the core city and the suburbs. However, suburbanization does not imply that the entire urban agglomeration is shrinking given that nearby towns and villages in the vicinity may actually increase in population and/or develop in economic terms (because of convenient location conditions for business). Moreover, postsocialist suburbanization generally does not entail physical degradation of city cores, abandoned buildings, brownfield sites, mass demolitions, or the formation of "perforated cities" typical of many shrinking cities in North America (cf. Jessen 2012). Lastly, it is worth noting that the notion of "postsocialist suburbanization" may be misleading in certain contexts, such as that of the Balkan countries, where the process of suburbanization, apart from its typical form associated with urban sprawl (i.e., growing suburbs "at the cost" of city cores), also includes the development of illegal suburban settlements with substandard housing conditions (Petrovic 2005). But unlike in Third World countries, such illegal settlements are shrinking together with city cores (see Antonić and Dukić 2018). Hence, patterns of suburbanization may differ significantly between individual postsocialist countries (contrasting examples are Serbia and Slovenia). These context-specific differences must be kept in mind while comparing shrinking cities.

Demographic trends are the next driver of shrinkage. They are similar in all postsocialist countries, despite occurring at different times. The main problem has been a dramatic decline in fertility rates, often termed a "demographic shock", which generally began earlier in the wealthier countries of the region. Combined with processes of suburbanization and migration, natural population decline has acted as a catalyst for the socioeconomic degradation of many shrinking cities.

International outmigration, which is one more cause of urban shrinkage in Central-Eastern and South-Eastern Europe, intensified significantly after opening the borders and labor markets of the EU to new member states, primarily owing to a sharp rise in economic migrants. However, in some countries of the Balkan Peninsula, such as Bosnia and Herzegovina or Serbia, the principal reason for migration was quite different; namely, it was frequently forced by the series of wars over the breakup of former Yugoslavia, which, in many cities, resulted in the destruction of dwellings, workplaces and physical infrastructure, the disintegration of social ties, and the intensification of ethnic conflicts.

Hence, apart from the aforementioned similarities, there are also many specificities (or national distinctions) behind the process of urban shrinkage in particular postsocialist countries. The following section analyzes these contextual differences with reference to the earlier subdivision of postsocialist Europe into two groups of countries, i.e., Central-Eastern and South-Eastern Europe. Special attention is paid to those countries and issues which are not discussed in the subsequent chapters, and the countries are described in alphabetical order.

Urban shrinkage in Central-Eastern Europe

In **Czechia** (formerly part of Czechoslovakia), shrinking cities are not a new phenomenon. In the postwar period of the 20th century, the growth of some cities was curbed due to population limits imposed by the communist authorities. The capital city of Prague could only be inhabited by 1,000,000 inhabitants and smaller towns had their limits, too. In the 1980s, a centralist system of settlement structure was introduced, under which some cities were intentionally stimulated while others were shrinking.

A new wave of shrinking cities began in Czechia in the late 1980s and early 1990s. The timing of changes in the urban structure was closely connected with changes in the economy. Former mining regions, such as the Ustecky and Moravskoslezsky districts, were most affected. The Czech government designed special development programs for those regions to prevent a deterioration of their situation, above all by advancing new employment opportunities (for details, see Rumpel and Slach 2014).

The process of gradual shrinkage can also be observed in cities that are not structurally weak. This mostly concerns medium-sized cities with populations between 100,000 and 500,000, including Brno, Olomouc, Pardubice, and Hradec Kralove. Relatively prosperous, these cities began shrinking because of demographic trends and outmigration to Prague and suburban areas (cf. Cermak 2005). This process was likewise observed in small towns located in peripheral regions. In some cases, former mercantile towns are now just larger villages since having lost their urban functions (Schmeidler 2009).

The population of **Hungary** has been declining since 1985, mainly due to negative demographic change (for details, see Tomay 2009, Pirisi and

Trócsanyi 2014). In this process of overall depopulation, a gradual spatial restructuring is taking place. Population growth is recorded in the central, urbanized region around the capital, while the population of the north-eastern as well as south-eastern and south-western parts of the country continues to decline. Throughout the 1980s, but especially after 1985, there were marked suburbanization processes around Budapest and other major cities, particularly in the west (e.g., Győr and Veszprem). While population growth in suburban settlements around Hungary's major cities had already arrived as a conspicuous trend in the 1960s, in those years, it was rather driven by an inflow of residents from rural regions.

Budapest has experienced a steady population decline since the beginning of the 1980s, though after 2000, depopulation slowed, and since 2008 there has even been slight growth. The capital city derives the greatest benefit from international migration. The urbanized belt around the city is growing steadily, and its territory is expanding, too. The core city is affected by several contradictory trends of growth and decline, prosperity and poverty, decay and regeneration (cf. Kovács 2006).

After the capital, with a population of 1,750,000 (and over 3,000,000 in the Budapest region), there is a considerable gap in Hungary's urban hierarchy. The next eight largest cities, each with a population of between 100,000 and 250,000 inhabitants, are similarly surrounded by suburban belts, usually undergoing growth, while their cores, after a period of steady population growth in the 1990s, have since fluctuated according to factors such as migratory dynamics, the economic health of local industries and flows of foreign direct investment (FDI).

As in other postsocialist countries, the most drastic examples of shrinkage are found in former heavy industry cities. One can distinguish two typical paths of population development using the cases of Miskolc and Győr (Figure 16.1). The former represents a heavy industrial city, where economic crisis and population decline started in the 1980s, and the process of shrinkage persists in spite of positive economic developments. The latter, conveniently located between Budapest and Vienna, is a multifunctional, export-oriented city with ties to the automobile sector as well as many R&D institutions and relatively high levels of FDI. Here, a period of population growth (briefer than in Miskolc) was followed by a period of population stabilization. The above two cities thereby serve as "model cases", illustrating the process of demographic shrinkage both in Hungary and in the entire region of Central-Eastern Europe.

Lithuania, located along the eastern border of the EU, is representative of post-Soviet Baltic states. It shares most characteristics of urban shrinkage with other postsocialist countries, i.e., low birth rates, population aging, high international outmigration, or deindustrialization. There are, however, some differences. Firstly, the density of urban settlements is smaller than in other countries of Central-Eastern Europe. Secondly, the number of shrinking cities and the scale of population decline is, relatively speaking, much

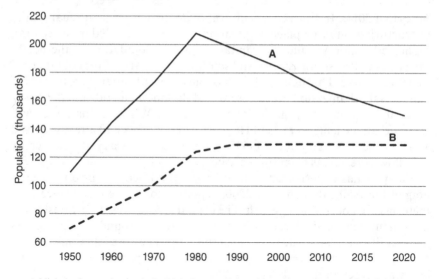

A Miskolc: former heavy industrial city – peripheral location – permanent shrinkage
B Győr: multifunctional city – convenient location for investments – population stabilized

Figure 16.1 Typical pathways of population development in different types of cities.

greater. Thirdly, the legacy of Soviet-type urban and economic development was more pronounced as, e.g., in Czechia, Hungary, or Poland, especially at the beginning of the postsocialist transformation. Some of its remnants are still present in the Lithuanian governance system (e.g., strong state control over publicly owned land). The collapse of socialism at the beginning of the 1990s was a shock for Lithuanian cities and towns. The former Soviet concepts of multimodal urban network (assuming growth limitations for major cities, in particular, Vilnius) became invalid and polarization trends emerged rapidly. This polarization is evidenced by the divergent development paths of three urban centers, including their suburbs (Vilnius, Kaunas, and Klaipeda), and the quick demographic and economic decline of peripheries (Ubarevičienė 2018). Small towns and second-tier cities (such as Šiauliai, Panevėžys, or Alytus) are particularly affected; the latter lost about one-third of their population during the postsocialist transformation period. The case of Šiauliai in this book illustrates how rapidly urban shrinkage befell a once booming socialist city, the consequences of such dramatic change and the policies adopted to cope with them.

Likewise, after 1989, **Poland** experienced a radical acceleration of depopulation and demographic change in many areas as a result of the collapse of state-run agriculture and industrial plants, as well as the failure of economic restructuring in monofunctional industrial centers. The increasingly evident polarization of the socio-demographic structure manifests itself in the concentration of population in a few metropolitan areas and the depopulation of remaining ones. With the exception of old industrial districts, most metropolitan areas, deemed "the winners" of the postsocialist transformation, have managed to adjust their economies to the new market conditions fairly quickly. The process of urban shrinkage largely affects cities and towns from the group of "the losers". They grapple with such problems as industrial decline, high unemployment rates, and an outflow of the population (especially young and well educated). These are, in particular, old industrial cities with the predominance of heavy, extractive, and textile industries. Most are located in southern Poland (Upper and Lower Silesia). Out of the 39 Polish cities with more than 100,000 inhabitants, a substantial population drop (more than 5% over the years 1990–2015) was recorded in 13, of which 6 are former mining cities that recorded population declines of greater than 10%.

In turn, when analyzing the biggest cities with well developing metropolitan functions, from the very start of the systemic transformation, one can observe a dynamic suburbanization process. A closer inspection of five major Polish cities (Warsaw, Krakow, Łódź, Wrocław, and Poznań) indicates that only Warsaw and Krakow were still growing in the 2000s within their city limits (though at a low rate). Considered along with their suburbs, those urban regions, apart from Łódź, grew steadily between 1995 and 2015. However, the population pattern in four metropolitan areas (except Warsaw) was marked by consistently declining cores and growing suburban zones.

The most important determinants of shrinkage in **Slovakia** are demographic processes, economic transformation, patterns of housing development, and suburbanization. Among Slovakia's 40 cities with a population of over 20,000, only four recorded population growth between 1995 and 2015 (Bardejov, Snina, and Trebišov in eastern Slovakia, and Pezinok in the Bratislava hinterland). Despite this general downward trend of the urban population within this group of Slovak cities, no city experienced a population decline exceeding 10% from the years 1995 to 2015. There were, however, 11 cities, including Bratislava, which declined in population between 5% and 10% over the same time.

The situation in Bratislava, the capital of the country, is specific. Despite certain signs of population shrinkage, Bratislava is not a typical shrinking city. As Bleha and Buček (2009) explain, in a certain group of Slovak cities, it is demographic development and especially the age structure that determines the basic characteristics of urban development. Since 1995, a decline in fertility combined with a net outmigration has resulted in a total annual decrease of between 2 to 4‰. The most important factor was substantial "pioneer suburbanization" (in Slovak terms), which occurred after 1995. Most rural settlements located in the capital's hinterland gained significantly from migration flows originating in Bratislava. One can observe a reversal in this trend after 2005: the decrease in population number stopped and population dynamics revived. This seems to be the result of a moderate decline in suburbanization, an increase in housing construction within the city limits, and a rise in births by women born in the 1970s.

More severe cases of shrinkage can be found when we focus on smaller Slovak towns with fewer than 20,000 inhabitants. To some extent, they can be considered "victims" of the postsocialist transformation. This is especially true of industrial centers that were challenged with the transformation of their local economies, many of which lost a significant number of jobs and recorded population declines exceeding 10% between 1995 and 2015. Examples of such towns include Zlaté Moravce, Štúrovo, Nová Bana, Liptovský Hrádok, and Kremnica. The dynamics of depopulation in Cierna nad Tisou, a former railway hub located on the border with Ukraine (previously the former Soviet Union), are unique. Here, shrinkage is connected to the partial decline of the town's gateway function. While some of the aforementioned towns suffered the most difficult period of their development during the 1990s, others struggled with shrinkage after the turn of the millennium.

Urban shrinkage in South-Eastern Europe

The development of cities in **Bosnia and Herzegovina** (B&H),[5] analyzed within the European urban context, has been very untypical due to the war following the breakup of Yugoslavia (1992–1995). Warfare, particularly ethnic cleansing, caused massive migrations of the population within and outside the borders of the country. Forced and free migrations intended to bring about ethnic and territorial homogenization were dominant during as well as following the conflict. During the war, around 2.2 million people, or more than half of the total population, left their homes, and the emigration rate reached 38.9% (European Commission Delegation to B&H 2005). In addition, one of the lowest fertility rates in the world (1.26) was recorded.

The violent disturbances which the war caused can be classified into three groups: demographic, socioeconomic, and administrative-territorial. Demographic disturbances primarily refer to a large number of war-related casualties and fertility decline. As a result of its specific circumstances, a post-war stage of urban development, characterized by a brief phase of accelerated population growth, was observed in B&H. Socioeconomic disturbances encompass the devastation of economic capacities and infrastructure, stagnation of the transformation process, closure of companies due to poorly conducted privatization, low foreign investment, increased delinquency and corruption rates, loss of highly educated professionals, increased social tensions, and inter-ethnic intolerance; all of which contribute to the negative international perception of B&H. Administrative-territorial disturbances are a consequence of the country's territorial rearrangement, disintegration of its functional economic regions and urban system, insufficient and inefficient public policies, and the slow progress of reforms. All the above disturbances either directly or indirectly affect the characteristics and dynamics of the modern urban transformation in B&H.

However, in some cities of B&H (notably the capital Sarajevo and Mostar), the physical destruction of their built environment is of particular importance. Their war experiences rekindled the debate on the notion of "urbicide" (first coined by Michael Moorcock in 1963; see also Sego 1992; Bogdanovic 1993; Covard 2004, 2008), i.e., physical, structural and symbolic forms of "violence against the city". Thus, B&H is illustrative of a country where urbicide as well as a prolonged phase of post-war urban recovery are "atypical" causes of urban shrinkage.

In the late 1970s and 1980s, cities of B&H were characterized by growing concentrations of central urban facilities and continuous population growth fueled by migration from peripheral areas. This process was closely associated with economic modernization, primarily industrialization. In general, until 1990, the urban population had been growing and urban shrinkage was observed in only five economically deprived municipalities, constituting 0.98% of the total population (Bosanko, Grahovo, Gacko, Kalinovik, and Ljubinje). The period after 1990 is characterized by a slowed pace of urbanization and stabilization of the urban population (39.2% in 1990, 39.3% in 2000, 39.2% in 2010, and 40.6% in 2018).

Recent predictions based on official planning documents expect turbulent urban dynamics in the future. Some of the largest cities are now exhibiting diverse patterns. The inner-city of Sarajevo, the capital, has begun shrinking while its periphery is growing. The city of Tuzla shows a similar trajectory, but its periphery is unstable. Mostar and its periphery are shrinking. Only Banjaluka, a second-tier city in the country, is growing. The emergence of new development poles is also evident (e.g., Bihać).

The shrinkage of **Bulgarian** cities should be regarded within the context of the country's steady population decline since 1990. Both diminishing birth rates and considerable emigration after 1990 are effects of the political, social, and economic processes of the transition period. Based on 2011 census data, 67% of the population decline in the recent decade was due to negative natural growth and 33% to emigration (NSI 2011). According to EU statistics, Bulgaria has at present the worst rate of natural decrease at −0.79% per year (EUROSTAT 2019). Four of six Bulgarian regions (NUTS 2 level) are among the ten regions facing the highest rates of shrinkage in Europe.

As of 2018, about 75% of the country's population reside in urban areas. The capital of Sofia is the only city with over 1 million residents; Plovdiv and Varna both have populations of about 330,000, and four cities, Burgas, Ruse, Stara Zagora, and Pleven, are home to between 100,000 and 200,000 inhabitants. Population growth was recorded in the 2011 census in only four urban districts: Sofia, Varna, Burgas, and Veliko Tarnovo. The remaining 22 urban districts shrunk at various rates compared to the 2001 census results (NSI 2011). The population in the urban districts of Vratsa and

Vidin decreased by more than 20%, while Gabrovo, Kyustendil, Lovech, Montana, and Silistra recorded population declines of between 15 and 20% (NRDS 2012). In total, more than 90% of cities and towns are shrinking. Ultimately, the demographic crisis facing Bulgaria could have long-term consequences for the country's settlement network or result in deep structural changes to urbanized territories.

The analysis of urban shrinkage in **North Macedonia** is not an easy task and, due to a lack of data, results may be ambiguous. As Yudah observes, "official population statistics are not just a little off—they are dramatically incorrect" (2020, p. 1). According to the State Statistical Office, the population of North Macedonia was approximately 2.08 million as of 2018. At the same time, the Office Director admits, "there are no more than 1.5 million people in the country, but no one can prove it" (Yudah 2020, p. 1). This discrepancy (one-fourth of the total population) stems from the fact that the country's most recent census took place in 2002. While a new census was scheduled for 2020, it had to be postponed on account of the coronavirus pandemic. Therefore, North Macedonia still uses population figures from the 2002 census as a baseline for all other data as well as for official demographic reports. In the meantime, "hundreds of thousands inhabitants of the country have emigrated—but are not registered as having done so, and no one knows how many they are" (Yudah 2020, p. 2). Taking into consideration the additional concern of manipulated population data resulting from ethnic tensions, a realistic estimation of North Macedonia's total population as well as its scale of urban shrinkage seems impossible. Thus, while the latest official figures suggest the cities and towns of North Macedonia are not shrinking, this is evidently far from the reality (this explains why a question mark is used for this country in Table 16.1). In spite of the above reservations, the case of North Macedonia seems worth discussing if only to raise awareness of the challenges faced by researchers and decision-makers concerned with urban shrinkage in some postsocialist countries.

The CIRES Synopsis Report (2014) indicates that, in the 20th century, the process of shrinkage was evident only in smaller Macedonian settlements and villages and was viewed as an inevitable consequence of the country's industrialization (cf. Siljanoska et al. 2012). The first signs that shrinking cities had become a reality despite forecasts of constant, albeit reduced growth, were revealed by the census of 2002. It showed clearly that the process of shrinkage was not restricted to cities of a certain size or location but that it affected various levels of the settlement system. For example, cases of shrinking cities include Veles, one of the country's larger formerly industrial cities with a population of 44,000, and Krushevo, one of the smallest cities with 5,000 inhabitants. The decline, although seemingly insignificant, clearly shows that the country is facing a new phenomenon that has to be addressed by both academia and the central and local governments. Moreover, it is crucial to note that, apart from the general decline in total population, the 2002 census revealed a slight increase in several cities of North Macedonia's least developed region, i.e., the Eastern region. However, it is very likely that the next census will report population losses in these cities. According to Yudah (2020), the capital of Skopje is currently the only city with a stable number of inhabitants.

The dynamics of urban shrinkage in **Romania,** as with the other countries under discussion in this chapter, cannot be understood without a brief review of its specific political, demographic and economic context. As a result of changes following the Second World War, Romanian society was profoundly affected in its composition and spatial behavior. During the postsocialist period, globalization, the collapse of industrial activities as well as internal and outmigration strongly affected the structure and spatial organization of cities. At present, Romania tops the list of European countries with the greatest number of shrinking cities. This is not only a consequence of the postsocialist transformation but also a result of earlier migration and population policies. Notable population growth, by more than a million, took place between 1977 and 1992 (Census data 1992) due, on the one hand, to heavy restrictions on external migration, and on the other, to the prohibition of abortion and birth control. The results of this sudden growth of newborns were unfavorable, both in the short- and long-term (Lataianu 2001; Flister 2013). This "artificial" boost was immediately followed by a sharp decline, however, due in large part to increasing mortality rates and outmigration, especially toward the end of the 1990s. Between 1990 and 2006, Romania lost more than 1.4 million inhabitants (Ghetau 2007). While the first two years of the country's postsocialist transition were marked by the exodus of the German minority from Transylvania and their return to Germany, the following years recorded increasing rates of labor migration, which exploded once Romania became part of the EU in 2007. Thus, according to the census, the population of Romania in 2002 was similar to that of 1977. Since 2002, the population has continued to decrease, and the latest forecasts predict a further decline of four million by 2050 (ENEPRI 2007).

According to the 2018 EUROSTAT Urban audit, a population growth analysis performed on 357 representative European and Turkish cities, four of the ten most severely shrinking cities were located in Romania (Bacău, Piatra Neamt, Târgu Mures, and Sibiu). However, small cities with under 20,000 inhabitants, former monoindustrial cities, nor the Danube harbors are mentioned in this report, despite being even more acutely affected by depopulation. The monoindustrial cities were greatly affected by the processes of postsocialist economic restructuring, followed by growing unemployment and poverty. The new investors are reluctant to clean the old industrial sites and prefer to build new facilities in the periphery, which deepens the decline of industrial cities. Today, former industrial cities, particularly small and medium-sized, are the most severe cases of shrinkage in Romania. Examples include Valea Jiului, Oltenia, Bălan, Moldova Nouă, Zimnicea, Călan-Victoria, and Făgăraș (cf. Constantinescu 2012). Another group of small shrinking cities is tourist resorts with poor infrastructure and lack of investment (Vaineasa, Căciulata, Herculane, Slanic, and Târgu Ocna). Their situation has gradually changed since Romania joined the EU. According to Wolff and Wiechmann (2018) more than 70 per cent of Romanian cities are shrinking. Hence, urban shrinkage is a crucial issue for the present and future urban policies, albeit one that is not always explicitly articulated. State of the art on shrinking cities in Romania is comprehensively described by Constantinescu (2018) and Eva, Cehan and Lazăr (2021).

In **Serbia**, the process of urban shrinkage in demographic terms already began in some parts of the country in the early 20th century. After the First World War, it was observed in a few cities in the northern part of the Republic of Serbia (Vojvodina), while Belgrade (the capital of Serbia), Nish, Kragujevac, and Krusevac increased in population. After the Second World War, many cities lost their populations, but Belgrade, as the capital of the former Yugoslavia, soon experienced steady population growth (mostly because of the increase in migration inflows). Belgrade doubled its population in the first postwar decade. The growth of population in settlements around Belgrade, i.e., suburbanization, was a trend since the late 1970s (earlier than in other cities of South-Eastern Europe). At the same time, the shrinkage process started in Serbian towns with population around 5,000.

A new wave of urban shrinkage in Serbia began in the early 1980s and has continued since. According to Antonić and Djukić (2018, p. 165), approximately 85% of 169 urban settlements declined in population between the last two censuses (2002–2011). The timing of changes in the urban structure is closely connected with those in the economy. The eastern and southeastern parts of Serbia, peripherally located, with obsolete, non-competitive economic base, have been the hardest hit: 19 cities from those regions have been shrinking since the 1980s (16 cities with a population of 5,000–10,000 and 3 with 15,000–50,000 inhabitants). On the other hand, relatively prosperous small peripheral cities in the north keep losing their inhabitants; not only because of demographic trends but also because of outmigration to Novi Sad and Belgrade.

Belgrade (especially its central districts) has been experiencing depopulation since 1990, although the trends are changing (now becoming more positive). It is hard to register population changes because the number of its inhabitants is very flexible. For instance, it is estimated that, in the 1990s, 200,000 war refugees settled in Belgrade, yet at the same time, thousands of young people left the city (there is no way of knowing if they did so for good or temporarily). Informal spatial development is another phenomenon worth emphasizing. It manifests itself, among other things, by illegal residential settlements (like slums) at the fringe of major urban centers, including Belgrade. This type of postsocialist suburbanization seems similar to that of the Third World. At present, shrinking illegal suburbs are one of the most acute problems of urban governance (cf. Antonić, Djukić 2018). However, Belgrade is still the largest city in Serbia and one of the biggest in the Balkan region with its 1.7 million inhabitants. There is a gap between Belgrade and three cities belonging to the next size group (Novi Sad, Nish, Kragujevac): Belgrade is 7 to 10 times more populous. Those cities are not shrinking because their population structures and diversified economies are favorable for development processes. Typological classification of Serbian cities has been recently performed by Djurkin, Antić and Budović (2021).

Slovenia is the most developed state among the postsocialist countries of South-Eastern Europe and the processes taking place here are the most similar to those in Western Europe. Slovenia witnessed accelerated urbanization in the 1960s and 1970s. Migrations from the countryside to cities/towns were substantial. After 1981, the growth of urban population stagnated to give way to the urbanization of suburban areas. Settling flows began to run in the opposite direction—from city centers to their edges and suburban villages. Already at the beginning of the 1990s, a third of the Slovenian population lived in suburban areas. For these areas, dispersed growth was a major characteristic. One can speak about a spatial reorganization of the population in the largest cities. The main demographic problems of these cities are a slowly decreasing population number, low fertility, and population aging.

The largest Slovenian cities, such as the capital city of Ljubljana, as well as Maribor, Celje, Koper, Kranj, Novo Mesto, and Nova Gorica, have been engines of robust economic, social and cultural development. However, as mentioned above, in recent decades, they have seen rapid development of urbanization on their outskirts. Consequently, these cities have been supplemented by suburb settlements forming a sort of new sub-centers. Several spatial and sociological studies (e.g., Robernik 2004, Sasek Divjak 2007) indicate an absence of a very high degree of urbanization in Slovenia in favor of suburbanization, which is illustrated by the fact that 63% of dwellings in Slovenia are family houses occupied by 1.3 million residents (Hočevar et al. 2005). The preference of Slovenes for living in houses of their own is reinforced by increasing mobility and the growth of individual standards of living in the last decade. The result has been a notable depopulation of city cores and an increase in the number of inhabitants in the suburbs, particularly where natural conditions are good and transport infrastructure well developed.

The changes and development of Slovenian cities represent a process of transition from a compact city toward a "regional" city. Since the 1980s, the number of city dwellers has been declining; this trend has not stopped. In the mid-1980s, the share of urban and rural population leveled up, and since 2010 "rural" population (living mostly in suburbs) has prevailed. This means that suburbanization is a driving force of urban shrinkage in this country.

Comparison and conclusions

The results of the above overview confirm the findings of Turok and Mykhnenko (2007) that national distinctions matter in explaining the patterns of urban shrinkage in postsocialist Europe, especially South-Eastern

Europe, and furthermore that the influence of such country-specific contextual factors on shrinkage makes generalizations and comparisons very difficult. The aforementioned difficulty notwithstanding, an attempt has been made to compare the countries of the region under analysis from the point of view of the scale and predominant determinants of urban shrinkage. The results of this highly generalized comparison are shown in Table 16.1, and the following conclusions can be formulated.

1 Looking at the scale, spatial pattern, and determinants of the process of urban shrinkage, the group of postsocialist countries of Central-Eastern Europe is more homogenous than that of South-Eastern Europe.
2 In most of the countries under discussion, demographic processes and the postsocialist transformation (including rapid deindustrialization) are crucial determinants of urban shrinkage. However, in Slovenia as well as in the largest cities of Czechia, Hungary, Poland, Serbia, and Slovakia, suburbanization is becoming a more significant driver of shrinkage. Importantly, in postsocialist Europe, unlike in the United States, suburbanization is associated primarily with demographic shrinkage of the main city (within its administrative boundaries) and residential spillover, and not so much with physical destruction of the built environment in the city core. In any case, while discussing shrinkage, it is important to distinguish between cities that are shrinking because of suburbanization and those which are shrinking due to the loss of their economic base.
3 International outmigration affects cities in both Central-Eastern and South-Eastern Europe. Immigration, although increasing in some regions, does not compensate for population loss.
4 Urbicide connected with war destruction is a specific cause of shrinkage in some Balkan countries, and particularly in Bosnia and Herzegovina.

It is striking that, in spite of its large scale and observed consequences, perceptions of urban shrinkage in postsocialist Europe remained vague and of minor importance for many years. While the phenomenon has been a subject of debate among small groups of experts and academics, it has attracted little attention from political circles or the general public. Politicians, mayors, and other leaders or local communities are especially reluctant to discuss shrinkage. They do not like the very notion of a "shrinking city" for two reasons: (1) the term is difficult to translate in many Slavonic languages; and (2) it sounds negative from a PR point of view. Thus, dominant themes are usually economic prosperity and possible constant growth of "their" cities. In North Macedonia, for instance, the central and local governments, even when confronted with statistical data that show the beginning of the tendency of shrinkage, choose to view it as a temporary setback on the road to growth, both in territorial as well as economic and social terms. One of the consequences of the poor perception of shrinkage is that urban planners are claimed to be ill-prepared for solving the problems that cities are confronted with (CIRES Synopsis Report 2014). In recent years, however, the awareness of central and municipal authorities has begun to change. Some of them have already taken measures to cope with the negative effects of the shrinkage process and have started looking for new development strategies (usually mitigation strategies) and forms of governance in the conditions of shrinkage (cf. Stryjakiewicz and Jaroszewska 2016). Such a change in policies oriented toward shrinking cities is particularly visible in the countries of Central-Eastern Europe, which are EU members since 2004.

A few common issues with managing urban shrinkage can be identified in postsocialist countries of Central-Eastern and South-Eastern Europe.

1 Dealing with demographic decline (e.g., incentives, such as allowances for the birth of a child, have proven unsuccessful so far).
2 Compensation for massive international outmigration: as Yudah (2019, p. 3) states, "only Poland has managed to significantly compensate large-scale emigration and low birth rates with the fortuitous immigration of more than a million Ukrainians").
3 Coping with negative effects of deindustrialization: there are both good and bad experiences related to economic, social, and spatial restructuring (some of which are presented in the following chapters); there are also emerging attempts at reindustrialization (however, it is too early to assess their results).
4 Dealing with "wild", uncontrolled suburbanization.
5 Integration of the national and local levels of governance (the latter seems too weak).
6 Necessity to supplement Western approaches to urban shrinkage with country-specific ones.

Although national specificities largely impede efforts to make generalizations about urban shrinkage, they also present an opportunity to merge the debate on shrinkage in the postsocialist realm with the broader, global discussion. The cross-national overview presented in this chapter confirms that urban shrinkage in the postsocialist realm cannot be seen as a process of convergence or catching up with western structures and processes. It has its own mechanisms and dynamics, still strongly rooted in the past. If we use one of the basic notions of the path dependence concept,[6] both the forced introduction of the state socialist system and its eventual demise represented critical junctures along the development paths of East-Central and South-Eastern European countries and cities. They constituted general preconditions (which I call "macro tracks") which took different forms in individual national and local contexts. One should agree with Haase et al. (2016b) that there is no single model of a postsocialist shrinking city (just as there is no single model of capitalism or socialism). What we are dealing with in this book is, in fact, "a mixture" of varieties of socialism (legacies of the past) and varieties of capitalism (the rapidly changing present) with

Table 16.1 Comparison of the scale and main determinants of urban shrinkage in selected postsocialist countries of Central-Eastern and South-Eastern Europe

Country	Population 2020 (millions)	Scale of shrinkage (in terms of population loss)*			Main determinants of urban shrinkage				
		Large	Moderate	Small	Demographic processes	Suburba-nization	Foreign outmigration	Postsocialist transformation (including deindustrialization)	Urbicide (war destruction)
Central-Eastern Europe									
Czechia	10.7		x		x	x		x	
Hungary	9.7		x		x	x		x	
Lithuania	2.8	x			x	x	x	x	
Poland	37.8			x	x	x	x	x	
Slovakia	5.5			x	x	x		x	
South-Eastern Europe									
Bosnia and Herzegovina	3.3	x			x		x	x	x
Bulgaria	6.9	x			x		x	x	
North Macedonia	2.1		?		x		x	x	
Romania	19.2	x				x	x	x	
Serbia	8.7	x			x	x	x	x	x
Slovenia	2.1		x		x			x	

Notes
* Scale of shrinkage:
Large: more than two-thirds of the total number of cities are shrinking.
Moderate: the share of shrinking cities is between 33.3 and 66.6 per cent.
Small: less than one-third of cities are shrinking.

Source: Own elaboration based on CIRES Synopsis Report (2014), Wolff and Wiechmann (2018), updated and extended, https://www.worldometers.info/world-population/; date of access: 20 October 2020.

varieties of urban shrinkage. The legacies of socialism are visible, e.g., in the structure and ownership of housing, social cultures (deficiency of entrepreneurship, the role of trade unions in state-run firms), governance systems and planning practices (weak power of local self-governments, failures in multilevel governance and participatory planning). The effects of new, capitalist development paths can be seen, among other things, in quickly growing spatial and social polarization, transformation of economic structures (privatization, deindustrialization, take-off of tertiary sector), labor markets (including unemployment) and housing markets (reduction of state rental sector, gentrification, emerging "gated communities"), openness to FDIs and supra-national institutions, free population movement, changing patterns of social reproduction (low fertility rate, population aging), and consumption and living preferences (suburbs). The interplay of the elements of "the old" and "the new" justifies the use of the label "postsocialist", in spite of some recent doubts expressed in the literature (cf. Müller 2019).

Varieties of urban shrinkage, summarized in this chapter at the national level and detailed in subsequent chapters at the local level, fall under two predominant types:

a shrinkage driven by deindustrialization of traditional (mining, metallurgy, or textile) industrial centers; and
b shrinkage driven by suburbanization.

While the former type encompasses all symptoms of shrinkage (demographic, economic, social, and spatial), the latter denotes primarily demographic decline in city cores alongside the simultaneous expansion of population, housing, and some economic activities in suburban areas. These two types of shrinkage demand different policies (so does the subtype of illegal suburbanization, as discussed above using the example of Serbia).

The experience of East-Central and South-Eastern Europe demonstrates the crucial role of institutional settings with regard to the scale, scope, and speed of urban shrinkage. This confirms the conclusion of Haase et al. (2016b, p. 307) that "both locally and nationally, policies and decision-making after the beginning of transition have proved to be crucial for the cities' trajectories". Of particular importance are state interventions, such as the creation of special economic zones, tax incentives, or national programs for shrinking cities and regions enabled to softer overcome a "shock" of deindustrialization at the beginning of postsocialist transformation (concrete examples are discussed in the following three chapters). On the other hand, national regulations and centralization of decision-making may hinder or slow down the implementation of some local actions or policies which are aimed at coping with shrinkage (e.g., city governments in Lithuania cannot dispose of publicly owned land without permission from central authorities).

However, the most important factor differentiating the trajectory and management of urban shrinkage in postsocialist Europe seems to be EU membership. The comparative national overview presented in this chapter suggests that the first postsocialist countries to become EU member states (since 2004) are far ahead of non-members. EU membership is one of the main reasons, together with legacies of the past, for the divergent development pathways of postsocialist shrinking cities, which are most pronounced in the neighboring ex-Yugoslavian countries (e.g., Slovenia vs. Bosnia and Herzegovina). The greatest positive impact of the EU relates to the exchange of experience and good planning practices, the development of a more stable institutional environment (which facilitates, e.g., attracting FDIs), and, last but not least, the financial support for shrinking cities under the EU's Cohesion Policy (which may be directed to the restructuring of urban economies or urban renewal projects). Moreover, in contrast to the priorities of some national and local governments, EU institutions put a lot of emphasis on sustainability, i.e., balancing economic, social, and environmental goals of policies to combat shrinkage and its effects.

Two common features characterizing the process of urban shrinkage stand out in all countries of postsocialist Europe, although at different scales: demography (low fertility rates, population aging) and massive international outmigration. With the exception of Poland, programs aiming to bolster populations by attracting immigrants have been largely inefficient to date. The scale of international outmigration significantly differentiates shrinking cities in East-Central and South-Eastern Europe not only from those in Western Europe but also from Russia and China, where outmigration is largely domestic. As far as perception, public discourse, and policymaking are concerned, one final common feature of shrinking cities in postsocialist Europe can be identified: urban shrinkage is very rarely perceived as a window of opportunity for the qualitative improvement of urban development, and, despite the writing on the wall, the mindset of growth-oriented planning continues to reign supreme.

Notes

1 A similar conclusion can be drawn from the study by Haase et al. (2016a).
2 This chapter is partly based on the unpublished synopsis report of the CIRES project (CIRES Synopsis Report 2014), which, however, has been thoroughly updated and adjusted to meet the requirements of the book and to offer the reader the most recent insight on the topic. The report was elaborated by the author of the present chapter on the basis of contributions by Nihad H. Čengić, Mirza Emirhafizović, (Bosnia and Herzegovina), Elena Dimitrova (Bulgaria), Karel Schmeidler (Czech Republic), Erzsebet Vajdovich Visy (Hungary), Vlatko P. Korobar, Jasmina Siljanoska (North Macedonia), Tadeusz Stryjakiewicz, Emilia Jaroszewska (Poland), Ilinca Păun Contantinescu (Romania), Aleksandra Krstić-Furundžić, Aleksandra Djukić (Serbia), Barbara Golicnik, and Mojca Šašek-Divjak (Slovenia).
3 One more problem making a cross-national comparative perspective difficult is mentioned by Martinez-Fernandez et al. (2016, p. 17). This is the lack of a standardized database at the local level, linked, among other things, to "the absence of a common definition of a city".

4 There were some exceptions, mainly in Poland. However, the development of small and medium-sized enterprises was limited in the regions with high intensity of urban shrinkage.
5 The sovereignty of this federal country was officially declared in 1992.
6 The foundations of this concept are presented in Arthur (1994) and Mahoney (2000), and new path dependencies in the context of postsocialist transformation are discussed in Sykora (2008).

References

Antonić, B. & Djukić, A. 2018. The Phenomenon of Shrinking Illegal Suburbs in Serbia: Can the Concept of Shrinking Cities be Useful for their Upgrading? *Habitat International*, 75, 161–170.

Arthur, B. 1994. *Increasing Returns and Path Dependence in the Economy*. Ann Arbor: University of Michigan Press.

Bleha, B. & Buček, J. 2010. Theoretical issues of local population and social policy in "shrinking" cities—some findings from Bratislava. In Kovács, Z. (ed.) *Challenges of Ageing in Villages and Cities: The Central European Experience* (pp. 110–131). Department of Economic and Social Geography, University of Szeged.

Bogdanovic, B. 1993. *Die Stadt und der Tod (City and Death)*. Klagenfurt-Salzburg: Wieser Verlag.

Cermak, Z. 2005. Migration and Suburbanization Processes in the Czech Republic. *Demography*, 47(7), 169–176.

CIRES Synopsis Report. 2014. *East-Central and South-Eastern Europe: Synopsis Report*. TU Dortmund & AMU University (typescript).

Constantinescu, I. P. 2012. Shrinking Cities in Romania: Former Mining Cities in Valea Jului. *Built Environment*, 38(2), 214–228.

Constantinescu, I. P. 2018. *Shrinking Cities in Romania: Research and Interventions*. DOM Publishers.

Covard, M. 2004. Urbicide in Bosnia. In Graham, S. (ed.) *Cities, War and Terrorism: Towards an Urban Geopolitics* (pp. 154–171). Blackwell.

Covard, M. 2008. Urbicide: The Politics of Urban Destruction. *Global Discourse*, 1(2), 186–189.

Djurkin, D., Antić, M. & Budović, A. 2021. Demographic and Economic Aspects of Urban Shrinkage in Serbia: Typology and Regional Differentiation. *Bulletin of the Serbian Geographical Society*, 101(2), 43–78.

ENEPRI. 2007. Balkandide. Research Report No. 40.

European Commission Delegation to B&H. 2005. Funkcionalni pregled sektora povratka u B&H. http://www.ceps.eu/ceps/download/1434 (Accessed 12 March 2012).

EUROSTAT. 2019. Database by themes. https://ec.europa/eu/Eurostat/data/database (Accessed 25 October 2020).

Eva, M., Cehan, A. & Lazăr, A. 2021. Patterns of Urban Shrinkage: A Systematic Analysis of Romanian Cities. *Sustainability*, 13(13), 7514.

Flister, L. D. 2013. Socioeconomic Consequences of Romania's Abortion Ban Under Causescu's Regime. *Journal of Alternative Perspectives in the Social Sciences*, 5, 294–322.

Gheatau, V. 2007. *Declinul demographic si viitorul populatiei Romaniei (Demographic Decline and the Future of Romania's Population)*. Centre of Demographic Research Vladimir Trebici. Alpha MDN, Bucharest.

Haase, A., Bernt, M., Großmann, K., Mykhnenko, V. & Rink, D. 2016a. Varieties of Shrinkage in European Cities. *European Urban and Regional Studies*, 23(1), 86–102.

Haase, A., Rink, D. & Großmann, K. 2016b. Shrinking cities in post-socialist Europe: What can we learn from their analysis for theory building today? *Geografiska Annaler, Series B Human Geography*, 98(4), 305–319.

Hočevar, M. M., Uršič, D. K. & Trček, F. 2005. Changing of the Slovene urban system: Specific socio-spatial trends and antiurban public values/attitudes. In Eckardt, F. (ed.) *Paths of Urban Transformation* (pp. 281–300). Frankfurt am Main: Peter Lang.

Jessen, J. 2012. Conceptualizing shrinking cities—a challenge for planning theory. In Piro, R. (ed.) *Parallel Patterns of Shrinking Cities and Urban Growth: Spatial Planning for Sustainable Development of City Regions and Rural Areas* (pp. 45–54), Routledge.

Kovács, Z. 2006. *Population and Housing Dynamics in Budapest Metropolitan Region after 1990*. Budapest: Geographical Research Institute of the Hungarian Academy of Sciences.

Lataianu, M. 2001. The 1966 law concerning the prohibition of abortion in Romania and its consequences: The fate of one generation. Conference Proceedings, 1–13.

Mahoney, J. 2000. Path Dependence in Historical Sociology. *Theory and Society*, 29, 507–548.

Martinez-Fernandez, C., Weyman, T., Fol, S., Audirac, I., Cunningham-Sabot, E. & Wiechmann, T., et al. 2016. Shrinking Cities in Australia, Japan, Europe and the USA: From a Global Process to Local Policy Responses. *Progress in Planning*, 105(3), 1–48.

Moorcock, M. 1963. Dead God's homecoming. *Science Fantasy*, 59.

Müller, M. 2019. Goodbye, Postsocialism!. *Europe-Asia Studies*, 71(4), 533–550.

NRDS. 2012. National Regional Development Strategy of the Republic of Bulgaria for the Period 2012–2022. *Ministry of Regional Development and Public Works*. Sofia. http://www.mrrb.government.bg (Accessed 2 September 2020).

NSI. 2011. National Statistical Institute of Bulgaria. Census 2011. Sofia. http://www.nsi.bg/census2011 (Accessed 25 October 2020).

Petrovic, M. 2005. *Cities after Socialism as a Research Issue: Discussion Papers (South East Europe series)*, 34, London School of Economics and Political Science.

Pirisi, G. & Trócsanyi, A. 2014. Shrinking Small Towns in Hungary: The Factors Behind the Urban Decline in "Small Scale". *Acta Geographica Universitatis Comeniane*, 58(2), 131–147.

Rebernik, D. 2004. *Recent Development of Slovene Towns: Social Structure and Transformation*. Ljubljana: Faculty of Arts, Department of Geography, University of Ljubljana.

Rumpel, P. & Slach, O. 2014. Shrinking cities in Central Europe. In Koutsky, J., Raška, P. Dostal, P. & Herrschel, T. (eds.) *Transitions in Regional Science – Regions in Transitons: Regional Research in Central Europe* (pp. 142–155). Prague: Wolters Kluwer.

Sasek Divjak, M. 2007. Planning for Sustainability in Slovenian Towns. *International Journal for Housing Science and its Applications*, 31(3), 205–214.

Schmeidler, K. 2009. Peripheral regions of European Union: Case study Czech Republic. In Ventura, P. & Tiboni M. (eds.) *Sustainable Developments Targets and Local Participation in Minor Deprived Communities: Sustainable Development Policies for Minor Deprived Urban Communities* (pp. 63–78). McGraw-Hill.

Sego, K. 1992. *Mostar'92 Urbicid*. Croatian Defense Council.

Siljanoska, J., Korobar, V. P. & Stefanovska, J. 2012. Causes, Consequences and Challenges of Shrinkage: The Case of Small Cities in a Transition Society. *Built Environment*, 38(2), 244–258.

Stryjakiewicz, T. & Jaroszewska, E. 2016. The Process of Shrinkage as a Challenge to Urban Governance. *Quaestiones Geographicae*, 35(2), 27–39.

Sykora, L. 2008. Revolutionary change, evolutionary adaptation and new path dependencies: Socialism, capitalism and transformations in urban spatial organisations. In Strubelt, W. & Gorzelak G. (eds.) *City and Region: Papers in Honour of Jiri Musil* (pp. 283–296). Opladen and Farmington Hills: Budrich Uni Press.

Tomay, K. 2009. *Demographic Challenges of the European and Hungarian Urban Areas*. Budapest: Falu Varos Regio.

Turok, I. & Mykhnenko, V. 2007. The Trajectories of European Cities, 1960–2005. *Cities*, 24(3), 165–182.

Ubarevičienė, R. 2018. City Systems in the Baltic States: The Soviet Legacy and Current Paths of Change. *Europa Regional*, 25(2), 15–29.

Wolff, M. & Wiechmann, T. 2018. Urban Growth and Decline: Europe's Shrinking Cities in a Comparative Perspective. *European Urban and Regional Studies*, 25(2), 122–139.

Yudah, T. 2019. *Bye-bye, Balkans: A region in critical demographic decline*. www.balkaninsight.com (Accessed 14 October 2019).

Yudah, T. 2020. *Wildly wrong: North Macedonia's population mystery*. www.balkaninsight.com (Accessed 14 May 2020).

17 Shrinking cities in Poland—recent trends of change and emerging policy responses

Tadeusz Stryjakiewicz and Emilia Jaroszewska

Introduction

The process of urban shrinkage can exhibit differing socioeconomic patterns depending on the spatial and temporal context. In each case, however, urban shrinkage leads to a decrease in the number of inhabitants. This process, already present in some cities (usually old industrial centers) of Western Europe and the United States (for details, see Pallagst, Fleschurz, Said 2017, Haase et al. 2017), has assumed especially large dimensions in postsocialist countries since 1990. These developments present new challenges for research and policy: the need, on the one hand, to identify the scale, rate, and forms of urban shrinkage and how these features differ spatially, and on the other, to reconsider the applicability of existing urban policies, which are usually designed to follow a growth paradigm and treat cities as "growth machines" (Logan, Molotch 1987).

In Central and Eastern Europe (CEE), the process of urban shrinkage has been influenced by great institutional changes that have significantly altered the situation of cities, diversifying them in terms of both demographic and economic development, which has been discussed in numerous studies (e.g., Bontje 2004; Franz 2004; Steinführer, Haase 2007; Großmann, Haase, Rink, Steinführer 2008; Wiechmann 2008; Wiechmann, Wolff 2013; Stryjakiewicz et al. 2012; Stryjakiewicz 2014; Wiechmann, Bontje 2015, Wolff, Wiechmann 2018). As follows from a comparative study of Europe's shrinking cities by Wolff and Wiechmann (2018), Poland belongs to the group of countries facing a moderate but growing scale of urban shrinkage.

Contrary to Western countries, contemporary problems stemming from shrinkage in Polish cities have largely occurred following the radical political and economic changes, or "shock therapy", which was declared in 1989 and implemented since 1990. New determinants of development, appearing at that time, contributed to the new division of cities into "winners" and "losers" of the postsocialist transformation (Parysek, Wdowicka 2002). While the beginning of the 1990s opened opportunities for dynamic development and growth of the former group, for the latter, it initiated the process of urban shrinkage. This was a direct outcome of the system

DOI: 10.4324/9780367815011-21

transformation, causing the fall of industrial plants and an economic crisis of entire cities. Hence, the mechanism of shrinkage consists of the combination of negative effects of the system transformation with the impact of worldwide processes, such as deindustrialization, globalization or demographic and behavioral changes.

However, apart from the above-mentioned general conditions of the process of shrinkage, the forms of its manifestation also depend strongly on local specificities, and in particular, on earlier development pathways of cities. Therefore, studies of the pattern and direction of evolution of social systems, the institutional context, as well as various events, choices, and decisions from the past, can provide deeper insight into the mechanism of shrinkage. This type of explanation relying on the interpretation of historical facts and employing a genetic approach is the essence of the concept known as path dependence (e.g., Arthur 1994; Boschma, Lambooy 1999; David 2001; Mahoney 2000; Gwosdz 2004, 2014). According to this approach, shrinkage can be understood as an outcome of a choice made at one time. It is then reinforced, reproduced, or transformed by successive chance events. As David (2000) claims, the concept combines isolated, unique occurrences with more general growth processes of dynamic structures. Thus, dealing with shrinkage requires a skillful combination of general rules of urban governance with "tailored" policies oriented to local specificities.

Taking the above into account, this chapter draws attention to the issues of the urban shrinkage process at national and local levels. The chapter is structured as follows: the first section presents the context of urban shrinkage in its demographic form as well as policy responses in Poland. The second offers a deeper chronological analysis of the most important post-World War II events, choices, and decisions, as well as their transformative effects on the Polish industrial shrinking city of Wałbrzych. Finally, future challenges and policy recommendations are formulated.

Urban shrinkage in Poland in its demographic dimension and policy responses

For three recent decades, Polish cities have been subject to many changes regarding material, functional and social aspects. The pace and direction of these changes were influenced by processes connected with political transformation, progressing internationalization of the economy, Poland's accession to the EU and new possibilities of financial support. In some cities, especially old industrial and small and medium-sized ones situated at peripheries, the changes contributed to the activation of negative processes leading to urban shrinkage in the long run.

Key drivers that have influenced the shrinkage of Polish cities include:

• demographic change (a decline in the birth rate, aging population);
• massive outmigration (intensified especially after the EU enlargement);

- transformation of the settlement system (metropolization, suburbanization); and,
- transformation of the economy (in particular its deindustrialization).

Demographic change is connected with the process known as the second demographic transition (Lesthaeghe, van de Kaa 1986). This process involves a unidirectional shift in many types of demographic behavior concerning marriage and replacement rates (Okólski 2005). Its effects include a change in lifestyle, the appearance of a new model of the family and household, a decline in birth rates, and an advancing aging of society. As Kurkiewicz (2010: 51) observes, the second demographic transition in Poland and other parts of CEE lags far behind Western European countries. The unfavorable demographic situation has its origins in the processes observed in the late 1980s and early 1990s. Since the beginning of the transformation, the downward trend in the number of births has been coupled with a simultaneous decrease in mortality and prolonged life expectancy.

Migration is another important component shaping local demographic structures. In particular, the departure of young people at reproductive age has serious consequences in the form of population structure disruptions, contributing to the acceleration of the aging process of a given community. Official statistics do not cover the entire migration process. This is due to incomplete registration and deregistration data, which do not record all events. Two main developments can be distinguished in internal migration, i.e., within Poland. First of all, there is a dominant shift toward a number of metropolitan areas, which are becoming places of concentration of the socioeconomic potential. This is paralleled by population deconcentration within these areas as a result of the suburbanization process. Secondly, the outflow of inhabitants from peripherally located small towns and rural areas (mainly eastern Poland) and from old industrial centers, observed for many years, continues. In turn, foreign outmigration increased, especially in the period after Poland's accession to the EU (an estimated two million people contributed to this outflow). Unfortunately, emigration is associated not only with quantitative but also qualitative shrinkage of population. The emigrants were mainly young people, often well-educated, which reduced local human capital resources.

Transformation of the settlement system is first of all linked to the metropolization process, i.e. a fast growth of metropolitan areas on the one hand and the suburbanization process on the other. The former process triggers a growing disproportions between cities of peripheral location and low competitive advantage and metropolitan areas (e.g., Warsaw, Poznań, Wrocław, Tricity, Szczecin, Krakow). The suburbanization process, in turn, triggered primarily by better conditions for the development of one-family housing in suburban zones, in the case of economically strong metropolitan centers (such as Poznań) leads to a loss of residents within their administrative boundaries to neighboring communes, functionally connected with a city (therefore we do not deal here with typical shrinkage). Although suburbanization causes a series of problems, the situation of the largest cities regarding the population number is often more favorable than available data indicates. Indeed, these cities are generally home to a substantial number of unregistered internal immigrants, including students and temporary workers, who constitute a significant group not mentioned in statistics. Unfortunately, the suburbanization process strengthens the scale of shrinkage in towns and former industrial cities, with no compensation of population outflow and a higher than in other cities natural population losses.

Economic transformation, caused mainly by the processes of deindustrialization and globalization, contributed to the economic crisis and, consequently, to the long-term shrinkage of old industrial cities, in particular the centers of coal mining and heavy industry (Bytom, Wałbrzych) as well as textile industry (Łódź). A return to the path of capitalist development following the period of the command economy was a particularly difficult experience for them. In the socialist period, due to industrial functions, these were the most privileged cities, to which the government development funds were directed more than to other cities. As a result of the postsocialist transformation, which began in 1990, this type of city had the greatest difficulty adapting to the new conditions. The problems resulted not only from the poor economic situation associated with the collapse of industrial plants and a dramatic increase in unemployment but also from the passive attitude of the inhabitants, who in socialist times commonly benefited from comprehensive social care (company flats, canteens, kindergartens, etc.) (Sagan 2000, p. 159). It was the inhabitants of these cities and, in particular, the low-skilled workers of closed-down industrial plants who became the biggest losers of the economic transformation (to be shown later in this chapter with a case study of Wałbrzych).

Forecasts for Poland's demographic future are alarming. According to Central Statistical Office (2014), the country can expect to face long-term population decline, falling from 38,411 million in 2018 to an estimated 33,951 million by 2050, with accelerating rates of depopulation over time. Important differences in demographic processes of urban and rural areas are clearly visible (Figure 17.1). However, it should be stressed that rural areas gain residents in particular as a result of the aforementioned suburbanization process. These are areas having the administrative status of a village, but in fact, they are part of the functional areas surrounding the largest cities.

According to Statistics Poland, the bulk of projected population decline will occur in the urban population (and more specifically, among people residing within the administrative boundaries of cities). By 2050, city dwellers will make up only 80% of their 2015 population. The decrease in urban residents will lead to unfavorable changes in age structure, including further aging of the population and declines in the number of women of childbearing age.

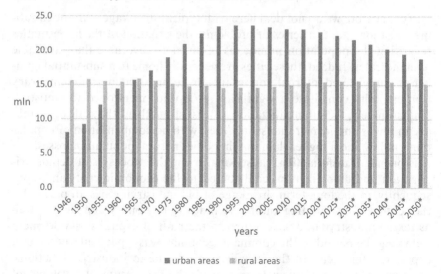

Figure 17.1 Change of urban and rural population in Poland, 1946–2050 (* forecast).

Source: Own compilation on the basis of Statistics Poland data.

Given that all of the above-mentioned processes are not evenly distributed in space, it is important to identify which cities have lost the most inhabitants. This study examined population changes in 829 Polish cities that existed in 1990, excluding the cities created after this time. The temporal scope of the analysis covers the years 1990–2015 as two and a half decades seem sufficient to illustrate the population changes which have occurred in Polish cities since the beginning of the systemic transformation.

The analysis of the annual population change over the above-mentioned period of 25 years was carried out on the basis of the methodology elaborated in the CIRES[1] project, see Wiechmann, Wolff 2013; Wolff and Wiechmann 2018) and uncovered the following types of cities: growing, stable and shrinking (Figure 17.2). Among the 829 cities, one can distinguish:

- 257 growing cities characterized by more than 0.15% of an annual increase in the population number over a 25-year period,
- 265 stable cities characterized by an annual change in the population number from –0.15% to 0.15% over a 25-year period,
- 307 shrinking cities characterized by more than 0.15% of an annual decline over a 25-year period.

Of the 307 shrinking cities, 28 are characterized by permanent shrinkage, i.e., in every five-year subperiod, the population fell at least –0.15%. There are several old industrial centers of coal mining and heavy industry (Katowice, Sosnowiec, Bytom, Zabrze, Ruda Śląska, Chorzów, and Wałbrzych) as well as of textile industry (Łódź) characterized by both significant absolute

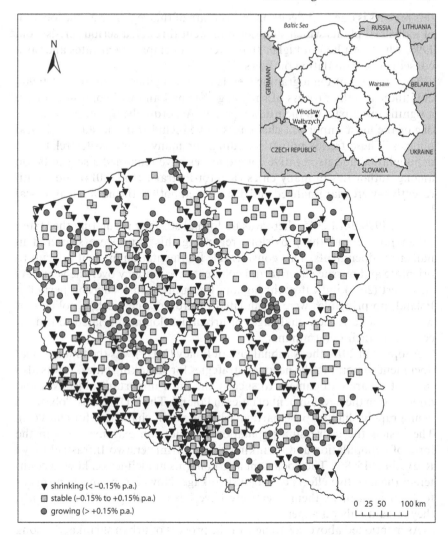

Figure 17.2 Shrinking, stable, and growing cities in Poland in demographic terms, 1990–2015.

Source: Own compilation on the basis of Statistics Poland data.

and relative losses of inhabitants. The shrinkage process causes a series of unfavorable social, economic, spatial, and also image-related phenomena, which, when left without intervention, lead to the intensification of future development problems.

As shown in Figure 17.2, urban shrinkage does not affect the whole country equally; huge regional disparities are apparent. The southwestern area of Poland (Upper and Lower Silesia) clearly stands out against other regions in terms of the intensification of the number of shrinking cities. The scale of this process is due to the economic structure of this region, e.g., the location of industrial plants which shut down or went through a serious crisis from 1990, which resulted in higher than average unemployment rates and, as a consequence, population outflows.

The analysis of demographic trends of Polish cities in the period 1990–2015 indicates the process of shrinkage has had and will continue to have a significant impact on their development. Accordingly, urban shrinkage is one of the most important challenges to be included in the state and local policies. These trends notwithstanding, for many years, issues related to depopulation and its negative consequences have remained a sort of taboo among authorities in many cities (Kantor-Pietraga et al. 2014) and—until recently—were marginalized and disregarded both at the national and local levels.

From 1990 until 2015, there was no formal strategy regulating the state's urban policy, and the provisions regarding this issue were scattered in numerous documents. This considerable fragmentation made it difficult to orientate and pursue effective and coherent urban policy over the years. As A. Billert (2012) indicates, throughout the entire transformation period in Poland, no principles were formulated regarding a new urban policy that would respond to new challenges of urban development related to radical changes in their determinants.

Adopted in 2015, the National Urban Policy (NUP) was the first national document to point to the need for strategies for shrinking cities. It was also the first separate document to specify the most important objectives and directions in the development of cities in Poland. The NUP states: "observed demographic changes pose particularly important challenges for cities (…). The first is the decreasing population in cities—in extreme cases in the form of depopulation and shrinking cities" (Ministerstwo Infrastruktury i Rozwoju 2015: 86). The document also contains guidelines on how to counteract the negative effects of urban shrinkage. However, no mechanisms or tools to implement them effectively have been developed. Consequently, they remain only on paper.

As mentioned above, awareness of the process of urban shrinkage among local politicians remains low; however, recent signs suggest this is slowly beginning to change. An example is the city of Wałbrzych, presented in the following section.

Shrinkage of the old industrial city of Wałbrzych in the light of the path dependence conception

The process of urban shrinkage under the unique conditions of the Polish postsocialist transformation is best exemplified by the city of Wałbrzych for several reasons. The combined effect of deindustrialization and the systemic transition has exacerbated the scale of shrinkage in Wałbrzych relative to other Polish cities and many "typical" cases of shrinkage in highly developed Western cities. Therefore the policies implemented to cope with the "shock of shrinkage" are also different and include such forms as, e.g., special economic zones (SEZ) and "top-down" revitalization programs.

Wałbrzych is situated in Lower Silesian Voivodeship in southwestern Poland, close to the borders of the Czech Republic and Germany. After Wrocław—the Voivodeship's capital—it ranks as the region's second center with 112,594 residents (as of 31 December 2018). By the mid-19th century, Wałbrzych had grown into an important industrial center. The chief industries of the city and the entire region were mining for hard coal and coke-making. In the wake of World War II, it was the biggest industrial center in the region. Apart from mining, there were also other industrial establishments, mostly coking plants, clothing, and textile plants, as well as glass and ceramics works. The transformation of the Polish economy initiated in 1990 and subsequent opening of Wałbrzych to global processes were the impetus of very deep changes in the economic structure of the city and its surrounding region, which ultimately served to reveal its obsolescence and make restructuring necessary. In 1990 the Lower Silesian Coal Basin, including four state-owned coal mines located in Nowa Ruda (1) and Wałbrzych (3), was faced with the decision of the Polish government of its liquidation. Never before had liquidation steps been taken at such a scale in Poland; Wałbrzych served as a sort of testing ground. Liquidation steps were taken at that time also in another, bigger mining region of Upper Silesia.[2] There, however, these activities were carried out gradually, and some of the mines there are still in operation today. This difference in the scale and pace of deindustrialization resulted, on the one hand, from better mining conditions in the Upper Silesian Coal Basin and, on the other hand, from strong pressure from mining trade unions and the local community rooted in the region, opposing the government's winding-up activities. The lower strength of the local community and its identification with the mining tradition in the Lower Silesian Coal Basin (which was part of the German state before World War II) made it easier for the government to carry out liquidation operations. In Wałbrzych, the social and economic costs of the rapid decision to liquidate the mining industry proved to be quite high. This decision led to a deep crisis and to the city's shrinkage since the early 1990s, which manifests itself in almost all aspects, i.e., demographic, economic, social, and spatial (both as to physical structure and aesthetic values).

Although the situation has improved in many aspects in recent years (e.g., creation of new jobs, decline in unemployment, revitalization of run-down areas, housing investments), population loss has followed a continuous trajectory since 1990 (Figure 17.3). In the years 1990–2015, the population decreased by 18.12%, making Wałbrzych one of the country's most drastic cases of urban shrinkage. According to the demographic forecast of Statistics Poland, the city will have fewer than 100,000 inhabitants (99,187) in 2030, a figure last seen in 1950/51. Looking further out, by 2050, the population is expected to drop to only 74,463, which means that—if the forecasts

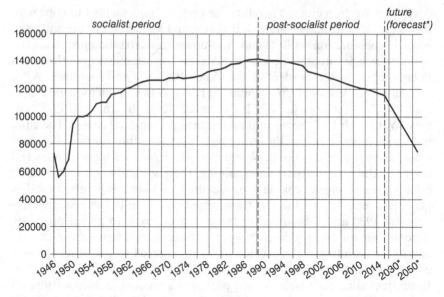

Figure 17.3 Population of Wałbrzych in the years 1945–2050 (* forecast).

Source: Own compilation on the basis of Statistics Poland data.

prove reliable—Wałbrzych will shrink by almost half (–47%) compared to its peak population of 141,504 inhabitants in 1989.

In order to better understand the process of shrinkage, apart from analyzing the city's current state of affairs, it is also very important to trace its historical context and evolution. Thus, we shall attempt to help explain the shrinkage of Wałbrzych on the basis of some elements of the concept of path dependence. In what follows, a chronological analysis of the most important events, choices, and decisions that have influenced changes in this city since the end of World War II is performed.

Under the term *development path*, the authors of this chapter understand the process of shaping the features and functions of a city. In turn, *path dependence* means a type of development, the evolutionary direction of which is formed by its history (Arthur 1994, Mahoney 2000, Gwosdz 2014). This dependence is especially apparent in the case of cities whose development was contingent on the monoculture of traditional industries, such as Wałbrzych. These cities may become "locked-in" on the path, i.e., a city's dependence on a failing industry and the subsequent disinterest or resistance of established institutional, social, and economic systems to change. The status quo can be overcome by a profound and quick change of the current structures and development forces, known as a *critical juncture*. As the critical juncture is reached, one of the key competences facilitating the creation of a new development path is the ability to manage change (Grabher 1993). However, the legacy inherited from the earlier development path makes the task extremely difficult and long-lasting, as seen in the case of Wałbrzych.

In the post-war history of the city, the following periods concerning the patterns of development pathways can be distinguished (Figure 17.4):

a   the period of the command (socialist) system (1945–1989);
b   the period of postsocialist neo-liberal transformation (1990–2003);
c   the first period of the EU membership with no significant actions against shrinkage (2004–2015);
d   the period of implementing National Revitalization Plan (after 2015).

In the next sections, the dynamics of change and policy responses will be discussed, following the above development paths of the city.

The years 1945–1989

A significant critical juncture in the history of Wałbrzych followed a new political division after the end of World War II in 1945. The city survived the war without being destroyed. Inhabited by German residents, Wałbrzych (called Waldenburg at the time) found itself within the Polish borders. In the first years after the war, the city's population was subject to considerable fluctuations due to the inflow of the Polish and displacement of the Germans. However, as early as 1947 and throughout the entire socialist period, population increases were recorded in Wałbrzych, driven in large part by the high birth rate. Immediately after the War, a record number of children were born thanks to the young age of the city's immigrant population.

With the imposition of the communist regime and the conditions of a socialist economy, Wałbrzych focused its development on hard coal mining and eventually became dependent on industrial monoculture. By 1975, aside from three porcelain factories, some minor clothing and food-industry plants, and public sector jobs in connection with the city's status as a regional capital, the economy of Wałbrzych was dependent on the mining of hard coal. Subsequent problems of the city were in part caused by the specific features of an industrial center, including the influx of low-skilled labor, as well as by other decisions and events resulting from the policy of the socialist state, such as the neglect of older, pre-war buildings. This negligence stemmed from, among other things, treating capitalist-era buildings as low-quality compared to newly constructed housing estates from so-called large plates.

The years 1990–2003

The next critical juncture in the development of Wałbrzych occurred in 1990, alongside the political changes and application of a "shock therapy" as a method of neo-liberal socioeconomic transformation. Dominated by industrial monoculture, Wałbrzych was facing a difficult, even—as some

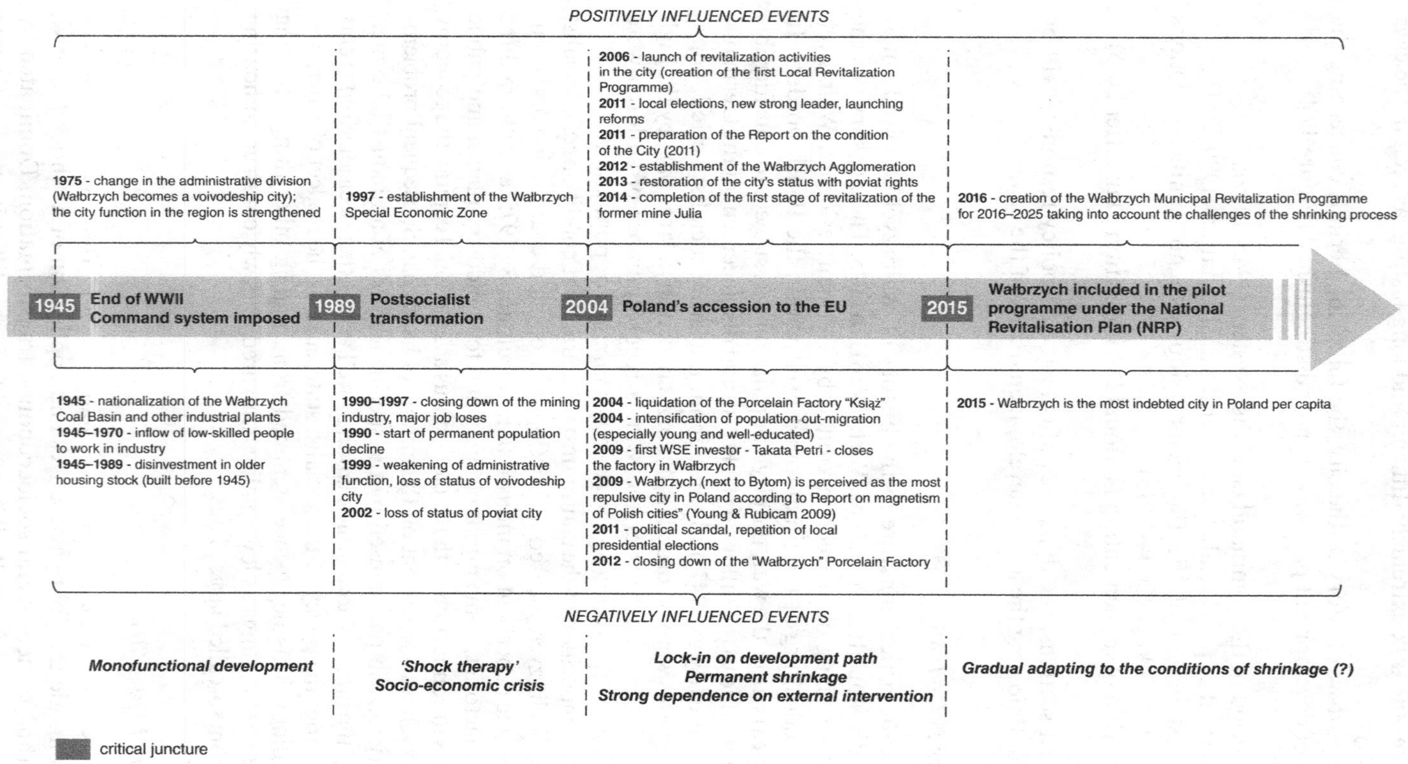

Figure 17.4 Development pathway of Wałbrzych after 1945.

Source: Authors' elaboration.

Table 17.1 Number of the unemployed and job offers in the Labour Office, 1990–2015 (data as of the end of December)

	*1990**	*1992**	*1997**	*2002***	*2005***	*2010***	*2015***
Number of unemployed registered	9,320	21,070	10,432	15,518	12,717	7,079	4,088
With right to unemployment benefit	–	–	3,490	3,024	2,056	1,658	601
Number of job offers	39	70	28	0	0	44	1,242

Notes
* Data refer to the former Wałbrzych voivodeship.
** Data refer to the city of Wałbrzych and Wałbrzych poviat.

Source: Own study based on the reports of the District Labour Office in Wałbrzych and the Poviat Labour Office in Wałbrzych.

authors indicate—tragic situation at the beginning of the 1990s (Skiba 1997; Gajda 2010; Rakowski 2009). As early as 1990, the problems of hidden unemployment arose. According to the District Labour Office in Wałbrzych, there were 9,320 people registered as unemployed and only 39 jobs offered in 1990 (Table 17.1). The heaviest blow for the local economy, however, was the governmental decision to liquidate three of Wałbrzych's coal mines on 29 November 1990.

The decision followed from the mines' unprofitability, a consequence of the increasingly high costs to excavate poorly accessible coal deposits, under-investment, technological backwardness, worldwide decline in demand for coal and growing external competitiveness. According to available sources,[3] in 1993, there were 7,251 miners employed in mines, in addition to approximately 20,000 others employed in positions connected in one way or another to the mining industry (about 50% of the city's working age population). The difficult situation was further aggravated by a generally bad economic condition of all industrial plants located in the city. Initial plans to relocate miners to other branches proved impossible. Consequently, many other industrial plants did not survive the transformation.

The closing down of mines caused a rapid increase in unemployment. Skiba (1997: 21) writes "the problem of fired miners cannot be described with the most accurate indices". The same author notices that these indices do not consider, e.g., persons on temporary mine leave and included in a periodic social protection. Moreover, many unemployed were not registered in the Labour Office. The dramatic situation was also reflected in the lack of entitlement to financial benefits for the great majority of registered unemployed and a lack of job offers (Table 17.1). It should be kept in mind that in those years, a migration balance remained negative, which influenced the lowered unemployment statistics.

According to official statistics, unemployment reached a record high in Wałbrzych in 2002, with an unemployment rate of 28.1%—one of the highest indices in the country. However, according to Rakowski (2009: 383), the real unemployment rate may have been as high as 50% in 2002–2006. The situation was worsened by the fact that the city was largely inhabited by low-skilled workers who had either elementary education or finished vocational mining schools. Finding a new, different job became a significant problem for these workers, and one of the main reasons for young people to leave the city. Indeed, since the beginning of the 1990s, economic migrations both within Poland and abroad have been among the main reasons for decreasing populations. Additionally, since 1991, the birth rate has been negative with the lowest coefficient of −6.7‰ (in 2015) in the city's history.

The closing down of the city's industry followed by a serious socioeconomic crisis together with a major population loss led the process of urban shrinkage to manifest itself in many ways. Subsequent political decisions, such as depriving Wałbrzych of its status of a voivodeship city[4] in the course of 1999 administrative reforms, and the subsequent removal of its poviat (district) status in 2003, have exacerbated the position of Wałbrzych against other large cities in the country, deepening its crisis.

Positive developments in this period include the creation of INVEST-PARK, or the Wałbrzych Special Economic Zone (WSEZ) in 1997 (three years after the Polish parliament had passed the Special Economic Zones Act). The creation of such zones, i.e., selected areas designated to run business activity on favorable terms (e.g., tax exemptions), aimed to reduce unemployment in areas that were most severely affected by the "shock" of transformation. Unfortunately, an increase in new jobs offered by the Wałbrzych Special Economic Zone occurred slowly in relation to the unemployed rate.

Despite high unemployment, WSEZ investors (representing mainly the automotive industry, and more precisely manufacturing car sub-assemblies, including Japan's Toyota) encountered many problems, especially in the initial stages, with finding not only highly skilled workers (e.g., managers with technical education and managerial staff) but also those with lower qualifications. This was mostly due to the monofunctional character of the region, which for years was oriented toward mining. When the situation in the labor market had changed, many retraining programs were organized to prepare people for other occupations.

The years 2004–2015

Poland's accession to the European Union can be regarded as another critical juncture in the city's development. On the one hand, new opportunities to obtain EU funds emerged, including funds for projects related to revitalization. On the other hand, the opening of the EU labor market to the Polish workforce influenced the outmigration to other EU countries, especially among the young and better-educated.

In 2004 the Local Programme for the Revitalization of Wałbrzych City for the years 2004–2006 was passed by the city authorities, followed in 2008 by a program covering 2008–2015. The programs prioritized downtown revitalization efforts and postponed the reclamation of former coal-mining areas. The first projects included the renovation of historic tenement houses and the modernization of public space. In 2008, Wałbrzych received EU funds for the revitalization of the former mine into the science and art center. The cost of the revitalization was a total of 52.5 million zlotys (ca. 12.5 million EUR), of which 35.7 million (ca. 8.5 million EUR) was funded by the EU. The first stage of revitalization was completed in 2014, although it was initially planned for 2012. During the process, however, the anticipated expenses were quite off the mark. Ultimately the project was over three times more costly than expected, reaching an estimated value of over 166 million zlotys (ca. 39.5 million EUR). The significant increase is explained by the unreliable inventory of the area, the need to find a new contractor as well as numerous late stage corrections and project changes. Despite many difficulties, Wałbrzych can be proud of its new complex, the Former Mine Science and Art Centre, the activities of which goes beyond that of a museum (e.g., in addition to its historical buildings and exhibits, visitors can listen to the history and culture of the mining as told by guides who are former miners). The revitalization process is also supported by numerous cultural projects, including joint transborder initiatives with neighboring communes in the Czech Republic. In 2012 city authorities created the association of communes, the Wałbrzych agglomeration, toward facilitating cooperation between the neighboring communes and to maximize the chances of receiving EU funds under Integrated Territorial Investments (so-called ITIs) in the EU 2014–2020 financial perspective.

Despite gradual improvements of Wałbrzych's economic and social situation, thanks to both revitalization projects and new WSEZ investments, the city continued losing inhabitants, which in the long run caused unfavorable consequences in the form of the disturbance of the age structure and intensification of ageing processes. As a result, Wałbrzych became "locked-in" on the path of permanent urban shrinkage. This "lock-in" was also strengthened by negative developments on the local political scene, culminating in 2010 with the need to repeat a local self-government election on account of electoral corruption. Although Wałbrzych later experienced some positive events following the adopted policy of "rebuilding" the city after 2011, effectively, the path taken remained largely unchanged. In the case of this city, while economic rebirth may take place, demographic rebirth with the co-occurrence of other unfavorable processes is unrealistic. The loss of population is already so significant and the negative demographic trends are so advanced that it will most probably not be possible to compensate for the resulting loss and Wałbrzych will continue to shrink in the future, although the rate of shrinking will probably decrease.

The problems connected with the closing down of mining led to the consolidation of the stereotype of Wałbrzych as a city of the unemployed, poor's shafts,[5] social pathologies and political corruption. This, in turn, was reflected in a negative image of the city, shaped not only by local and national media but also outside the country, which can be defined as the *mindware* result of the city's shrinkage (Hospers 2012, 2014a,b). The image they create is of a fallen city with no future. In particular, poor's shafts attracted national as well as foreign media, including Austrian *News,* Dutch *de Volkskrant* and American *New York Times.*

The city's unfavorable image was also reinforced by various rankings related to cities' attractiveness, which tended to rank Wałbrzych among the worst. For instance, according to "Magnetism of Polish cities" (Young, Rubicam 2009), Wałbrzych was recognized as the most revolting place in Poland (next to Bytom and Ruda Śląska). In 2014, a ranking carried out by the country's weekly *Polityka* concerning the quality of urban life in the largest 66 cities in the country had Wałbrzych in the last position (Kowanda 2014), whereas in a ranking published by the online news service *Wirtualna Polska* it achieved the status of the worst city to live in Poland in 2015. While the methods and criteria of the aforementioned rankings may raise doubts; their results undoubtedly reinforce Wałbrzych's negative image.

Since 2015

The most recent critical juncture in the construction of the city's future seems to be the decision of the national government to include Wałbrzych—along with Łódź and Bytom—as part of a pilot program of the National Revitalization Plan (NRP) to develop a model solution for revitalization. Thanks to this decision, the city received special funds for revitalization. Accordingly, in 2016 a new 2016–2025 Commune Revitalization Programme (CRP) for Wałbrzych City was drawn up by local authorities; the city's first official document to use the term "shrinking city". These developments appear to mark an important step toward undertaking real measures to address the challenges of urban shrinkage. Indeed, the attitudes of local authorities seem to have shifted, as demonstrated by the official recognition that "the decreasing population of Wałbrzych causes ever higher operation costs due to a low intensity of land-use and high expenditure on infrastructure. The revitalization measures have to counteract the main problem, which is the shrinkage of Wałbrzych" (Urząd Miejski w Wałbrzychu 2016: 17).

Recent positive changes to help support the image of Wałbrzych as a city that "rises from its knees" or even "has been reborn". Quite unexpectedly, the increased interest in the city among tourists and the media has grown due to the revival of legends about a Nazi German "golden train", rumored to be buried in an underground tunnel somewhere in Lower Silesia at the end of World War II in January 1945. A wealth of treasure and military secrets are said to be hidden inside its armored carriers. The search in 2015/16 for the infamous "golden train" helped to improve perceptions of the city. Although the train was not found, Wałbrzych, earlier characterized as nothing more than a poor, dilapidated mining city, suddenly became a city of secrets and treasures. At this moment, Wałbrzych seems to stand a chance of entering the path of adaptation to the conditions of shrinkage. Still, answers to the questions of whether this will happen and, if so, whether the city will be able to mitigate the negative effects of shrinkage by becoming a smaller but better place to live remain to be known.

Taking into account demographic forecasts, a departure from the path of permanent shrinkage, as mentioned earlier, will be practically impossible. Wałbrzych will continue to shrink in light of/a variety of historical, demographic, social, economic, spatial, and political factors. The adaptation to the conditions of shrinkage and the creation of a long-term development vision for a smaller but more resident-friendly city seems to be key for its future.

Concluding remarks

There is no doubt that urban shrinkage will be a major challenge for future urban policies, especially in postsocialist cities. Hence a discussion about a strategy to counteract the detrimental effects of this process seems highly topical. The course of the urbanization process in Poland changed fundamentally after 1989, primarily due to the political and economic transformation. Many cities have experienced a reversal of their earlier development trends; especially affected are those that used to rely on traditional industries. Today most of them are shrinking cities. The scale and intensification of the process of urban shrinkage since the beginning of the transformation require adopting measures to mitigate its unfavorable after-effects. This is true now more than ever considering current forecasts for further urban population losses and aging, which, when left uncontrolled, can cause a series of problems in urban functioning. Of key importance for the future of many Polish cities will be the creation of long-term development visions under the conditions of shrinkage. This calls for an increasing awareness and knowledge of the processes among central authorities responsible for the implementation of the adopted National Urban Policy and local decision-makers. Especially at the local level, it is very important to accept the shrinking process and to notice its positive implications, such as lower population density, larger housing supply, better access to some services, wider areas of greenery, or improvements in local environmental quality. Therefore, shrinking cities should implement strategies aimed at planning for shrinkage.

In order to understand the nature of urban shrinkage in old industrial cities, like Wałbrzych, various internal and external drivers of local development need to be analyzed. Some drivers, however, such as deindustrialization or negative demographic changes, play a leading role and influence the emergence of others, e.g., an unfavorable image that is difficult to change. The case of Wałbrzych illustrates that the adopted regeneration strategies evolve over time, from those initially aiming to improve the labor market to more complex approaches whereby the main focus is on the qualitative development of the city. An important element of the revitalization strategy is the use of industrial heritage, including the development of new cultural and creative spaces. This, in turn, facilitates the tourist movement. In Wałbrzych, the former Mine Science and Art Centre has become one of the main tourist attractions of the city. It also serves as the new (due to its function) and old (because of its mining past) symbol of Wałbrzych as well as an important meeting spot, with the capacity to foster a new local identity while preserving local cultural heritage. Although the Centre has not solved the city's main problems, it contributes to strengthening its cultural potential and public image. Moreover, the implemented program of the city's revitalization has presented Wałbrzych with an opportunity to enter the path of adaptation to the process of urban shrinkage.

According to the latest demographic projections, Wałbrzych will continue shrinking into the foreseeable future. Its adaptation to the process of shrinkage and a long-term development vision for a smaller but more residentially friendly city seems to be of key importance for its future. Building awareness of the city's problems among local politicians and inhabitants will be a significant step toward this aim. This also applies to other shrinking cities across Poland. In their future development strategies, they should adapt the experiences of Western shrinking cities (e.g., those located in the Ruhr coal mining basin) on the one hand and utilize their own unique local resources on the other. Moreover, the coordination of national, regional, and local revitalization programs is key to their successful implementation and capacity to produce positive long-term outcomes in shrinking cities.

Notes

1 *Cities Regrowing Smaller. Fostering Knowledge on Regeneration Strategies in Shrinking Cities across Europe* (CIRES)—a project implemented under the European Union's COST Action (European Cooperation in Science and Technology) in the years 2009–2013.

2 Employment in coal mines in the Upper Silesia region was 388,000 in 1990 and 82,000 in 2018.

3 The program of the liquidation and restructuring of employment in the mine "Julia" for 1993–1995. Wałbrzych 1993 (after Urbański 2004).

4 Between 1975 and 1998, Poland was divided into 49 voivodeships (regional units), with Wałbrzych voivodeship being one of them. As of 1997, Wałbrzych voivodeship was 41,168 km² in size and had 735,300 inhabitants. As a result of the administrative reform of Poland carried out in 1999, smaller voivodeships were liquidated and 16 new voivodeships, 308 *poviats*, and 2,489 *gminas* were created. Wałbrzych *poviat* was established as a local unit of territorial division and at the same time one of 308 units of this rank in Poland. The *poviat* area is much smaller than the former voivodeship and covers 430 km²; in 2000, it had a population of 186,000 (including the city of Wałbrzych).

5 Poor's shafts refer to illegal coal mining extraction (Rakowski 2009; Gajda 2010; Krygowska 2016). As many as 2,000 people were estimated to perform such work at the beginning of the first decade of the 21st century. This life-threatening way of mining, performed in poor conditions mainly by unemployed former miners, proved fatal for some of them.

References

Arthur, B. 1994. *Increasing Returns and Path Dependence in the Economy*. Ann Arbor: University of Michigan Press.

Billert, A. 2012. Założenia, modele i planowanie polityki rozwoju miast. Próba konfrontacji dwóch światów jednej Unii Europejskiej (Assumptions, models and planning of urban development policiy). In Derejski, K., Kubera, J., Lisiecki, S. & Macyra R. (eds.) *Deklinacja odnowy miast. Z dyskusji nad rewitalizacją w Polsce* (pp. 21–53). Poznań: Wydawnictwo Naukowe Wydziału Nauk Społecznych Uniwersytetu im. Adama Mickiewicza w Poznaniu.

Bontje, M. 2004. Facing the Challenge of Shrinking Cities in East Germany: The Case of Leipzig. *GeoJournal*, 61(1), 13–21.

Boschma, R. A. & Lambooy, J. G. 1999. Evolutionary Economics and Economic Geography. *Journal of Evolutionary Economics*, 9, 411–429.

Central Statistical Office. 2014. Population projection 2014-2050. Warszawa.

David, P. A. 2001. Path dependence, its critics and the quest for "historical economics". In Garrouste, P. & Ioannides, S. (eds.) *Evolution and Path Dependence in Economic Ideas: Past and Present* (pp. 15–40). Cheltenham: Elgar.

Franz, P. 2004. Shrinking Cities—Shrinking Economy? The Case of East Germany. https://difu.de/publikationen/shrinking-cities-shrinking-economythe-case-of-east.html (15 June 2012).

Gajda, M. 2010. Wałbrzych naznaczony węglem. Od etosu górnika do stygmatu bieda szybownika (Wałbrzych marked with coal. From the miner's ethos to the stigma of the poor's shaft worker). In Kamińska K. (ed.) *Miejskie wojny. Edukacyjne dyskursy przestrzeni (Urban Wars: Educational Discourses of Space)* (pp. 87–102). Wrocław: Oficyna Wydawnicza ATUT.

Grabher, G. 1993. The weakness of strong ties: The lock-in of regional development in the Ruhr area. In Grabher, G. (ed.) *The Embedded Firm: On the Socioeconomics of Industrial Networks* (pp. 255–277). London: Routledge.

Großmann, K., Haase, A., Rink, D. & Steinführer, A. 2008. Urban shrinkage in East Central Europe? Benefits and limits of a cross-national transfer of research approaches. In Nowak, M. & Nowosielski, M. (eds.) *Declining Cities/Developing Cities: Polish and German Perspectives* (pp. 77–99). Poznań: Instytut Zachodni.

Gwosdz, K. 2004. *Ewolucja rangi miejscowości w konurbacji przemysłowej. Przypadek Górnego Śląska (1830–2000) (Evolution of Locality Ranks in an Industrial Conurbation: The Case of Upper Silesia (1830–2000))*. Kraków: Instytut Geografii i Gospodarki Przestrzennej Uniwersytetu Jagiellońskiego.

Gwosdz, K. 2014. *Pomiędzy starą a nową ścieżką rozwojową. Mechanizmy ewolucji struktury gospodarczej i przestrzennej regionu tradycyjnego przemysłu na przykładzie konurbacji katowickiej po 1989 roku. (Between the Old and the New Development Path. Mechanism of Evolution of Economic and Spatial Structure of a Traditional Industrial Region: The Case of the Katowice Conurbation after 1989)*. Kraków: Instytut Geografii i Gospodarki Przestrzennej Uniwersytetu Jagiellońskiego.

Haase, A., et al. 2017. Representing Urban Shrinkage: The Importance of Discourse as a Frame for Understanding Conditions and Policy. *Cities*, 69, 95–101. https://doi.org/10.1016/j.cities.2016.09.007

Hospers, G. J. 2012. Urban shrinkage and the need for civil engagement. In Haase, A., Hospers, G. J., Pekelsma, S. & Rink D. (eds.) *Shrinking Areas: Front Runners in Innovative Citizen Participation*. The Hague: EUKN.

Hospers, G.J. 2014a. Policy responses to urban shrinkage: From growth thinking to civic engagement. *European Planning Studies* 22(7): 1507–1523.

Hospers, G. J. 2014b. Urban shrinkage in the EU. In Richardson, H.W. & Nam C.W. (eds.) *Shrinking Cities: A Global Perspective*. Abingdon: Routledge.

Kantor-Pietraga, I., Krzysztofik, R., Runge, J. & Spórna, T. 2014. Problemy zarządzania miastem kurczącym się na przykładzie Bytomia (Problems with governing the shrinking city on the example of Bytom). In Markowski, T. & Stawasz, D. (eds.) *Społeczna odpowiedzialność w procesach zarządzania funkcjonalnymi obszarami miejskimi (Social Responsibility in the Governance of Functional Urban Areas)*. Biuletyn KPZK PAN 253, 162–175.

Kowanda, C. 2014. W pogoni za stolicą. Ranking jakości miejskiego życia *(Chasing the Capital. Ranking of Urban Life Quality)* Polityka, Wydanie specjalne Niezbędnik Inteligenta *(Special Issue The Essentials of an Intellectual)*, 10, 71–75.

Krygowska, N. 2016. Biedaszyby jako obszar formowania się subkultury pogórniczej (Poor's shafts as an area of the formation of post-mining subculture). In Filimowska, A. & Krygowska, N. (eds.) *Wałbrzych miasto poszukiwaczy (Wałbrzych—The City of Explorers)* (pp. 45–60). Kraków: Wydawnictwo AGH.

Kurkiewicz, J. (ed.) 2010. *Procesy demograficzne i metody ich analizy (Demographical Processes and Methods of Their Analysis)*. Kraków: Wydawnictwo Uniwersytetu Ekonomicznego w Krakowie.

Lesthaeghe, R. & van de Kaa, D. 1986. Twee Demografische Transities? (Two demographic transitions?). In Lesthaeghe, R. & van de Kaa, D. (eds.) *Bevolking: Groei en Krimp. Van LoghumSlaterus*. Deventer.

Logan, J. & Molotch, H. 1987. *Urban Fortune: The Political Economy of Place*. Berkeley: University of California Press.

Mahoney, J. 2000. Path Dependence in Historical Sociology. *Theory and Society* 29, 507–548.

Ministerstwo Infrastruktury i Rozwoju (Ministry of Infrastructure and Development). 2015. Krajowa Polityka Miejska 2023 (National Urban Policy 2023). Warszawa.

Okólski, M. 2005. *Demografia. Podstawowe pojęcia, procesy i teorie w encyklopedycznym zarysie (Demographics: Basic Concepts, Processes and Theories in an Encyclopedic Outline)*. Warszawa: Wydawnictwo Naukowe Scholar.

Pallagst, K., Fleschurz, R. & Said, S. 2017. What Drives Planning in a Shrinking City? Tales from two German and two American Cases. *Town Planning Review* (Special issue on shrinking cities), 88(1), 15–28. https://doi.org/10.3828/tpr.2017.3

Parysek, J. J. & Wdowicka, M. 2002. Polish Socio-economic Transformation. Winners and Losers at the Local Level. *European Urban and Regional Studies*, 9, 60–72.

Rakowski, T. 2009. *Łowcy, zbieracze, praktycy niemocy (Hunters, Gatherers, Practitioners of Powerlessness)*. Gdańsk: Wydawnictwo Słowo/Obraz Terytoria.

Sagan, I. 2000. *Miasto. Scena konfliktów i współpracy. Rozwój miast w świetle koncepcji reżimu miejskiego (City. The Arena of Conficts and Collaboration. The Development of Cities in the Light of the Concept of Urban Regime)*. Gdańsk: Wydawnictwo Uniwersytetu Gdańskiego.

Skiba, L. (ed.) 1997. *Zagłębie węglowe w obliczu restrukturalizacji (Coal Basin in the Face of Restructuring)*. Wrocław: Wydawnictwo Silesia.

Steinführer, A. & Haase, A. 2007. Demographic Change as Future Challenge for Cities in East Central Europe. *Geographiska Annaler B*, 89(2), 183–195.

Stryjakiewicz, T. (ed.) 2014. *Kurczenie się miast w Europie Środkowo-Wschodniej (Urban Shrinkage in East-Central Europe)*. Poznań: Bogucki Wydawnictwo Naukowe.

Stryjakiewicz, T., Ciesiółka, P. & Jaroszewska, E. 2012. Urban Shrinkage and the Post-socialist Transformation: The Case of Poland. *Built Environment*, 38(2), 197–213.

Urbański, M. 2004. *Restrukturyzacja gospodarki regionu wałbrzyskiego: instrumenty, cele, środki (Restructuring of the Economy of the Wałbrzych Region: Instruments, Goals, Means)*. MSc dissertation Uniwersytet Ekonomiczny w Poznaniu.

Urząd Miejski w Wałbrzychu (Wałbrzych City Office). 2016. *Gminny Program Rewitalizacji Miasta Wałbrzycha na lata 2016–2025 (The Commune Revitalization Programme for Wałbrzych city for the years 2016–2025)*. Urząd Miejski, Wałbrzych.

Wiechmann, T. 2008. Conversion strategies under uncertainty in post-socialist shrinking cities—the example of Dresden in Eastern Germany. In: Pallagst, K. et al. (eds.) *The Future of Shrinking Cities: Problems, Patterns and Strategies of Urban Transformation in a Global Context*. Berkeley: IURD.

Wiechmann, T. & Bontje, M. 2015. Responding to Tough Times: Policy and Planning Strategies in Shrinking Cities. *European Planning Studies*, 23(1), 1–11.

Wiechmann, T. & Wolff, M. 2013. Urban shrinkage in a spatial perspective: Operationalization of shrinking cities in Europe 1990–2010. AESOP-ACSP Joint Congress, 15–19.07.2013. Dublin.

Wolff, M. & Wiechmann, T. 2018. Urban Growth and Decline: Europe's Shrinking Cities in a Comparative Perspective 1990–2010. *European Urban and Regional Studies*, 25(2), 1–18. https://doi.org/10.1177/0969776417694680

Young & Rubicam. 2009. Magnetyzm polskich miast, http://wiadomosci.gazeta.pl/wiadomosci/1,114873,7253954,Bytom__Walbrzych__Ruda_Slaska___najbardziej_odpychajace.html (Accessed 10 December 2010).

18 Coping with shrinkage in old and young mining cities in Slovakia
The cases of Banská Štiavnica and Prievidza

Ján Buček, Branislav Bleha, and Marek Richter

Introduction

Although urban shrinkage is a global phenomenon (Pallagst, Wiechmann and Martinez-Fernandez 2013, Richardson and Nam 2014), East-Central Europe is often identified as a region with a large share of shrinking cities (Turok and Mykhnenko 2007, Stryjakiewicz 2014, Haase, Rink and Grossmann 2016). Numerous cities, also in Slovakia, are losing population with multiple consequences. This is especially the case of lower-ranked cities in peripheral locations with industrial and mining traditions. This process has attracted the attention of scholars in Slovakia primarily during the last decade. In addition to the more general focus on urban population development and shrinkage (Šprocha et al. 2017), attention had been paid to planning in shrinking cities (Buček and Bleha 2013) and the specific demographic conditions and policies in Bratislava (Bleha and Buček 2015), and Banská Štiavnica (Buček and Bleha 2014). Nevertheless, more extensive insight into shrinkage processes, as well as policies adopted in such cities are missing in Slovakia. The term "shrinkage" is not widely used (no equivalent exists in Slovak) and only a minor number of cities face a significant impact of shrinkage on their functioning (Buček and Bleha 2013).

The geographical focus of this study is on the cities of Banská Štiavnica and Prievidza. They can be considered as mining cities (Martinez-Fernandez et al. 2012) due to their dependence or development based on mining (silver and gold in Banská Štiavnica, brown coal in Prievidza) and linked industries. Both are typical in that they have a non-mining future. While in Banská Štiavnica mining activities ceased, Prievidza is in the last phase of extraction, with the expected closure of mines in 2023. Differences in the mining period, population size, location conditions, scale, and importance within the Slovak economy offer excellent potential to study the reasons and effects of shrinkage, changing development priorities, and their implementation from a comparative perspective. We also intend to contribute to the debate on the emergence of shrinkage, as well as mitigation and adaptation approaches to shrinkage in cities facing mining-induced decline under postsocialist conditions. The key research questions include the following:

DOI: 10.4324/9780367815011-22

(1) Are these cities able to cope with shrinkage alone? and (2) Are more sophisticated approaches applied in the case of a "young" shrinking city compared to a long-term shrinking city? Although many experiences can be unique and hardly transferrable, they can provide useful knowledge from an international perspective. We used data on population, social and economic development provided by the Statistical Office of the Slovak Republic, in addition to governmental planning documents, company information (Trend 2019), and reports prepared by advisory and non-governmental bodies.

Theoretical framework

To understand shrinkage in the selected cities, we must consider the unique processes of postsocialist transition, especially in its early period. We can mention the introduction of a market economy (including privatization and the collapse of less competitive local businesses), a reduction of public funds transfers to cities, and an absence of selected public policies (e.g., in housing). Additionally, the state-building process in Slovakia (a new state since 1993), including public administration reorganization and progress in decentralization, is important. As a result, in addition to population loss caused by birth rate decline and out-migration, typical signs of shrinkage such as changing needs for public services, deindustrialization, unemployment growth, unused and abandoned premises (Haase, Rink and Grossmann 2016; Stryjakiewicz 2014) appeared in Slovakia.

Urban shrinkage substantially affects local public sector operations. It leads to changed local services and infrastructure demands accompanied by inevitable managerial and local finance implications (Wolf and Amirkhanyan 2010). Among the typical consequences of shrinkage, we can mention infrastructure overcapacity, housing market imbalance and decreasing local property values, school closures, increased spending on social and health care, fewer public service users (e.g., of public transport), and a decline in local public sector jobs. These challenges are often multiplied by decreasing revenues (from taxes and fees for services), which further threaten the provision of local services and new investments. Not surprisingly, financial issues are crucial in the debate concerning the right-sizing of cities facing shrinkage (Hummel 2015). The adaptation of local public sector operations to a modified financial framework is significant in the battle against the effects of shrinkage. This depends, however, on country specific local finance frameworks. Primarily it is a question of local government powers in relation to local budget revenues. Under postsocialist conditions, we can also observe the limited scope of fiscal decentralization (e.g., as a portion of local government budgets compared to total public budgets), more equalizing approaches, and fewer direct links of the local tax system to the local economy.

The non-renewability of resources places postsocialist mining cities at particular risk. This vulnerability has been exacerbated by the fact that socialist era mining developed outside global markets. The new economic framework, uncompetitive production, long-term underinvestment, difficulties with labor safety standards, and rising environmental standards have placed this sector and many mining cities under multifaceted pressure, including Ostrava in the Czech Republic (Rumpel and Slach 2012) and Walbrzych in Poland (Jaroszewska 2019). Smaller, less efficient mines, with lower quality coal or ores, located at the periphery and dependent on state support, faced closure or a substantial reduction of operations. Some mining operations have survived under the threat of social and political tensions thanks to close production linkages that are not easy to replace (energy production and distribution systems are often defined as national economic interests) or due to the large number and traditionally high status of miners within the workforce.

The twilight of mining has seriously threatened the future of many local communities. However, for those mining areas facing the exhaustion of resources, we can accurately predict the decline of mining according to the exhaustion of resources, technological limits, and market conditions. These areas offer the opportunity to prepare economic restructuring and regeneration strategies when the first signs of irreversible changes and shrinkage are known. They can mitigate the effects of diminishing mining activities and assist in the adaptation to a future "non-mining existence". Typical responses include re-industrialization (new segments of industry, including culture and creative industries), deindustrialization and a shift to services (e.g., tourism), improvements to the physical environment (e.g., housing renewal), environmental sustainability, and the protection of public sector employment. Responses are usually context-specific and dependent on external actors, including central state and foreign investors (Rink et al. 2012, Pallagst et al. 2013).

We decided to follow the institutionalist and governance concepts and their application for the successful adaptation to shrinkage (Dale 2002). They reflect the important role of context (historical, political), institutional influences (relevant organizational base and the more general regime, rules, and regulations), multiple actors involved, as well as the symbolic position of mining as a traditional sector of the economy. The specific multi-level governance system established in Eastern German mining regions after 1990 (Harfst and Wirth 2011) and the governance approach to the economic regeneration of the shrinking city of Ostrava (Rumpel and Slach 2012) are good examples. Institutional development and a coordinated effort between a plethora of actors/stakeholders' capacities and resources (with a significant role played by external actors) are essential. It is reflected in various partnership-based strategies, policies and planning documents, and implementation projects. They mostly focus on local economic development measures, the labor market, and responses in the field of "public consumption" (social services, education, housing, infrastructure), including its financing.

When focusing on urban shrinkage and governance in the postsocialist Slovak context, we must also consider public administration and territorial organization changes. The existing dual public administration model is based on separate lines of state administration and self-government and the related division of powers. The revised territorial division consists of eight self-governing regions, 79 districts, and 2,890 local self-governments (141 have the status of cities). The state administration below the central level represents district offices representing central state ministries and agencies. They have a dominant role in selected fields (e.g., social affairs or the labor market). Both case study cities are district centers (Banská Štiavnica district—16,134, Prievidza district—134,546 inhabitants; 2018).

Local self-governments, introduced in 1990, developed as the main power center at the local level. The term local self-government is used instead of local government to signify democratic changes, autonomy, link to administrative tradition, and the differences compared to the previous socialist regime. Local self-governments have wide-ranging powers in managing all local affairs and the provision of public services, including, e.g., development planning and regulation, primary education, local roads, and waste management. Although they have limited autonomous revenue-raising powers, stable tax revenues are in general guaranteed by legislation and they may access other sources such as intergovernmental grants. Due to the decentralization processes introduced in 2001, selected powers (e.g., in secondary education, regional roads, regional development coordination) were obtained by regional self-government. This means that more levels and lines of public administration are active on each city's territory according to their powers, and their cooperation is often inevitable.

Brief city profiles

Banská Štiavnica

Banská Štiavnica is located in a mountainous region in the southern part of central Slovakia. At present, its population is slightly above 10,000 (2018). Nevertheless, in 1782, it was the second-largest city in the territory of present-day Slovakia with more than 22,000 inhabitants. Thanks to silver and gold mining, the city grew. Natural resource exploitation supported its development into a vibrant historical city with an architecturally rich built environment. The decrease in mining efficiency, problems with groundwater, and increasing international competition caused a substantial reduction in mining activities, especially since the 19th century, led to long-term economic, social as well as population decline. This trend was partly mitigated by attempts to locate new activities since the 19th century. Socialist industrialization after 1948, later accompanied by an effort to transform the city into a service center (secondary education, tourism), played a major role. Depopulation tendencies were also mitigated by new socialist mass housing developments that attracted new inhabitants. Since the change of regime in 1989, the city has had to cope with shrinkage-related pressures caused by the postsocialist transition including population decline, a collapsing local economy, and a poorly maintained yet valuable historical city center. The city's long-term shrinkage was accompanied by the loss of its economic and administrative position in the broader region. The development core shifted from the traditional, resource-rich mountain areas of the region (with Banská Štiavnica at its center) to the more accessible Hron river valley and the city of Žiar nad Hronom, which grew thanks to the location of an aluminum factory during socialist industrialization. Banská Štiavnica decayed into a peripheral, second-rate urban center, with its own small functional region based on less intensive commuting flows and the provision of services to its hinterland.

Prievidza

Prievidza, located in the western part of central Slovakia in the upper Nitra River basin, is surrounded by mountains and beyond the main railroad and highway lines. It is a medium-sized city by Slovak standards with a population greater than 45,000 as of 2018. It boasts of a compact urban environment with the neighboring small spa town of Bojnice. During most of its history, the city served as a center for small local services and a market for the surrounding agricultural region, with minor factories (food, wood, and construction materials processing). In 1930, it had only 5,608 inhabitants. Although brown coal resources had been identified by the mid-19th century, its extraction was limited partly due to its poor quality. The first phase of larger-scale mining started after the city's connection to the railroad network at the end of the 19th century. Eventually, three mines (Handlová—1909, Nováky—1939, Cígeľ—1962) were established. Coal production peaked in the 1970s and 1980s at about 5 million tons per year (Office of the Government 2018) as traditional energy resources in Slovakia became increasingly scarce. The primary consumers of local coal (Szőllős 1993) have been the neighboring power plants in Nováky and Handlová (total installed capacity almost 600 MW in 1990), along with the chemical industry in Nováky (plastic and carbide production). The construction material industry based on waste by-products generated by the Nováky power plant also contributed to the local industries. Thus, Prievidza developed into internally integrated region of mining, energy production, and processing industry, considered the fourth most important industrial area of the country, with approximately 23,000 employees in 1980 (Mládek 1990). The city had been planned principally as a housing and service center for the neighboring mining and industrial areas. It led to rapid growth during the second half of the 20th century till 1989. Nevertheless, the city faces the decline of an inherited economic structure, which has led to intensive shrinkage in the last two decades.

Latest population and socioeconomic change

Historical context is crucial to understanding the developmental trajectories of both cities, whose historical backgrounds could not be more different. As Banská Štiavnica was the third-largest city in the Hungarian part of the Austro-Hungarian Empire in the 18th century, Prievidza was a very small town (Figure 18.1). The socialist period changed everything. Prievidza was planned as one of the development cores with intensive industrialization and immigration. It was one of the leaders in relative population growth and experienced a five-fold increase during the socialist era. As a result, it was one of 11 cities with more than 50,000 inhabitants in Slovakia. On the other hand, Banská Štiavnica represents an entirely different story. The city had been systematically losing population since its peak in the 18th century, although less dramatically during the last 150 years. The periods during and immediately after WWI and WWII (including the emigration of part of the local population) also had a negative influence. After World War II, population development became more stable, and the population slightly increased in the later phase of the socialist period. The demographic scissors in the growth of Prievidza and the stagnation of Banská Štiavnica during the socialist era say much about the diverging social and economic trajectories of these cities. Most Slovak mining cities lost their position, while others gained at the expense of these historically formed cities. From the demographic point of view, this stagnation was a combination of lower

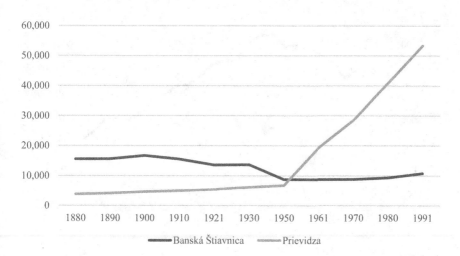

Figure 18.1 Population change in 1880–1991 (according to historical census data).

Sources: Federal Statistical Office (1978) and Statistical Office of the Slovak Republic (2003).

natural increase, negative net migration, and unfavorable trends concerning the age structure.

The fall of the Iron Curtain triggered the deterioration of the demographic situation for Banská Štiavnica and was also a negative turning point for Prievidza. Annual population loss became pronounced around 2000. Prievidza lost its position among Slovak cities with more than 50,000 inhabitants, and Banská Štiavnica's population fell below 10,000. A decomposition of the overall trend to the natural increase and migratory component highlights some differences between the cities in terms of population loss. In Banská Štiavnica, the natural increase was negative from 1995 until 2017; however, the slight increase in 2018 does not necessarily indicate the start of a positive trend. The balance in the number of births and deaths remained positive up to 2013 in the case of Prievidza, mainly due to a younger age structure. However, the city has been losing this advantage continuously. The out-migration component has played a critical role in population stagnation and loss, although the period after 1993 has been volatile in terms of trend curves. Prievidza's population decline is a result of higher emigration throughout the entire period. Banská Štiavnica faced better times before 2003, but since then, annual decreases have prevailed. Population loss is induced more by the migratory component in the case of Prievidza since its annual decrease of about 10 persons per 1,000 is one of the highest negative values among Slovak cities.

Population aging is the most specific feature of the population trend in both cities (Figure 18.2). The aging index (65 years and above per 100 population 14 years and below) of the population of Banská Štiavnica rose from 45 to 120 between 1996 and 2018, and from 29 to 152 for the same period for Prievidza. The increase in the senior population of the Slovak Republic

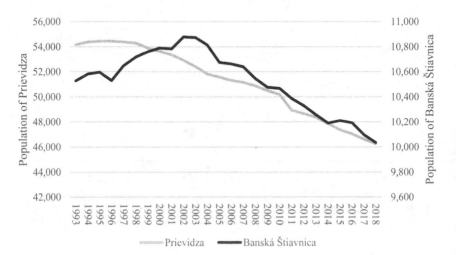

Figure 18.2 Population change in the postsocialist period.

Source: Statistical Office of the Slovak Republic (2019).

was 50 persons (to 102 in 2018) during this period. Prievidza is an "older" city than Banská Štiavnica, as expressed by the aging index, and one of the "oldest" cities in Slovakia. Such a high level of aging is a combination of the outflow of the population in the reproductive age (direct aging) with a lower number of births as a secondary effect. Together with this, large cohorts that were born during the socialist era and moved to the city are aging. This is due to a shift from the productive to post-productive age (aging from the middle) as well as increasing life expectancies (aging from the top). Prievidza is facing some suburban trends like other cities, but it is also losing its population to other regions, mainly Western Slovakia and the capital city of Bratislava. Since this migration relates to younger and middle-aged inhabitants, the impact on aging is clear.

Both cities have suffered from the strong impact of postsocialist transformation, accompanied by reduced employment opportunities and the slow expansion of new businesses. While in the case of Banská Štiavnica, shrinkage has been visible for two centuries (with a new phase occurring after state socialism), in the case of Prievidza, major changes only began after 2000. Although there is an absence of detailed social and economic data at the city-level, district-level data is available. The population of the district of Prievidza in 2018 was 95.5% of its population in 2000, while the population of the district of Banská Štiavnica in 2018 was 95% of its population in 2000. The changing situation is reflected in the higher level of unemployment, which culminated in 2012 at 13% in Prievidza district, and at 18.5% in Banská Štiavnica district. Thanks to the positive influence of the economic cycle, the employment situation subsequently improved. The unemployment rate in Prievidza district has remained close to the national average for a long time (e.g., it was 5.2% in Prievidza district compared to the national average of 5.9% in December 2017) and reflects stability partly guaranteed by the continued role of the local economy in the national energy production system and its ability to attract some new investors. The social situation and attractiveness of both cities are reflected in the average monthly salary of their inhabitants, which is below the national average in both cases (2017 average salary in Slovakia—1095 EUR, in Prievidza district—993 EUR, in Banská Štiavnica district—690 EUR). In Banská Štiavnica, salaries are stagnating and uncompetitive. The change of salary level in Prievidza is parallel to the nationwide trend, although on a lower level. While access to jobs has improved, average salaries remain low compared to better-performing Slovak regions due to employment in less knowledge-intensive and low-paid sectors of the economy.

Effects of and adaptation to shrinkage

The postsocialist transformation and its accompanying reforms had many new and not always positive effects for cities. Many anti-shrinkage activities were adopted under the pressure of changing local economic and social situations. Major responses have occurred during the last two decades thanks to market economy consolidation, the clarification of the role of local self-government and the state, decentralization, and better access to international experience and resources.

Postsocialist transition and anti-shrinkage activities in Banská Štiavnica

Due to long-term shrinkage, initiatives addressing depopulation have been implemented in Banská Štiavnica since the 19th century. These measures had the greatest mitigating effect during the socialist regime (1948–1989) but were not sustainable after 1989. Earlier mitigation activities included industrialization (establishment of a tobacco factory in the 19th century; new wood processing and furniture, engineering, textile/clothing industry; completion of a standard gauge railroad after 1945; an attempt to revitalize mining and ore processing since the 1970s), heritage revitalization (renovation of historical buildings after 1978; initiation of massive national media attention to valuable local heritage since the 1980s), quaternary sector development (general and vocational secondary education; tertiary education—field centers of some universities; scientific and cultural institutions location—Slovak Academy of Sciences branch, Mining Museum, State Mining Archive); and, improved living conditions (public housing construction to attract new inhabitants and to resettle inhabitants who previously lived in the neglected historical city center and new public facilities in 1980s).

The postsocialist transition has posed a serious challenge to the local society. Many larger industrial employers collapsed (mining activities ceased in 1993). The local tobacco factory was closed a few years after its privatization by the multinational company Reemtsma (later Imperial Tobacco). The opening of global markets led to the collapse of the non-competitive local textile and clothing industries. Local engineering factories closed or experienced diminished production. Other local industries, such as wood processing and construction, also faced problems. Only residual minor employers remained active in these traditional sectors. This contributed to higher unemployment, migration, and abandoned premises.

However, the postsocialist transition also offered many new opportunities. Numerous initiatives addressed the rising pressure of shrinkage. Adopted strategies (City of Banská Štiavnica 2006) emphasized the need to change the development base from mining and industry to tourism, small and medium businesses, and public sector activities (state administration, education, health, and social services). In particular, tourism-oriented activities moved to the forefront of local development. The city's rich mining history, attractive buildings, and surrounding natural beauty, together with its status as a UNESCO World Heritage Site since 1993, is exploited as in other former mining cities around the world (Pretes 2002, Dale 2002). The first Geopark in Slovakia, which combines mining monuments with the unique environment of the most massive Carpathian stratovolcano, was established here. A new winter tourism center on the outskirts of the city (Salamandra Ski Resort) has been operating since 2010. Other locally specific, mild adventure tourism products are under preparation (City of Banská Štiavnica 2019). The number and variety of events are increasing each year. Positive trends are documented by increases in the number of beds in accommodation facilities from 665 in 2001 to 1,182 in 2017 (data for the Banská Štiavnica district, Statistical Office of the Slovak Republic 2019).

Ongoing shrinkage mitigating efforts include small and medium business development (besides tourism services, in industries with local traditions); physical environment revitalization (historical buildings in the city core, socialist housing estates, public spaces), accessibility improvements (sustaining and improving the quality of roads), public transport (rail and bus access and service), retaining public sector functions and employment (district level offices; nationwide institutions in mining, water management, science; secondary and minor university education).

Postsocialist transition and anti-shrinkage activities in Prievidza

The Prievidza city-region has remained strongly dependent on its traditional coal-energy base (City of Prievidza 2016) and only partially diversified its local economy. The transition period affected the local economy seriously but not across all sectors. Coal production gradually decreased and the mining workforce fell below 3,500 employees in 2018 compared to 9,000 in the mid-1990s (employment peaked at an estimated 10,000–13,000 employees in the 1970s), and the Cígeľ mine was closed in 2017. The power plant in Nováky closed part of its power generators, and employment has declined to less than 250 employees (2018), compared to 1,700 during the socialist era (Slovenské elektrárne 2013, Office of the Government 2018). The leading local chemical enterprise in Nováky (Fortischem) is in operation after a partial modernization with less than 1,000 employees (2017), compared to 2,500 in the mid-1990s. Two companies (Xella and Porfix) are active in construction material production based on porous concrete and enjoy a stable position on the international market. The privatization of the traditional local food company by Nestlé in 1992 and the expansion of its culinary production (almost 800 employees in 2017, Trend 2019) has also improved the local economy. Employment in many other local companies decreased during the transition as a reaction to the necessary adaptation to market conditions as well as technological modernization.

After decades of rapid, post-WWII growth, the city has had to cope with long-term shrinkage. Suitable general conditions, including a consolidated national economy, EU membership, easily accessible expertise, and increased capacities of the local self-government have played a decisive role in such efforts. Nevertheless, the phasing-out of the local mining-energy system may prolong shrinkage and have more damaging effects on unemployment and out-migration. In order to adapt to the "post-mining" era, the main focus is directed to local economy restructuring and diversification. In particular, links to the expanding automotive industry in western Slovakia, with companies like Brose, Il Jin, and GeWiS have been sought-after. A set of small and mid-sized innovative companies in light-sport aircraft production (Aerospool), construction products (INCON), plastics (JP Plast) are also operating. However, some new foreign investors (e.g., Yazaki—wires, Geox—shoemaking) have left the city for other locations with a cheaper labor force. Mining and industrial heritage will be added to already operating tourism industry sites (the spa, castle museum, and zoo in the adjacent town of Bojnice).

Local economy reorientation also requires the large-scale adaptation of the transport, technical and business infrastructure (with an industrial park and business incubator; new research and development facilities; connection to the national highway network; modernization of the local airport, and the adjustment of the inherited local infrastructure previously serving mining and energy e.g., a new source of heating to replace centralized heating based on local coal). It also requires a reorientation of higher education, including the conversion of schools focused on mining and chemistry to better meet current needs (secondary schools, undergraduate level university education). Prievidza, as the second-largest urban center of the administrative region (the eastern part of the Trenčiansky region), would like to strengthen its role as an administrative and service center. The post-mining period will also require special environmental, social affairs, and labor market measures (e.g., focusing on the conversion of derelict land and mining areas; improving social and health services focusing primarily on the elderly; requalification opportunities). Expanding foreign investments (e.g., expansion of companies in the automotive industry) and increasing the use of the local airport (better access to business travelers and tourists, increased popularity as glider airport, providing more services) demonstrate the development potential of this region.

Governance and local finance in shrinking cities

Governance

Both cities are coping with shrinkage thanks to a practical application of the local and multi-level governance approach. Shrinkage mitigation activities have been organized in partnerships and in coordination with many stakeholders, with city governments playing a crucial role. Both cities, together with other local self-governments in the region, have called attention to the shrinking and threatened economic base. Other entities have participated in the preparation and implementation of development initiatives and plans (e.g., master plans, development strategies, community plans, proposing measures and projects).

While miners (represented by labor unions) are among the key stake-holders, they oppose the changes discussed herein. Compared to the closure of mines in Banská Štiavnica in the early transition period, miners in Prievidza had been able to achieve better conditions. Local branches, as well as central representatives of labor unions, have exerted moderate pressure (petitions, negotiations) regarding the suitable mitigation of mine closures and their impact on miners (My Horná Nitra-SME 2018). Thanks to a well-developed social dialogue adopted at all levels, a suitable "exit" strategy and special measures were developed to guarantee their livelihood. This includes specific legislation (Act 385/2019 Compensation Benefits for Miners) that provides generous financial assistance for laid-off miners.

Despite the absence of a national policy addressing shrinking cities, both Banská Štiavnica and Prievidza have enjoyed extensive attention from the national and regional governments. Banská Štiavnica enjoys the support of central state institutions thanks to individual legislation—Act No. 100/2002. It resulted in financing development projects, the location of its institutions, and subsidizing of historical heritage reconstructions. In the case of Prievidza, shrinkage has been countered by ongoing attention to its mining and energy systems. Considerable state resources have been allocated to address the survival of local mines and energy production each year, including more than 100 million EUR per year in the last few years. It is combining three approaches: hidden subsidies in electricity prices (approximately 60% of the annual sum), tax exemptions on coal used for electricity production (approximately 30%), and direct subsidies provided by the Ministry of the Economy to the mining company (Haluš 2011). Recently, the central government (PwC and Office of the Vice-Premier for Investment and Informatization, 2019) has actively participated in formulating specific measures addressing the problems of the Prievidza region. Both cities also benefitted from national development policies addressing Slovak regions in general. Regional self-government is involved thanks to its responsibility for secondary education, transport infrastructure, and its coordinating role in regional development and tourism.

Progress in anti-shrinkage activities has also been achieved thanks to cooperation with the private and nonprofit sectors. In addition to creating jobs, private businesses have participated in the preparation of development documents and offered their own projects for future expansion (PwC and Office of the Vice-Premier for Investment and Informatization 2019). Their involvement in the historical heritage revitalization of Banská Štiavnica and the modernization of the local airport in Prievidza is especially noteworthy. Numerous NGOs, associations, and inter-sectoral institutions focus on environmental improvement, heritage protection, social care (e.g., addressing the elderly and disabled), and enrichening cultural life. Selected local activities are supported by large national NGOs such as Pontis and Ekopolis. The NGOs have also been actively calling for the closure of the mining–energy chain in the Prievidza region and supporting the revival of traditions in

Banská Štiavnica. Intersectoral regional tourism organizations are active in both city regions (marketing, coordination in tourism development, including the presentation of coal mining traditions). The business and innovation center (BIC-TI), as well as the Regional Development Agency based on partnership principles, operate in Prievidza.

Attracting external partners with relevant expertise and funding capacities is an important aspect in coping with shrinkage. EU institutions and specialized agencies are crucial for Prievidza. In the case of Banská Štiavnica, its status as a World Heritage Site has had a substantial mobilizing effect. Various external consultancy and advisory bodies provide specialized know-how. They have participated mostly in the elaboration and implementation of various development strategies. For example, Baten&Partners Slovakia participated in a project focusing on the integrated urban development of historic cities in Banská Štiavnica. PricewaterhouseCoopers (PwC and Office of the Vice-Premier for Investment and Informatization 2019) prepared an action plan addressing the transformation of the Prievidza mining region. These cities frequently use the expertise of Slovak universities and research centers, as well as the experiences of their sister cities.

Local finance

Slovak public finances have less extensive fiscal decentralization. Shrinkage-related processes further reduce the financial capacity of city self-governments. Nevertheless, fiscal pressure has not dramatically threatened cities' current budgets. However, more costly measures and development initiatives depend on the ability to develop multi-source financing schemes that include other stakeholders (regional self-government, state, private sector, EU).

The impact of shrinkage on city self-government finances seems less serious due to the reduced scope of powers allocated to local governments, as well as the Slovak local finance system's equalizing/redistributing features. The primary sources of local self-governments revenue in Slovakia include personal income tax (PIT, as shared tax—70% of the total national yield goes to local self-governments), property tax, local fees, and subsidies from the state budget (to selected powers). Pure local income is only comprised of property tax and selected local user fees. Due to population decline, the most sensitive issues for city governments in financial terms are the national formulas of PIT redistribution according to the population number and the subsidies for primary education based on the number of pupils. Cities are not directly financially dependent on local economic situations (e.g., in their main tax revenues), with their minor impact on property tax yield. They also have a minor role in funding costly public policies addressing unemployment, social care, infrastructure networks (provide regional or nationwide companies), and housing. It reduces pressure for more intensive debate on rightsizing and this model gives city representatives a solid basis for day-to-day operation. During the positive part of the economic cycle, accompanied by rising salaries and low unemployment, local self-governments have a guaranteed income increase. Nevertheless, cities tend to initiate decisions forced by financial scarcity and inefficiency caused by shrinkage.

Prievidza recently lost a large portion of its budget due to its depopulation and relegation to a lower category of cities with less than 50,000 inhabitants since 2012 (more than 500,000 EUR per year). Both cities face decreasing usage of their facilities due to shrinkage. The most typical are school closures caused by the lower number of students. Between 2005 and 2010, two primary public schools in Prievidza and one primary public school in Banská Štiavnica were closed. The lack of demand places pressure on some public services provided locally in Banská Štiavnica (e.g., specialized health services) that face obstacles related to decreasing the scale of their operation. Both cities expect costs associated with care for the elderly and citizens with disabilities (e.g., former miners) to continue rising. Community plans for social services (e.g., City of Prievidza 2018) anticipate new facilities and new service users. These requirements can only be satisfied by the coordinated efforts of all levels of government, insurance companies (social, health), and citizens.

The existing local finance model and impacts of shrinkage have also reduced governments' fiscal capacity to invest in larger-scale measures. They are strongly dependent on external sources of loans or financial transfers provided by higher levels of government, among other non-governmental entities. In 2010, Prievidza almost faced excessive debt proceedings due to its large portion of the debt relative to incomes. The city borrowed money to finance a new industrial park development; however, a lack of capital prevented more comprehensive plans for derelict areas and brown-field regeneration, as well as the conversion of abandoned buildings for new purposes. Thus, the role of the national government is crucial in managing and financing large-scale measures.

Discussion

Banská Štiavnica and Prievidza are among the most typical cases of shrinkage in Slovakia influenced by various phases of their mining and industrial development. Despite the complicated postsocialist era, they have exerted considerable effort to deal with shrinkage. They must form new development bases, stabilize their populations, maintain a suitable physical structure, and find and adapt their functioning to the new conditions. Their approach has been largely based on economic restructuring and diversification. However, overcoming shrinkage is far from complete. As in the case of mining cities in general, various activities (land reclamation, redevelopment, energy source replacements) will be needed for decades. It is clear that the cities alone will not be able to cope with shrinkage. While Banská Štiavnica is already less dependent on external assistance, Prievidza and its phasing-out mining industrial base will be dependent on massive external support for at least the next five to seven years.

The experience of both cities shows that postsocialist development has various local phases. While Banská Štiavnica was faced with large-scale changes immediately after 1989, the core of Prievidza's local economy remains in the final stage of postsocialist transformation and at-risk of shrinkage. In the case of coal mining, the transition has been more careful due to the importance of coal for energy production, as well as the position of miners at the forefront of the working class and their links to influential leftist political parties. This, in turn, meant significant delays to the closure of mines and local transformation. In the case of Prievidza, it also led to a sort of "agony". The inhabitants did not trust the future of mining (e.g., local mines had to attract miners from outside the region), and they searched for other opportunities, including out-migration. Nevertheless, the diminishing number of miners (e.g., compared to other postsocialist countries), their "dwindling" respect within the postsocialist realm, rising environmental awareness, the public's declining acceptance of huge state subsidies for energy prices and a "compensation agreement" with miners brought about a substantial change in the approach to mining.

Both cities offer useful approaches to their future. It seems, however, that overall the response is better in those cities that enjoy a stronger position within the national economy and the settlement and have a generally recognized role in national historical heritage, with multiple values to protect. The ability to declare at least interim urgency (e.g., by local self-governments, NGOs) increases awareness and the mobilization of support. The danger of deep socioeconomic problems, environmental threats, and the loss of national heritage can motivate important entities to act thoughtfully. The successful placement of shrinkage-related problems in the broader political environment (national, international) is essential. Both cities, in cooperation with national authorities, have been able to shift their problems to the international framework, which has improved their chances for more efficient transformation (Banská Štiavnica as a UNESCO World Heritage Site; the Prievidza region on the EU's coal mining and energy transformation agenda, Donnari et al. 2018).

Banská Štiavnica and Prievidza adopted a mixture of local and multi-level governance approaches in addressing their prospect. Under the leadership of active local self-governments, both cities developed extensive local cooperation practices among various actors, including citizens. The ideas of governance and intersectoral partnership also expanded into the field of planning. After two decades of local planning experiences (regularly updated master plans, strategic development plans, community plans), the latest documents are more realistic about future population growth and better identify local situations. They are more interconnected, search for new endogenous potential, and focus on possible measures and implementation. The local effort is crucial but insufficient. The scale and complexity of shrinkage require coordinated participation across levels of government and between all involved stakeholders (public, private, non-governmental). Local planning outcomes have been assessed and incorporated into regional and national policies (economic, energy, transport, regional development, social). They include the implementation of measures that respect cities' needs and interests (e.g., motorway access, new central heating source, new social facilities, allowances to miners, new administrative and public sector functions, and motivations to investors).

The impact of international experiences, policy transfer, and inspiration are well visible, especially in directing economic restructuring and diversification. Like other mining cities (Dale 2002, Pretes 2002; Ballesteros and Ramirez 2007, Harfst and Wirth 2011, Martinez-Fernandez et al. 2012, Marais et al. 2018, Jaroszewska 2019), they turned to services, tourism and culture (museums, cultural events), and re-industrialization focusing more on traditional non-mining industries. They also support residential functions (modernization, affordable housing), provision of public services to their hinterlands, good commuting conditions for residents, and accessibility for business activity. The two case study cities also confirm the importance of size, location, and local condition awareness. There are limited chances to imitate successful strategies of large re-growing cities (Wiechmann and Pallagst 2012, Rink et al. 2012), for example, by focusing on the development of extensive cultural and creative sectors, knowledge-intensive economic activities, large manufacturers, or desirable office locations. However, Prievidza has managed to attract new investors into its new industrial park thanks to an initiative oriented on global industrial production networks, which was already operating in Slovakia and neighboring states (primarily in the automotive industry). Additionally, in light of growing levels of media attention to their transformations, the image of both cities has improved. Despite these activities, outcomes have been limited to the mitigation of shrinkage.

Conclusion

The two Slovak cases document potential approaches to urban shrinkage in smaller countries, with a national urban system dominated by smaller cities (only 10 out of 140 cities in Slovakia have more than 50,000 inhabitants) and relatively few cities characterized by acute shrinkage. In addition to intense local efforts, the approach allows strategic external entities (primarily the national government) to pay more individual and systematic attention to selected cities facing such problems (individual legislation, the concentration of funding), despite the absence of national policies focusing on shrinkage. The inclusion of higher levels of decision-making bodies (e.g., the relevant ministries), in addition to taking leadership on particular issues (planning, funding large investments, mitigating the social and labor market impact) is extremely important. They have sufficient power to influence legislation, financial allocations, and attract the participation of top-level stakeholders, for example, large corporations. Despite the important role of local resources, stable, massive, and longer-term financial support from national and international sources is also crucial.

Banská Štiavnica, which has been coping with shrinkage for a long time, is closer to being considered as "shrunk" in a new sustainable equilibrium (with minor oscillations, e.g., in population number). The national and regional support framework has been stable without any new strong impulses since the turn of the century. Restructuring and adaptation processes have been issues for decades, with extensive experience and a strong mobilization of local capacities based on deeply rooted local identity of citizens. It has consolidated its local economy in tourism, public services, small and medium business in various sectors, coupled with a flourishing cultural and social life based on traditions. It is also an attractive location for residents commuting to work outside the city. Renovated historic buildings, improvements of public spaces, and the modernization of socialist housing estates have contributed to the stabilization of the local community.

Prievidza, as a young shrinking city, faces greater uncertainty. Its most acute challenge is connected to the end of mining and the conversion of related industries. Despite the active participation of external actors and promised support, the outcome is less clear and the threat of further population declines remains. Nevertheless, the final decision concerning the closure of mines in 2023 and intensive work on a clearer vision for this urban region are positive steps. For one, debates about "when it will happen" have ended. This step has facilitated new and more positive expectations among the local population and political elite. Furthermore, investors are thinking about the workforce available from the mining and traditional industries. Due to the combined efforts of several levels of the government, the EU, and the business community, the prospects are better than they were a few years ago. The Prievidza case also documents the substantial shift to a more elaborated, "ex-ante" prepared conversion (PWC and Office of the Vice-Premier for Investment and Informatization, 2019) of the mining region compared to Banská Štiavnica. In the later stage of the transition, more intensive international links and shared experience contribute to this. Prievidza, as a higher ranked and larger city region, with a more important economic position and better location, also has better preconditions to mobilize external support compared to Banská Štiavnica at present. It may yet mitigate shrinkage or even stabilize its population in years to come.

Acknowledgments

This chapter had been prepared with the support of Slovak Research and Development Agency (APVV) project No. 17-0079.

References

Ballesteros, E. R. & Ramirez, M. H. 2007. Identity and Community—Reflections on the Development of Mining Heritage Tourism in Southern Spain. *Tourism Management*, 28, 677–687.

Bleha, B. & Buček, J. 2015. The Decade of Shrinking in Bratislava at the Turn of Centuries: An Attempt for Explanation. *Acta Geographica Universitatis Comenianae*, 59(2), 161–172.

Buček, J. & Bleha, B. 2013. Urban Shrinkage as a Challenge to Local Development Planning in Slovakia. *Moravian Geographical Reports*, 21(1), 2–15.

Buček, J. & Bleha B. 2014. Bańska Szczawnica—przeciwdziałanie negatywnym skutkom kurczenia się; lokalne inicjatywy a wsparcie zewnętrzne (Fighting the adverse effects of shrinkage in Banská Štiavnica: Local initiatives and external support). In Stryjakiewicz, T. (ed.) *Kurczenie sie miast w Europie środkowo-wschodniej* (pp. 39–50). Poznan: BWN.

City of Banská Štiavnica. 2006. *Program hospodárskeho, sociálneho, environmentálneho rozvoja mesta Banská Štiavnica (Program of economic, social and environmental development of Banská Štiavnica)*. Banská Štiavnica: City Office and Ekotrust.

City of Banská Štiavnica. 2019. *Strategický dokument rozvoja zážitkového turizmu lokality Banská Štiavnica a okolie (Strategic document of adventure tourism development in Banská Štiavnica and its surrounding)*. Banská Štiavnica: City Office and Geotour.

City of Prievidza. 2016. *Program rozvoja mesta Prievidza 2016–2023 (City of Prievidza Development Program 2016–2023)*. Prievidza: City Office.

City of Prievidza. 2018. *Komunitný plán sociálnych služieb 2018–2022 (The Community Plan of Social Services 2018–2022)*. Prievidza: City Office.

Dale, B. 2002. An Institutionalist Approach to Local Restructuring: The Case of Four Norwegian Mining Towns. *European Urban and Regional Studies*, 9(1), 5–20.

Donnari, E. et al. 2018. Socio-economic transformation in coal transition regions: Analysis and proposed approach: Pilot case in Upper Nitra, Slovakia. *JRC Science for Policy Report*. Luxembourg: Publications Office of the European Union.

Federal Statistical Office. 1978. *Retrospektivní lexikón obcí Československé socialistické republiky 1850–1970 (Retrospective lexicon of communities in the Czechoslovak Socialist Republic 1850–1970)*. Praha: Federální statistický úřad.

Haase, A., Rink, D. & Grossmann, K. 2016. Shrinking Cities in Post-Socialist Europe: What Can We Learn from Their Analysis for Theory Building Today? *Geografiska Annaler: Series B, Human Geography*, 98(4), 305–319.

Haluš, M. 2011. Podpora na baníka predstavuje dvojnásobok jeho hrubej mzdy (Aid to a Miner is Twice his Gross Salary). Institute of Finance Policy: Commentary, 9, Bratislava: Ministry of Finance of the Slovak Republic.

Harfst, J. & Wirth, P. 2011. Structural Change in Former Mining Regions: Problems, Potentials and Capacities in Multi-level-Governance Systems. *Procedia: Social and Behavioral Sciences* 14, 167–176.

Hummel, D. 2015. Rightsizing Cities: A Look at Five Cities. *Public Budgeting & Finance*, 35(2), 1–18.

Jaroszewska, E. 2019. Urban Shrinkage and Regeneration of an Old Industrial City: The Case of Wałbrzych in Poland. *Quaestiones Geographicae*, 38(2), 75–90.

Marais, L., McKenzie, F. H., Deacon, L., Nel, E., van Rooyen, D. & Cloete, J. 2018. The Changing Nature of Mining Towns: Reflections from Australia, Canada and South Africa. *Land Use Policy*, 76, 779–788.

Martinez-Fernandez, C., Wu, C-T., Schatz, L.K., Taira, N. & Vargas-Hernández, J.G. 2012. The Shrinking Mining City: Urban Dynamics and Contested Territory. *International Journal of Urban and Regional Research*, 36(2), 245–260.

Mládek, J. 1990. *Teritoriálne priemyselné útvary Slovenska (The Territorial Industrial Systems of Slovakia)*. Bratislava: Univerzita Komenského.

My Horná Nitra-SME. 2018. Banskí odborári spustili petíciu, žiadajú zachovať ťažbu na hornej Nitre (Mining trade unionist started petition to sustain mining in Upper Nitra). Available at https://myhornanitra.sme.sk/c/20811458/ (Accessed 25 April 2020).

Office of the Government. 2018. Problematika transformácie regiónu horná Nitra v súvislosti s návrhom všeobecného hospodárskeho záujmu na zabezpečenie bezpečnosti dodávok elektriny (Upper Nitra region transformation issues—public service obligations in general economic interest to protect electricity delivery security). Bratislava, Available at https://rokovania.gov.sk/RVL/Resolution/17440/1 (Accessed 18 July 2019).

Pallagst, K., Wiechmann, T. & Martinez-Fernandez, C. (eds.) 2014. *Shrinking Cities: International Perspectives and Policy Implications.* London: Routledge.

PWC and Office of the Vice-Premier for Investment and Informatization. 2019. *Transformácia uhoľného regiónu horná Nitra—Akčný plán (Action plan for Transformation of the Upper Nitra Coal Region).* Bratislava: PwC EU Services EESV and Office of the Government.

Pretes, M. 2002. Touring Mines and Mining Tourism. *Annals of Tourism Research,* 29(2), 439–456.

Richardson, H. W. & Nam, C. W. (eds.) 2014. *Shrinking Cities: A Global Perspective.* London: Routledge.

Rink, D., Haase, A., Grossmann, K., Couch, C. & Cocks, M. 2012. From Long-Term Shrinkage to Re-growth?: The Urban Development Trajectories of Liverpool and Leipzig. *Built Environment,* 38(2), 162–178.

Rumpel, P. & Slach, O. 2012. Je Ostrava "smršťujícím se městem"? (Is Ostrava a Shrinking City?). *Sociologický časopis/Czech Sociological Review,* 48(5), 859–878.

Slovenské elektrárne. 2013. *Elektrárne Nováky—60. rokov (Power station Nováky—60. years).* Bratislava: Slovenské elektrárne.

Šprocha, B., Bleha, B., Vaňo, B. & Buček, J. 2017. *Perspektívy, riziká a výzvy demografického vývoja najväčších miest Slovenska (Perspectives, Risks and Challenges of the Largest Slovak Cities Demographic Development).* Bratislava: Infostat.

Statistical Office of the Slovak Republic. 2003. Historický lexikón obcí Slovenskej republiky 1970 – 2001 *(Historical lexicon of municipalities in the Slovak Republic 1970–2001).* Bratislava: Statistical Office of the Slovak Republic.

Statistical Office of the Slovak Republic. 2019. Data packages—population, social statistics, tourism statistics. Available at: datacube www.statistics.sk/ (Accessed 17 July 2019).

Stryjakiewicz, T. (ed.) 2014. *Kurczenie sie miast w Europie środkowo-wschodniej (Urban Shrinkage in East-Central Europe).* Poznan: Bogucki Wydawnictwo Naukowe.

Szőlős, J. 1993. Analýza funkčnej a priestorovej štruktúry hnedouhoľného energetického reťazca Hornej Nitry (Analysis of Functional and Spatial Structure of Soft Coal Energy Chain in the Upper Nitra Region). *Geografický časopis,* 45(1), 29–40.

Trend. 2019. Databáza firiem (Companies database). Available at: https://www.etrend.sk/databaza-firiem (Accessed 19 July 2019).

Turok, I. & Mykhnenko, V. 2007. The Trajectories of European Cities, 1960–2005. *Cities,* 24(3), 165–182.

Wiechmann, T. & Pallagst, K. M. 2012. Urban Shrinkage in Germany and the USA: A Comparison of Transformation Patterns and Local Strategies. *International Journal of Urban and Regional Research,* 36(2), 261–280.

Wolf, D. A. & Amirkhanyan, A. A. 2010. Demographic Change and Its Public Sector Consequences. *Public Administration Review,* 70, 12–23.

19 Why is Ostrava in Czechia still shrinking?

Petr Rumpel and Ondřej Slach

Introduction

The transformation and adaptation processes in Central and Eastern European Countries (CEECs) started after the wave of democratic revolutions in 1989, and the 1990s led to significant political and economic changes at different spatial scales (European, national, regional, local), as well as to and spatial differentiation and polarization of countries into winners and losers (Heidenreich 2003). Some regions and cities performed better during this transformation, including capital cities with a more diverse economic base, as is the case of Prague, the capital of Czechia. Others, however, especially industrial cities with an outdated economic base (such as former coal mining cities), had a less successful transformation, and their former attractiveness was lost. Ostrava, as the third biggest city in Czechia, is the regional capital of the Moravian-Silesian region (one of 14 administrative regions of Czechia), which has been losing economic attractiveness and population since 1990 because of its former dependence on coal mining and iron and steel production and processing. Such losing cities can be termed "shrinking cities", as urban shrinkage is a pathway of urban development characterized by declining numbers of inhabitants. A declining population is subsequent to macro-processes in the economic, social, and political systems or subsequent to natural disasters (Haase et al. 2014). Theoretically, such shrinkage can be explained as a result of different but strongly interconnected processes, such as spatially uneven economic development and peripheralization, second demographic transition, and suburbanization (Couch et al. 2007, Martinez-Fernandez et al. 2012).

Urban shrinkage is specific to a given context, as each city has its characteristics, distinct features derived from its specific historical, political, economic, and social conditions. On the other hand, broader or global phenomena can influence the development trajectory of a city much beyond the local context (Turok and Mykhnenko 2007, Reckien and Martinez-Fernandez 2011, Großmann et al. 2013). For the development of Central European cities, such phenomena include globalization (and global competition in traditional manufacturing), European integration and enlargements

DOI: 10.4324/9780367815011-23

(e.g., application of EU law and policies), a transformation of a political and economic system, restructuring, or economic crises. Furthermore, a different temporal and spatial context of urban shrinkage occurs in the postsocialist CEECs compared to the Western member states of the EU (Mykhnenko and Turok 2008, Buček and Bleha 2013, Stryjakiewicz and Jaroszewska 2016).

Our aim is to analyze the trajectory and causes of the population and economic shrinkage in Ostrava during the period from 1990 to 2019. We will describe the consequences of this urban shrinkage for the policymaking and governance structures as well as the processes and projects of economic and urban regeneration. We intend to explore and determine whether Ostrava will be a shrinking city in terms of demographic and economic development. We use the emerging regrowth theory (e.g., Power et al. 2010) as a comparative contrasting framework for the assessment of policy initiatives dealing with demographic and economic shrinkage and influencing the desirable (by policymakers) population and economic regrowth. The general research questions are:

What are the primary or secondary causes of the urban shrinkage in Ostrava in the period 1990–2020? What does the trajectory of the urban shrinkage look like?

What factors—external and internal—are influencing Ostrava's shrinkage?

What have we learned from regrowth theories for the policy initiatives facing population and economic shrinkage? What are the factors and contexts of population and economic regrowth?

What are the governance strategies and measures in the shrinking city of Ostrava dealing with shrinkage and struggling for population and/or economic regrowth? Were there any strategies or policy measures that "successfully" tackled shrinkage during certain policymaking stages, and what were those?

Does urban governance in Ostrava execute appropriate strategies and measures to turn the shrinkage trajectory toward regrowth? Is Ostrava still a shrinking city losing population, or has it made the turn toward regrowth in the field of population and/or economic regrowth?

What were the successful/failed/missing policy initiatives, strategies, measures dealing with shrinkage and working toward population and/or economic shrinkage?

What will the future population and economic development trajectory of Ostrava be like according to our findings like?

Drawing on the conceptual and analytical framework of urban shrinkage and urban regrowth, a mixed-methods research design including both quantitative and qualitative approaches and involving stakeholder interviews was applied to the case study.

It is difficult to politically "promote" the concept of shrinkage, especially in European countries where growth is considered to be one key measure of political success (Haase et al. 2016). Although linked to many negative consequences, policymakers can respond to shrinkage using different strategies. Hospers (2014) identified four strategies to handle shrinkage. Firstly, if the shrinkage is insignificant, it can be ignored by politicians, resulting in the so-called strategy of trivializing shrinkage. Secondly, if depopulation is only a temporary problem, it can be tackled by attracting new people and businesses. This strategy of countering shrinkage is adopted through policy and measures aimed at fostering urban growth, including building new residential areas and landmarks, along with place marketing. Thirdly, if the shrinkage is highly significant, it should be accepted as normality (in contrast to desirable growth). While accepting shrinkage, policies should aim to mitigate the negative effects of shrinkage and retain residents by improving quality of life, renovating the housing stock, and providing more green space (Hospers 2014). Fourthly, shrinkage can be added to the political agenda, linked to discourse on economic decline and the necessity to combat unemployment, to undesirable drops in birth rates and a surplus of kindergartens and schools, to aging and the necessary pension and health care reforms, to unwanted commercial and residential vacancies, to emerging unattractive vacant lots and brownfields, to social segregation and exclusion, and others (Rink et al. 2010, Reckien and Martinez-Fernandez 2011). This represents a strategy of utilizing shrinkage (Hospers 2014) in which policymakers take advantage of shrinkage by, for instance, demolishing housing blocks to lay out green space for urban farming. Hospers (2014) suggests that accepting shrinkage by improving the quality of life for the city's existing residents is the most suitable and sustainable strategy. Dealing with shrinkage is a complex urban governance process that requires a mental transformation from growth to shrinkage as well as regional rather than local thinking (Hartt 2018). Similarly, Pallagst et al. (2017) deal with changes in planning systems in cities in Germany and the United States during post-industrial transformations and the consequent population shrinkage; they present three planning strategies for shrinking cities: expansion strategy and economic growth projects; maintenance strategy and raising attractiveness and inner-city development; planning for decline and right-sizing with comprehensive planning strategies (Schilling et al. 2008).

Some cities across the world and in Europe have made the turn from long-term population and economic shrinkage to becoming regrowing cities. There is no commonly recognized definition of a regrowing city. However, regrowing cities mainly refer to cities whose population or economy begins growing after a long-term decline (see Power and Katz 2016). Power and Katz (2016, pp. 288–289) and Haase et al. (2021) emphasize the following for a regrowing city: a sustained and coordinated public sector leadership; massive external funding and investment in economic redevelopment; provision of affordable, attractive housing, amenities, and infrastructure; economic regeneration and the creation of new jobs; revitalization of the inner city; control of suburbanization in cooperation with suburban authorities, and strong inward-oriented planning policies. Another comparative study of post-industrial European and US cities that successfully made the turn toward new growth (Carter et al. 2016) highlighted that regrowth is a long-term process. Carter et al. (2016, pp. 237–240) add several lessons such as the recommendation to have a long-term vision, and the courage to take risks, a metropolitan focus instead of mere urban, the use of public-private partnerships and diversification of the economy, and new education strategy as a must.

The above-mentioned theory derived from empirical evidence in regrowing cities in Western Europe and the United States can become the point of departure for the formulation of strategies dealing with shrinkage and struggling for regrowth, at least for cities in the CEECs, which are new EU members with strong cohesion policies and efforts. As scientists, we are aware that there can be further population and/or economic shrinkage or there can emerge a decoupling of population and economic development. Then, policymakers should rather get used to population shrinkage and consider it as a chance for qualitative changes in urban space and economy. On the one hand, we do not have to follow Western regrowth models and experiences but can also take into consideration also national and local specificities. On the other hand, we can get inspired by policy measures derived from Western regrowth theory. Nevertheless, the quite different macroregional and national contextual conditions and specific factors of shrinkage and regrowth in the CEECs as countries with a communist past must be taken into account when formulating strategies dealing with shrinkage or struggling for regrowth, especially traditional industrial ones. A one-size-fits-all approach using the body of knowledge on regrowth in Western Europe and the United States will not work. According to Rink et al. (2012), there is an ambivalence of those success factors which lead to population and economic regrowth, which were identified at the time of research (e.g., Power and Katz 2016, Carter et al. 2016). What today may support regrowth, may tomorrow lead to new problems and hinder regrowth. Haase et al. (2021) conclude that several factors are responsible for regrowth, but it is not just their presence or absence that is the key determinant. Much more important is their interplay and contingency, the combinations in which they are present or absent, and the impacts of contextual conditions on those factors.

Urban development trajectory and causes of shrinkage in Ostrava and its region

The development trajectory of Ostrava consists of more than 160 years of economic and population growth based on hard coal mining, followed by 30 years of economic and population decline. After the discovery (1763) and extensive utilization of hard coal and coke (1834) for the iron and steel industries, as well as for the heavy machinery and chemical industries, Ostrava grew dynamically until 1989. The industrialization process and the increasing number of jobs generated growth of physical structures and the population between 1830 and 1990. In this period, industrialization induced and pulled urbanization and thus population growth. Particularly in the communist period (1948–1989), Ostrava was named "The Steel Heart of the Republic", and its development was promoted according to communist economic ideologies based on the growth of power industries and heavy industries within a centrally planned economy, completely nationalized by the Communist party. Within this period, Ostrava was quite an attractive city and had the highest level of wages across Czechia.

Czechia has a centralized public administration system combined with regional and local/municipal self-government. The most important things for city and regional development are the structural conditions defined by law, national macroeconomic policy, and structural development programs at the national level. The process of Europeanization (adaptation to and adoption of the law and policies of the European Union) in the CEECs since their entry into the EU in 2004 is particularly important. The EU membership and EU structural funding conditions have a profound influence on decision making about development priorities at both regional and local levels (Rink et al. 2014).

Ostrava's development has been dependent on the contextual and external development conditions within Czechia. Simplifying the development pattern of Czechia between 1990 and 2019, the dynamic population and economic growth of Prague (the capital city) with its adjacent municipalities in Central Bohemia emerged (David et al. 2013). On the other hand, long-term population shrinkage emerged in old industrial and mining regions, such as the agglomeration of North-West Bohemia (with the regional capital Ústí nad Labem) and the Moravian-Silesian region with the regional capital Ostrava (Maier and Franke 2015). Certainly, rural border areas and internal peripheries inadequately equipped with infrastructure have also been shrinking (Havlíček et al. 2008, Bernard and Šimon 2017), as has happened throughout Europe.

The Moravian-Silesian region (established in 2000) is a shrinking region as well, currently having 1,200,533 inhabitants—which is 65,000 fewer inhabitants compared to 2000. The main reason for the demographic shrinkage at the regional level is out-migration to other regions, amounting to 7,439 persons in 2018 (compared to an in-migration of only 6,114 persons). The secondary reason for the population shrinkage in the region is the negative natural balance, with 1,262 more deaths than births in 2018. Furthermore, mining cities located eastward from Ostrava—Orlová or Karviná—are typical shrinking cities with negative consequences for further development (Orlová had 38,364 inhabitants in 1990 and 28,852 in 2019; Karviná had a peak of 81,882 inhabitants in 1977 and 52,820 in 2019). The situation of these two shrinking cities is even worse than that of Ostrava, which, being the regional capital, has a more diversified economic structure, along with more and better amenities.

Ostrava experienced its peak of population growth in 1990 with 331,219 inhabitants, declining by 13.7% to 285,904 inhabitants in 2020 (Ministry of the Interior of the Czech Republic). Despite short periods of positive deviations in 2006 and 2007 due to slightly higher birth rates than death rates, the trend of slight long-term population decline in Ostrava between 1990 and 2020 is similar to the trends in other old industrial cities in Europe.

Today's Ostrava can be characterized as a postsocialist, old industrial city facing a negative image of a working-class city, air pollution, and social exclusion—socio-economic group segregation in most damaged parts of the city (Slach et al. 2019). However, Ostrava's development trajectory in the last 30 years can be summarized as "everything has improved, but nothing is definitely quite well". Compared to other old industrial cities (Power et al. 2010), Ostrava's transition to a post-industrial city can be considered somewhat successful, as it managed to attract investors, create jobs, and bring basic changes in its economic and urban structures, as well as providing new amenities and cultural and sports events.

The main causes of urban shrinkage in Ostrava are:

1 Deindustrialization, economic transformation, and restructuring, especially concerning mining and iron and steel industries since 1990. The loss of relative economic attractiveness (loss of relatively well-paid jobs in mining) led to job-related out-migration of young, well-educated people to other Czech regions or abroad.

2 Suburbanization—the movement of people from the inner city, or other neglected and unattractive housing estates and neighborhoods, to "villages" on the fringes of Ostrava or beyond its administrative borders, e.g., the town-district Krásné Pole, which is not a part of a compact town, increased its population by 63.3% (adding 1,062 inhabitants) between 1991 and 2020. Such population growth and development of physical structures have appeared all over the fringes of Ostrava in suburban and peri-urban areas. In the end, the Ostrava agglomeration, containing 12 other municipalities, gained population due to its higher attractiveness to suburbanites. In the period 2000–2020, the very general pattern of settlement system development in the Moravian-Silesian region with its core in the Ostrava agglomeration has been as follows: the inner city of Ostrava (and other postsocialist and industrial cities in the region) declined, and the suburban zone slightly grew because of favorable conditions for suburbanization at all levels of the territorial administration. But the Ostrava agglomeration and the whole Moravian-Silesian region declined mainly because of out-migration (approximately 700 young people leave Ostrava yearly) and demographic changes.

3 Demographic changes and the rapid drop in birth rates across Czechia as a natural adaptation to the second demographic transition of low

birth rates and prolonged life expectancy in developed countries. Ostrava has experienced declining birth rates since 1994, with a negative natural change of—328 inhabitants in 2018 (3,163 births and 3,491 deaths). Furthermore, with an elderly rate of 14.5% in 2007 and 19.9% in 2018, Ostrava is an aging city. The aging index (ratio of population over 65 years and population of 0-14 years) was 55.5 in 1991, and 132.7 in 2018 in Ostrava (similar to the aging index in Czechia). The average age in Ostrava was 38.5 years in 2000, and 42.8 years in 2018 (comparable to that recorded in Czechia). Additionally, considering the rising life expectancy from 1990 to 2010, the trend of absolute aging in Ostrava and the Czech Republic can be clearly understood. Low fertility rates and rising life expectancy determine this relative population aging, which has actually been experienced in almost all well-developed countries in the world. However, immigration from abroad has grown since the 2000s, as 11,070 foreigners lived in Ostrava in 2019, compared with 7,339 in 2005. Nevertheless, immigration—which is economically motivated and depends on the current economic performance and job opportunities available in Ostrava—cannot compensate for mid-term population losses caused by out-migration and higher death rates.

Stages of policymaking addressing shrinkage in Ostrava

The last 30 years of development in Ostrava can be subdivided into three stages of policymaking and governance, each addressing shrinkage in different ways (see Table 19.1). The first stage is that of trivializing shrinkage (1990–2003); the second stage involved both ignoring and countering shrinkage (2004–2014); the third and current stage is observing shrinkage without fully accepting or countering it (from 2015 onwards). The particular stages have specific approaches toward shrinkage and regrowth, and quite different institutional arrangements, governance structures, and policy initiatives were applied. We describe and assess these institutional changes and policy initiatives in the particular stages in the framework of emerging regrowth theory. We are aware of the limited knowledge transfer possibilities during policy learning from Western countries and significantly different institutional contextual conditions in the postsocialist CEECs, such as in Czechia, as well as the weaker role of the public sector and state as a consequence of transformation in the 1990s (Maier 2012, Pavlínek 2017).

From the perspective of regrowth theories, Ostrava is not a regrowing city and its strategy is to deal with shrinkage and stop it. It is not necessary to follow the regrowth path but rather to use the Western experience in order to further improve urban structures and the economy. Nevertheless, the city has improved and is further developing, but with a rather controversial development (Slach et al. 2019). In comparison with Western-European cities, such as Leipzig and Liverpool, that turned from economic and population shrinkage toward regrowth (Rink et al.2012), Ostrava has developed in

Table 19.1 Stages of policymaking addressing shrinkage and/or supporting regrowth

Period	1990–2003	2004–2014	2015+
Perception/ strategy	*Ignoring/ trivializing*	*Ignoring/ countering*	*Observing without full acceptance/ countering*
Instruments of regrowth	0	+	++
Metropolitan scale in focus	0	0	+
Long-term vision on rebuilding and transforming the city	0	0	+
Being bold (courageous) and ready to take risks	0	+	++
Increasing the use of public-private partnerships	0	++	++
A strong collective or individual leadership	+	+	++
Diversification of the economy	+	++	++
Investment in education	+	++	++
Developing sustainably	0	0	+
Good planning and urban design	0	0	+
New housing provision	0	+	+

Notes:
0—no activity; +—some activity; ++—intensive activity.

Source: Own data collected by authors via interviews.

totally different external conditions in the last 30 years—driven by a strong Prague centralism, prevailing neoliberal belief in free-market forces (see Sýkora and Bouzarovski 2012), and economic development, and by rejecting the pro-active role of the public sector, rejecting immigration, and by still accommodating "minigarchs" in cities and clientelism. Only by radically changing the external and internal conditions of development in Ostrava and the Moravian-Silesian region can the negative population trends be changed. However, this radical change is highly improbable, as will be explained hereafter.

Ignoring/trivializing shrinkage 1990–2003

After the velvet revolution in former Czechoslovakia, which overthrew the communist government in 1989, a process of political, economic, and social transformation began, strongly impacting the previously favored mining and industrial cities in Czechia, such as the cities in the Ostrava agglomeration.

New political elites came into power at the national and local levels in 1990, imposing a transformation process. The new reformist government and the newly established—and thus weak and inexperienced—public administration aimed to create a free market economy by setting the precondition of privatizing fully nationalized property. Politicians believed in neoliberalism as the prevailing economic and political ideology; a negative attitude toward the pro-active role of the public sector in urban development remained, as many politicians consider that the public sector can pose a "restriction to entrepreneurial, private interests". The activity of the public sector was overall seriously limited due to the lack of experience and knowledge of urban development within the new conditions of the market economy, along with a lack of finances (Rumpel and Slach 2012a). Together with the economic transformation, the ongoing restructuring and deindustrialization led to a loss of economic attractiveness and an increase in unemployment. In 1994, all the mines in the territory of Ostrava were closed, as were some other large ironworks, coke plants, and chemical factories. Consequently, significant job-related out-migration was recorded. In 1998, an economic depression started, and the public sector became more active in emulating Western policies with measures to handle restructuring and deindustrialization (see Drahokoupil 2012). In the adaptation process to a Western social structure, and as birth rates dropped, demographic changes started taking place. A significant population decline was recorded primarily in mining and industrial cities—such as the cities in the Ostrava agglomeration, including Ostrava itself. Politicians then believed that nearly all the problems faced in these cities would be solved through the free market economy, through privatization, and other changes in the economic base that were meant to lead to economic growth. The serious population decline was not taken into account, rather being completely ignored, and the phenomenon of shrinkage was not a part of the political agenda. The main problem considered was unemployment; but, with the main focus on economic restructuring and job creation, addressing the population decline did not suit the marketing plans and strategies of the city.

In 1997, the first development strategy for Ostrava was elaborated through the cooperation of the public sector, the private sector, and NGOs. The main recommendation was to establish a Department for Economic Development, meant to establish industrial and business zones and attract foreign investors. From 1998, the most important policy was attracting FDIs into the Ostrava region, which was suffering from unemployment and related phenomena such as social exclusion and indebtedness of the poorest. Furthermore, from the 1990s, Ostrava's local authority pursued an "external low road strategy" (see also Cooke 1995) based predominately on low wages and low-cost inputs and subsidies. This, however, attracted FDIs into the region, and a significant increase in jobs in IT industries and services took place. The economic development was rather controversial in Ostrava at this stage. On the one hand, Ostrava's economic base has been reindustrialized and did not follow the post-industrial route of similar Western European cities; on the other hand, there has been a considerable increase in jobs, albeit relatively low paid, or at least worse paid than in Prague or Brno. Nevertheless, the support offered to establish the University of Ostrava (in 1991), and new faculties at the Technical University of Ostrava was important in the period, which contributed to the diversification of the labor pool in the Ostrava region. The decontamination of the Karolina brownfield (a former coke plant near the city center) was also a major milestone.

To summarize, there are several reasons why Ostrava did not make the turn toward regrowth during this stage. First, the economic development of the Ostrava city region was controversial. In the transition from an old industrial city to a post-industrial city, the public and political discourse in Ostrava was primarily concerned with reindustrialization and the role of manufacturing instead of economic diversity (Rumpel and Waack 2004). Ostrava presented itself as an entrepreneurial city and beginning in the 1990s, it pursued an "external low road strategy" based on low wages, subsidies, and low costs of inputs, which, however, attracted FDI into the Ostrava region.

Ignoring/countering shrinkage 2004–2014

Czechia became a member state of the EU in 2004, thus being part of the EU common market with the possibility to get grants from EU structural funds. Such a favorable situation led to an increase in FDIs in Ostrava, including the establishment of the ICT company Tieto (2004) and the Korean automotive company Hyundai (2006). Such foreign companies chose the Ostrava city region as a low-cost location with an available and skilled labor pool. Approximately 40,000 new jobs were created in the 2000s, and, in this sense, Ostrava can be considered successful (Rumpel et al. 2013). However, during 2008–2013, Czechia, and especially the Ostrava region, were impacted by the global economic crisis (Ženka et al. 2019). EU grants were massively used for urban regeneration projects. One major brownfield regeneration project was that of the "Lower Vitkovice Area" in Ostrava, formerly an ironworks with a mine and coke plant (1828–1998), shut down entirely in 1998. After 2004, with strong political support from the Ostrava government and Moravia Silesia government, which were similar to public-private partnership (PPP), the new owner of the Lower Vitkovice Area obtained conversion grants to an industrial heritage site, with dedicated spaces for museums and conference halls (Bosák et al. 2020). Another important project was the privately financed mixed-use development of "New Karolina" on the brownfield site of a former coke plant in the city center (Rumpel and Slach 2012b). Other projects of significance concerned investments in tertiary education and R&D—with a new medical faculty and a supercomputing center (IT4I) established. The former mayor of Ostrava (2009) stated that he pursued "only economic growth to rapidly induce economic revival and to later induce population revival". The mayor's leadership toward economic regeneration was overall positively received, as it attracted a significant inflow of FDIs into the Ostrava agglomeration. Entering the EU led to favorable market conditions and the economy of the Ostrava region was diversified, as many jobs were created in new branches, such as in the automotive industry. However, some challenges were not addressed appropriately such as air pollution; the public initiative toward new, attractive, and affordable housing; much needed metropolitan planning (see Slach et al. 2015), comprehensive urban planning, and better urban design (especially concerning city center development); the need for sustainable planning toward a greener city, while a clear long-term vision for a "new" Ostrava as a post-mining city was missing.

Contrary to regrowth theory, no sprawl containment measures were imposed in Czechia. The process of suburbanization (supported by benevolent zoning regulations for land-use policy in many municipalities and by the easy availability of mortgages) accelerated in 2004 (in the second stage of shrinkage) and intensified after 2015 (in the third stage of shrinkage). A necessary and massive investment from the public sector in new, attractive, but affordable housing and in social infrastructure and amenities which could contribute to the crucial turn toward regrowth is not yet on record in Ostrava.

Observing without acceptance/countering 2015+

Institutional changes have taken place in recent years (2015–2019), leading to re-constellations in the networks of stakeholders and decision-makers responsible for the local and regional development in the Ostrava and Moravian Silesian regions. For example, in 2017, the new publicly sponsored agency, Moravian Silesian Innovation Centre (MSIC), was established and started its activity of supporting SMEs (small and medium-sized enterprises), entrepreneurship, and innovation in the Ostrava region. Also, in 2017, the former Regional Development Agency Ostrava (RDA) was re-branded as the Moravian Silesian Investment and Development Agency (MSID) to play a more important role in the transformation of post-mining cities and brownfield redevelopments. In 2019, MAPPA (City Atelier of Spatial Planning and Architecture) was established as a new agency in Ostrava to prepare urban development projects and cultivate public spaces. Moreover, the authorities of both the Ostrava and Moravian Silesian regions offer support to regional universities and major regeneration projects in the city and in the regions, together with improved and increased activity of the public sector in the Ostrava city and region in the last five years. Nonetheless, the rapid positive impact of these new supporting agencies cannot be expected in determining population and economic regrowth. These new agencies do not have any strong political position and or support, they are just gathering the necessary body of knowledge for appropriate action, and the implementation of regrowth policy and strategy will take time.

Owing to the successful economic regeneration policy adopted in the previous stages and following the structural change, the number of jobs in new companies and branches increased and the economic base became more diversified in Ostrava. The unemployment rate was approximately 5% in November 2019, which is not high in comparison with other old industrial European cities. Despite this fact, the Prague and Brno city regions still have more diverse economies and offer more attractive job opportunities and higher salaries for both a skilled workforce and graduates.

Conclusions

Ostrava's population declined by 13% (approximately 40,000 inhabitants) between 1990 and 2019, making Ostrava thus far a shrinking city in terms of population. The population shrinkage can be classified as slight, without strong negative impacts. There is an interesting phenomenon of the decoupling of economic development with relatively successful results in terms of job creation and restructuring toward a more diversified economy on the one hand and still slightly negative demographic development on the other. Nevertheless, some successful projects started enhancing the attractiveness of the city, especially after the waves of economic growth (in 2004–2008 and 2013–2019) that occurred based on the favorable conditions of EU membership with strong cohesion policy support. In general, however, the weak population decline demonstrates that, in comparison with its mining period (1830–1990), Ostrava and the whole region have lost their former high economic attractiveness, which was based on mining and the industrial economic structure before 1989.

It is thus important to deal with the causes and consequences of this light population shrinkage. On the one hand, Ostrava's policymakers have tried to copy and implement policies and measures derived from the experiences of Western European shrinking and regrowing cities in terms of economic and urban development. On the other hand, we are aware of the differences in factors and contexts in CEECs' shrinking cities, which hinder the use of Western knowledge on regrowing cities in the contexts of CEECs' shrinking cities. First, the policymakers should get used to slight population shrinkage and consider it as a chance for qualitative changes in urban space and the economy. Some measures and tools used in regrowing Western cities became an inspiration for cities experiencing shrinkage to improve their economic base and urban attractiveness.

The causes and consequences of population shrinkage were not considered of importance for policymaking or strategy formulation on the political agenda of Ostrava's public authorities for a long time. That is because of the slight population shrinkage and because of the consensus of Ostrava's politicians that economic restructuring and job creation are the right and the only way toward regrowth. The economic redevelopment was considered by policymakers to be a panacea for all urban illnesses. Overall, Ostrava managed to attract investors, create jobs, and implement basic changes of economic and urban structures, which, however, were insufficient in contributing to population regrowth in the city of Ostrava.

Across the analyzed 30-year period of Ostrava's regeneration efforts, policymakers mainly placed emphasis on economic regeneration via reindustrialization, manufacturing development, and service sector development, especially ICT. Simplifying and generalizing the situation throughout this period, politicians considered technical education and technical skills for manufacturing development to be the most appropriate solutions for modernizing the Ostrava region. Only since 2015 (the latest stage of shrinkage) has the emphasis started to be placed more clearly on innovation and diversification by implementing a regional innovation strategy. It is further necessary to switch from "low road" to "high road" economic development through innovation, quality products, and services, along with the high-added value. The case study of Ostrava as still a slightly shrinking, not demographically regrowing city shows that relatively successful economic diversification and economic regrowth in the neoliberal policymaking framework is not enough to change the demographic trajectory toward regrowth.

Massive external support by the national scale and public investment into modern affordable housing, infrastructure, and amenities are particularly important conditions for population regrowth. This means that economic recovery itself does not necessarily lead to new population growth as long as an appropriate policy for the delivery of affordable housing in the inner city or control of suburbanization is missing. On the contrary, suburbanization and sprawl are, in fact, politically supported and wanted by policymakers and political leaders in Czechia and Ostrava region. Middle-class families prefer to move into the suburbs, where they build single-family houses, thus becoming a cause of shrinkage in Ostrava.

Development strategy has been aimed chiefly at attracting investors and FDIs into Ostrava and its region. There have been some policy initiatives aimed at improving living environments for the existing population, such as brownfield regeneration, e.g., the New Karolina development or Lower Vitkovice Area (industrial heritage used recently for cultural functions), the regeneration of socialist prefabricated housing estates, and the development of new faculties and universities. However, there has been no support for building new modern affordable housing via new smart neighborhoods in the walkable vicinity of the city center and no sprawl containment measures. The recommended focus by regrowth theorists on the provision of new affordable, more attractive housing for both the local population and newcomers has been ignored for a long time. The absence of a housing policy working toward attractive, affordable housing in Ostrava is and will be one of the factors of the lack of population regrowth.

The projects of urban regeneration and brownfield decontamination in Ostrava were very significant. The Lower Vitkovice Area is considered to be a successful project of industrial heritage conversion from iron production to attractive culture, education, and sports facilities. More controversial was the New Karolina mixed-use development on the former coke plant brownfield. Furthermore, the opening of Forum as a new retail center is considered to be a cause of retail decline and vacancies in the former city center around the main city square (Slach et al. 2020).

Currently, Ostrava is struggling to be recognized as a university city and as a center of high-quality education and research in Czechia and abroad. Despite all the support from the regional and local governments and of the efforts of university leaders, many students from the Ostrava region have chosen other universities instead, the leading cause of this being the negative image of Ostrava as an industrial city. The out-migration of young people (they leave in order to study or work in the capital city of Prague or in Brno, with more diverse and better-paid job opportunities for students and graduates as well) is an essential factor of the population shrinkage.

The other main challenges in the current discourse are how to address the environmental quality in the city, notably air pollution (Nováček et al. 2019, Balcar and Šulák, 2021). Other problems include the intensification of social polarization, segregation of the poorest social groups, and the creation of socially excluded localities in particular parts of the city, which has an adverse impact on the city's image.

Haase et al. (2021) have concluded that several factors are responsible for the regrowth of the comparable case study cities of Liverpool and Leipzig, but it is not just their presence or absence that is the key determinant. Much more important are their interplay and contingency, the combinations in which they are present or absent, and the impacts of contextual conditions on those factors. As Haase et al. (2021) emphasized, regrowth required a strong pro-active role from the public sector, the state, and the local and regional public authorities in the Western cities of Liverpool and Leipzig. The public sector has to be the main actor and leader, not only in terms of a massive concentration of investment in economic development and job creation but also in the field of affordable new housing, an attractive inner-city, amenities, and infrastructure as well. Ostrava has long disregarded the necessity of creating an attractive inner-city of attractive, affordable housing provided by the public sector and well-designed public space in the city center. Ostrava is a good (not an excellent) place to work, but it is still not good enough to live in. The long-term "shallow" implementation or use of knowledge concerning regrowth is a reason for the lack of population regrowth. Ostrava's leaders will have to be more courageous in the implementation of major development projects, will have to take risks and apply strategic long-term pro-active thinking instead of mere problem-solving.

In Ostrava, "everything has improved, but nothing is definitely quite well". In terms of population development, Ostrava is a slightly shrinking city, but in terms of economic and urban development, we can consider it to be a regrowing city. Due to all the above-mentioned facts and their explanations,

Ostrava will continue to be a shrinking city in terms of population. On the other hand, we can see many positive changes in the institutional, urban, and economic structures. It should be stressed that Ostrava, compared to similar other old industrial cities in Central Eastern Europe, has been successful in transitioning from an industrial mining city to a post-industrial city with a stronger role in retail, education, and services.

Funding

This project has received funding from the Czech Grant Agency, Project No. GACR 18-11299S.

References

Balcar, J. & Šulák, J. 2021. Urban Environmental Quality and Out-migration Intentions. *The Annals of Regional Science*, 66(3), 579–607.

Bernard, J. & Šimon, M. 2017. Vnitřní periferie v Česku: Multidimenzionalita sociálního vyloučení ve venkovských oblastech. (Inner Peripheries in Czechia: Multidimensionality of Social Exclusion in Rural Areas). *Sociologický časopis/ Czech Sociological Review*, 53(1), 3–28.

Bosák, V., Slach, O., Nováček, A. & Krtička, L. 2020. Temporary Use and Brownfield Regeneration in Post-socialist Context: From Bottom-up Governance to Artists Exploitation. *European Planning Studies*, 28(3), 604–626.

Buček, J. & Bleha, B. 2013. Urban Shrinkage as a Challenge to Local Development Planning in Slovakia. *Moravian Geographical Reports*, 21(1), 2–15.

Carter, D. K. (ed.) 2016. *Remaking Post-industrial Cities: Lessons from North America and Europe*. London: Routledge.

Cooke, P. (ed.) 1995. *The Rise of The Rustbelt*. London: UCL Press.

Couch, C., Karecha, J., Nuissl, H. & Rink, D. 2007. Decline and Sprawl: An Evolving Type of Urban Development—Observed in Liverpool and Leipzig. *European Planning Studies*, 13(1), 117–136.

David, Q., Peeters, D., Van Hamme, G. and Vandermotten, C. 2013. Is bigger better? Economic performances of European cities, 1960–2009. *Cities*, 35, 237–254.

Drahokoupil, J. 2012. Beyond Lock-in versus Evolution, towards Punctuated Co-Evolution: On Ron Martin's "Rethinking Regional Path Dependence". *International Journal of Urban and Regional Research*, 36(1), 166–171.

Großmann, K., Bontje, M., Haase, A. & Mykhnenko, V. 2013. Shrinking Cities: Notes for the Further Research Agenda. *Cities*, 35, 221–225.

Haase, A., Bontje, M., Couch, C., Marcinczak, S., Rink, D., Rumpel, P. & Wolff, M. 2021. Factors Driving the Regrowth of European cities and the Role of Local and Contextual Impacts: A Contrasting Analysis of Regrowing and Shrinking cities. *Cities*, 108, 102942.

Haase, A., Rink, D. & Großmann, K. 2016. Shrinking Cities in Post-Socialist Europe: What Can We Learn from Their Analysis for Theory Building Today? *Geografiska Annaler, Series B, Human Geography*, 98(4), 305–319.

Haase, A., Rink, D., Großmann, K., Bernt, M. & Mykhnenko, V. 2014. Conceptualizing Urban Shrinkage. *Environment and Planning A*, 46, 1519–1534.

Hartt, M. 2018. Shifting Perceptions in Shrinking Cities: The Influence of Governance, Time and Geography on Local (In) action. *International Planning Studies*, 25(2), 150–165.

Havlíček, T., Chromý, P., Jančák, V. & Marada, M. 2008. Innere und äussere Peripherie am Beispiel Tschechiens (Inner and External Peripheries of Czechia). *Mitteilungen der Österreichischen Geographischen Gesellschaft*, 150, 299–316.

Heidenreich, M. 2003. Regional Inequalities in the Enlarged Europe. *Journal of European Social Policy*, 13 (4), 313–333.

Hospers, G. J., 2014. Policy Responses to Urban Shrinkage: From Growth Thinking to Civic Engagement. *European Planning Studies*, 22(7), 1507–1523.

Maier, K. 2012. Europeanization and Changing Planning in East-Central Europe: An Easterner's View. *Planning Practice and Research*, 27(1), 137–154.

Maier, K. & Franke, D. 2015. Trendy prostorové sociálně-ekonomické polarizace v Česku 2001–2011 (Trends of Spatial Socio-economic Polarization in Czechia 2001–2011). *Czech Sociological Review*, 51(1), 89–124.

Martinez-Fernandez, C., Audirac, I., Fol, S. & Cunnigham-Sabot, E. 2012. Shrinking Cities: Urban Challenges of Globalization. *International Journal of Urban and Regional Research*, 36(2), 213–225.

Mykhnenko, V. & Turok, I. 2008. East European Cities—Patterns of Growth and Decline, 1960–2005. *International Planning Studies*, 13(4), 311–342.

Nováček, A., Slach, O. & Schachlová, N. 2019. *Šetření kvality života včetně push a pull faktorů z pohledu obyvatel centra Ostravy* (*Research into Quality of Life: Push and Pull Factors from the Perspective of Residents in the Inner City of Ostrava*). University of Ostrava.

Pallagst, K., Fleschurz, R. & Said, S. 2017. What Drives Planning in a Shrinking City? Tales from Two German and Two American Cases. *Town Planning Review*, 88(1), 15–28.

Pavlínek, P. (ed.) 2017. *Dependent Growth: Foreign Investment and the Development of the Automotive Industry in East-Central Europe*. Cham: Springer.

Power, A. & Katz, B. (eds.) 2016. *Cities for a Small Continent*. Bristol: Bristol Policy Press.

Power, A., Plöger, J. & Winkler, A. (eds.) 2010. *Phoenix Cities: The Fall and Rise of Great Industrial Cities*. 1st ed. Bristol: Bristol University Press.

Reckien, D. & Martinez-Fernandez, C. 2011. Why Do Cities Shrink? *European Planning Studies*, 19(8), 1375–1397.

Rink, D., Couch, C., Haase, A., Krzysztofik, R., Nadolu, B. & Rumpel, P. 2014. The Governance of Urban Shrinkage in Cities of Post-socialist Europe: Policies, Strategies and Actors. *Urban Research and Practice*, 7(3), 258–277.

Rink, D., Haase, A., Grossmann, K., Couch, C. & Cocks, M. 2012. From Long-Term Shrinkage to Regrowth? The Urban Development Trajectories of Liverpool and Leipzig. *Built Environment*, 38(2), 162–178.

Rumpel, P. & Slach, O. (eds.) 2012a. *Governance of Shrinkage of the City of Ostrava*. Praha: European Science and Art Publishing.

Rumpel, P. & Slach, O. 2012b. Je Ostrava "smršťujícím se městem"? (Is Ostrava a Shrinking City?) *Czech Sociological Review*, 48(5), 859–878.

Rumpel, P., Slach, O. & Koutský, J. 2013. Shrinking Cities and Governance of Economic Regeneration: The Case of Ostrava. *E+M Ekonomie a Management*, 16(2), 113–127.

Rumpel, P. & Waack, C. 2004. Die Mährisch–Schlesische Region. Perspektiven für die tschechische Altindustrieregion im Europa der Regionen. *Geographische Rundschau*, 56(4), 53–59.

Schilling, J., Logan, J. & Tomaney, J. 2008. Greening the Rust Belt: A Green Infrastructure Model for Right Sizing America's Shrinking Cities. *Journal of the American Planning Association*, 74(4), 451–466.

Slach, O., Bosák, V., Krtička, L., Nováček, A. & Rumpel, P. 2019. Urban Shrinkage and Sustainability: Assessing the Nexus between Population Density, Urban Structures and Urban Sustainability. *Sustainability*, 11(15), 4142.

Slach, O., Nováček, A., Bosák, V. & Krtička, L. 2020. Mega-retail-led Regeneration in the Shrinking City: Panacea or Placebo?. *Cities*, 104, 102799.

Slach, O., Nováček, A. & Rumpel, P. 2015. Metropolitní governance v mezinárodní perspektivě. Metropolitan Governance in International Perspective. In Ježek, J. (ed.) *Strategické plánování obcí, měst a regionů. Vybrané problémy, výzvy a možnosti řešení (Strategic Planning of Municipalities, Cities and Regions: Selected Problems, Challenges and Solutions)* (pp. 27–55). Praha: Wolters Kluwer.

Stryjakiewicz, T. & Jaroszewska, E. 2016. The Process of Shrinkage as a Challenge to Urban Governance. *Quaestiones Geographicae*, 35(2), 27–37.

Sýkora, L. & Bouzarovski, S. 2012. Multiple transformations: Conceptualising the post-communist urban transition. *Urban Studies*, 49(1), 43–60.

Turok, I. & Mykhnenko, V. 2007. The Trajectories of European Cities, 1960–2005. *Cities*, 24(3), 165–182.

Ženka, J., Slach, O. & Pavlík, A. 2019. Economic Resilience of Metropolitan, Old Industrial, and Rural Regions in two Subsequent Recessionary Shocks. *European Planning Studies*, 27(11), 2288–2311.

20 Drivers, consequences, and governance of urban shrinkage in Lithuania

The case of Šiauliai

Gintarė Pociūtė-Sereikienė and Donatas Burneika

Introduction

Political and economic changes which struck Central and Eastern European (CEE) countries at the end of the last century triggered significant variations in their respective urban systems (Krišjāne 2001, Mykhnenko and Turok 2008, Burneika 2012, Ubareničienė 2018). The key features of the postsocialist period in these countries were the redistribution of inhabitants within their territories, growth of metropolitan centers (Sýkora and Bouzarovski 2012), and rapid population decline elsewhere (Hospers 2012, Grossmann *et al.* 2013, Pociūtė-Sereikienė 2019). Population decline most strongly affected peripheral territories characterized by small towns and industrial cities that were developed during the Soviet era (Raagmaa 1996, Nagy and Turnock 1998, Stryjakiewicz *et al.* 2012, Lang *et al.* 2015). The small Baltic country of Lithuania, located along the eastern border of the European Union (EU), is no exception as it experienced similar trends of population redistribution, followed by sharp population declines (Ubarevičienė 2018).

Though all CEE countries suffer from similar trends, each country has its own specificities as inherited urban systems along with other local factors were different. Šiauliai, one of several industrial centers of the Lithuanian multimodal urban network, was purposefully developed during the Soviet era, which lasted from the end of the Second World War until 1990 and corresponded with a major wave of industrialization and urbanization. The idea of a multimodal urban network (Šešelgis 1996) was based on the limitation of the growth of former major centers, especially the capital of Vilnius. Industry and most public services were evenly distributed throughout Lithuania across ten (and later six) cities, which served as major regional centers. Since the early 1990s, Lithuania has seen a rapid transformation of its settlement system, which nevertheless remains much more polycentric than in other CEE countries (Vanagas *et al.* 2002, Ubarevičienė 2018).

This chapter seeks to conceptualize the phenomenon of shrinkage in Lithuania in the global context using the case of Šiauliai. The authors

DOI: 10.4324/9780367815011-24

present key empirical evidence of the shrinkage of Šiauliai, including related consequences and attitudes of local actors. Šiauliai was selected as a case study because, firstly, it is the largest non-metropolitan Lithuanian city, and secondly, its population has declined by 31 percent since 1992. While it was a major industrial center under state socialism, its industrial base was one of the most damaged sectors of post-Soviet development. Moreover, although it is formally recognized as a large city (under Lithuanian law more than 100,000 residents), Šiauliai has neither benefitted from any major scheme of regional development nor special support measures as in some smaller areas (e.g., financial support for farmers, tourism, small local entrepreneurs in rural areas and so on). While three major Lithuanian cities continue to act as interregional centers, concentrating jobs and population (Burneika 2019), Šiauliai seems to be losing its status as a regional center. We argue that Šiauliai still has grounds to be considered a strong center of northwestern Lithuania in light of its university, international airport (mostly used for military NATO air police purposes at present but working to expand to civil use), railway and road connections, and longstanding Free Economic Zones. These factors all contribute to making Šiauliai a lively city with some visibility in local and foreign media; however, these advantages over other non-metropolitan Lithuanian cities have not prevented its rapid urban shrinkage.

Background to the research: theoretical discussions on urban shrinkage, methodology and data

Theoretical discussions

While countries follow different paths of urban shrinkage, population decline is the main feature across all cases (Oswalt and Rieniets 2006, Rink *et al.* 2010, Reckien and Martinez-Fernandez 2011, Li and Mykhnenko 2018). There is no widely accepted definition to explain all cases of shrinkage. Instead, the literature contains several explanations. According to Audirac (2014, p. 43), Grossmann *et al.* (2013, p. 221), and a number of others, shrinkage should be viewed as a multidimensional, process-based phenomenon that is highly dependent on historical background.

Researchers have attempted to explain urban shrinkage using the theories and models of "life cycle development" or "delayed process of adjustment" (Dietzsch 2009). Meanwhile, some scholars (Rink *et al.* 2010, Haase *et al.* 2013, p. 89) have presented multi-theoretical understandings of shrinkage involving a combination of explanations rooted in "stage" or "life-cycle" theories, "uneven development" or "accumulation of capital" concepts, discussions on "post-suburbias", changing territorial divisions of labor, or even findings based on the "second demographic transition" or "fourth urban revolution" (Soja 2000).

The drivers of shrinkage are diverse. Scholars point out that shrinkage is influenced by economic decline and job-related out-migration, demographic change, suburbanization, structural upheaval, political changes, resettlement, and environmental disasters, among other reasons (Dietzsch 2009, Rink *et al.* 2010, Haase *et al.* 2013, Pallagst *et al.* 2014). The drivers of shrinkage are often found in combinations of two or more (Wiechmann and Bontje 2015).

The direct and indirect consequences of urban shrinkage are wide-ranging and vary from case to case (Haase *et al.* 2013). Direct consequences include those influenced by depopulation, such as the under-use of infrastructure or housing vacancies, and those affected by deindustrialization, such as the emergence of brownfield sites. Indirect consequences, on the other hand, "are defined as a combined product of feedback loops" (Haase *et al.* 2014, p. 1524).

Researchers have identified the main tools used by national and local authorities (in some cases with the help of entrepreneurs, public agencies, and other institutions) to manage urban shrinkage. In the majority of analyzed cases, economic development, especially foreign investment, is believed to be the most effective tool for attracting new residents (Stryjakiewicz *et al.* 2012, Cortese *et al.* 2013, Pallagst *et al.* 2017). Another essential tool to manage shrinkage is rethinking the city's development path: for example, strengthening the image of universities (Pallagst *et al.* 2017), establishing hi-tech hubs (Stryjakiewicz *et al.* 2012), restructuring the local economy and finding new niches (Leetmaa *et al.* 2015), developing green areas and social infrastructure (Fol 2012, Stryjakiewicz and Jaroszewska 2016). The literature also notes the positive effects of urban renewal in shrinking cities (Cortese *et al.* 2013). Economic development, renovations, renewal of engineering, and social infrastructure in postsocialist countries of the EU have often been funded by the EU (Stryjakiewicz *et al.* 2012, Wolff and Wiechmann 2018). Furthermore, active civic engagement is also identified as one of the essential keys to cope with shrinkage (Hospers 2012, Leetmaa *et al.* 2015).

Methodology

The research in this article is based on the heuristic model of urban shrinkage (Haase *et al.* 2014), previously applied to shrinking cities in several European countries, including Poland, the Czech Republic, Ukraine, Romania, (Eastern) Germany, Italy, and the United Kingdom. The model assumes that population loss is the key indicator of urban shrinkage. The main idea of the model is to view shrinkage from a broad perspective and consider not only the reasons (drivers) for depopulation but also the impacts (consequences) as well as responses (governance) (Haase *et al.* 2014).

Data

The research is based on quantitative and qualitative data analyses. Following the heuristic model, the first analysis was quantitative in nature and focused on uncovering the drivers and consequences of the shrinkage. Sources of data for the analysis include official statistics (Statistics Lithuania 2019) on demographic and macroeconomic indicators. In order to gain an understanding of the development trajectories of the city during the last 30 years, 25 statistical indicators were analyzed, from which the main indicators explaining the shrinkage were selected.

The qualitative part of the research is based on semi-structured interviews about the attitudes of the local authorities, stakeholders, and active citizens toward shrinkage. The main criterion behind the selection of respondents was expertise (leading position, active involvement in activities, work experience). The formulation of questions was guided using previous research (Rink *et al.* 2009). In total, ten interviews were conducted, including five with local authorities (municipality representatives), three with citizens involved in various institutions, one with an entrepreneur who is the leading person of the chamber of commerce, industry, and crafts, and one with the deputy to Parliament responsible for presenting Šiauliai affairs in the sessions. The interviews usually lasted around two hours.

Šiauliai and geography of population decline in Lithuania

Population shifts in Lithuania that began in the 1990s occurred more rapidly and with greater intensity than in other CEE countries (Ubarevičienė 2018). However, the causes of depopulation in Lithuania are similar to other CEE countries and include high emigration, low birth rates, and an aging population (Haase *et al.* 2014, Smętkowski 2017, Daugirdas and Pociūtė-Sereikienė 2018, Pociūtė-Sereikienė 2019). The result of this rapid depopulation is that the Lithuanian population decreased by about 25 percent in the last 25 years. Currently, a great majority (86.9 percent) of Lithuania's 2.8 million residents are Lithuanian, not unlike during the socialist period (76.4 percent in 1989) (Statistics Lithuania 2019). As of 2019, the biggest minority groups were Russians and Poles, which made up 12.4 and 6.0 percent of the population, respectively. On the other hand, the distribution of ethnic minorities is extremely uneven as they comprise roughly one-third of residents in Vilnius and Klaipėda and less than 5 percent in the other large cities, Šiauliai included.

Although Lithuania is a small country, a total of 103 cities/towns are officially recognized (Figure 20.1). The smallest urban settlement has around 500 residents and the largest, Vilnius, counts 550,000 inhabitants. Deindustrialization occurred throughout Lithuania and was the main cause of persistent population decline in all cities, except Vilnius, where the number of residents remained stable. Nearly all rural areas also faced drastic depopulation, with the exception of suburban areas near the largest cities of Vilnius, Kaunas, and Klaipėda (Ubarevičienė 2018).

Šiauliai's development trajectory is similar to that of other Lithuanian industrial regional centers; for instance, the fifth-largest city in Lithuania, Panevėžys, has also lost 33 percent of its inhabitants since 1992, and the sixth-largest city,

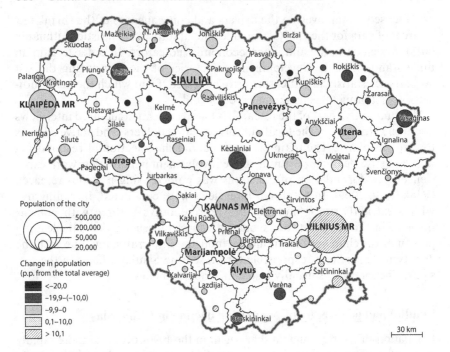

Figure 20.1 Population development in Lithuanian cities 1990–2018.

The case study area (city of Šiauliai) is underlined.

Source: Graphics prepared by R. Ubarevičienė and D. Burneika (adapted from Burneika 2019, p. 49).

Alytus, has lost 33.4 percent. Shrinkage of other small peripheral towns is particularly noticeable and can be explained by the decline in rural jobs in agriculture. It should also be emphasized that the processes of shrinkage were driven by foreign emigration, with the whole country losing approximately one-fourth of its population since 1991. Unlike in Western countries or even former East Germany, the urban shrinkage of Lithuania should be perceived at an international scale. Second-tier cities like Šiauliai experienced population declines not because of the redistribution of residents and jobs inside Lithuania but first and foremost because of out-migration to other parts of Europe.

Šiauliai, the fourth largest city in Lithuania (after Vilnius, Kaunas, and Klaipėda) with 100,100 inhabitants as of 2019, is located in northern Lithuania, approximately 50 kilometers from the Latvian border. Šiauliai became an important urban center because of its geographical position in the 16th century, but the city's historical development was very turbulent. It was heavily impacted during World War I (65 percent of buildings were destroyed; Baliutavičienė and Baliutavičius 1999) and World War II (85 percent of the city was destroyed along with the majority of population, which was primarily of Jewish origin; Sireika 2007).

Consequently, the city's period of stable growth under state socialism is often perceived as its "golden age". Indeed, during the socialist period, Šiauliai developed into an economically strong industrial and cultural regional center (Vanagas *et al.* 2002) with well-developed social infrastructure and a constantly growing population (Baliutavičienė and Baliutavičius 1999). The majority of inhabitants were recruited by the state to work in city's manufacturing sector (especially food, machinery and metal, clothing and textile industry) (Sireika 2007, p. 172), which made products for export all around the Soviet Union. It was a period of relative prosperity, marked by an increasing number of houses, schools, kindergartens, entertainment services, and green areas (Sireika 2007). From 1959 to the end of the Soviet era in 1990, the population increased from roughly 58,600 to 145,500. The population peaked in 1992, and since then, the city has faced persistent population decline (Figure 20.2).

Drivers and consequences of shrinkage in Šiauliai

The shrinkage of Šiauliai might be explained by general macro-level political, economic, and socio-demographic changes that have taken place in Lithuania and across CEE. Though no detailed studies exist, micro-level factors related to the roles performed by local actors, such as local government leaders, entrepreneurs, or politicians also appear to have played a role. However, the overwhelming similarity of trends throughout Lithuania suggests the impact of micro-level factors on shrinkage was limited.

There is little doubt that the macro-level triggers of shrinkage were related to the inability of the city's economy to adapt to the changing political and economic systems of Lithuania and Europe as well as growing global competition. Most Soviet era industries were extremely energy inefficient, had low levels of productivity, and produced low-quality goods or machinery. Under the conditions of a competitive global economy, the majority of factories had no future. In fact, Šiauliai suffered from deindustrialization in a similar way to other European cities at the end of the 20th century (Hall 1998); however, the pace of deindustrialization was much faster and its scope much wider due to the fact that the industry was orientated toward the East, and meeting the needs of the Soviet Union. During the socialist period, Lithuania, together with other Soviet republics (and similar to Latvia and Estonia), implemented plans imposed by the Soviet Union, which meant its direction of industrialization and spatial development were strictly planned and "laid down from the top" (Vanagas *et al.* 2002). The central Soviet government decided the location of factories in specific cities, which agricultural goods ought to be grown in specific rural areas, and even the migration patterns of inhabitants. Generally, the bulk of goods produced in one Soviet republic was distributed all over the Soviet Union. After the collapse of the Soviet Union, however, most of the previous distribution links were terminated, resulting in the closure of numerous factories.

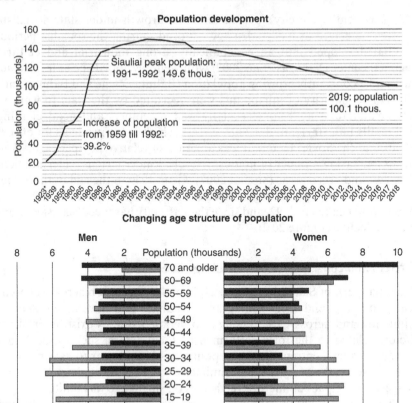

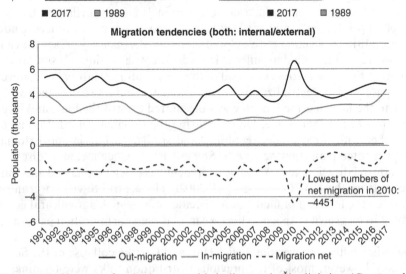

Figure 20.2 Changes of main demographic indicators in Šiauliai city. *(Continued)*

Source: Authors' own calculations based on data from Statistics Lithuania (2019).

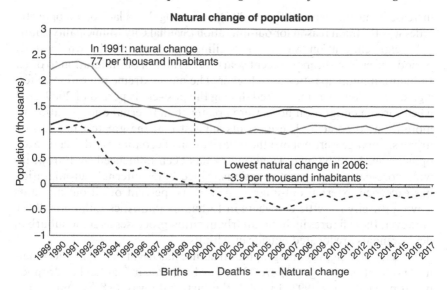

Figure 20.2 (Continued)

Factory closures led to drastic increases in the unemployment rate, which jumped from 1.3 percent in 1992 to 16.1 percent in 2000. Though the unemployment rate decreased to just 3.7 percent reached in 2007, this can be explained by out-migration. In recent years, the unemployment rate has remained around 5 percent, but this is only partly related to the growth of jobs in Šiauliai. Out-migration has helped to mitigate unemployment, but it has also skewed the age structure of the population (Figure 20.2). The working age population decreased by one-third (from 94,400 to 60,400). The largest age group are those over 55 years, which translates to an aging labor force that is not the most mobile or adaptable to changes. Extrapolating from the present trend, further population declines to 94,600 residents in 2022 and 87,400 in 2027. Meanwhile, the elderly population is increasing: it is estimated that by 2022, 22.9 percent of residents will be older than 65, and by 2027, 25.4 percent. In general, the age structure of Lithuania's population has radically changed during the last 30 years and the country is now facing a major population aging trend (Jasilionis *et al.* 2015). The index of aging (the population aged 65 and older per 100 children aged under 15) has almost doubled from 71 in 2001 to 131 in 2019 (the same period in Šiauliai from 61 to 134) (Statistics Lithuania 2019).

Due to economic decline, net migration has been negative since the early 1990s, but the greatest gap between in-migration and out-migration was during the first few years of the 21st century. From 2001 to 2005, almost 2.5 times more residents left the city than moved to it (Statistics Lithuania 2019). The out-migration wave was selective, as most migrants were 20 to 40 years old. The second wave took place from 2009 to 2011 and immediately followed a drastic increase in unemployment (Figure 20.2), indicating that a lack of job opportunities was the main reason for out-migration (Šiauliai City Municipality 2016).

Foreign emigration was the prevailing trend during the whole study period, except for the most recent years when rates of out-migration to other Lithuanian municipalities were higher. The most extreme migration to foreign destinations was monitored during the post-crisis period of 2009–2011 when foreign emigration flows were greater than internal out-migration by more than 30 percent. These results suggest that the greatest impacts on urban shrinkage were not just the polarization of economic and social development inside Lithuania. The country has been influenced by European-wide core-periphery transformations. Since 2016, internal out-migration has exceeded external out-migration, with 66 percent of all out-migrants leaving the city to settle elsewhere in Lithuania (Statistics Lithuania 2019). However, these figures include suburbanization processes as some suburban developments are located outside city limits.

The city's changing demographic structure has negatively influenced natural population change (Figure 20.2). The birth rate in Šiauliai has dropped by 30 percent since 1992. In 1992, the birth rate was 15.8 live births per 1,000 people, whereas, in 2017, there were 11.0 births. The number of children in Šiauliai more than halved (from 33,700 to 15,600) between 1989 and 2017. In the same period, the number of residents aged 60 or older increased from 17,400 to 25,900 (Statistics Lithuania 2019). The decreasing number of pupils has led to school closures and corresponding job losses in public services. Moreover, the decreasing number of consumers has caused problems for local government and businesses. For example, due to increasing maintenance costs of infrastructure, the municipality struggles to ensure the convenience of and accessibility to public services, namely public transportation and education. (Šiauliai City Municipality 2016, p. 21).

Deindustrialization also impacted the cityscape. After the closure of industrial enterprises, buildings were left abandoned. Select parts of former factories were restructured and reused, but others became huge brownfield sites and, in turn, a burden on the local government. Moreover, the owners of many abandoned buildings are not willing to invest in their renovation, instead preferring to leave them as "ghosts" of the city.

All of the above inevitably exerted a negative influence on the image of the city and its ability to attract investors and residents. On the other hand, shrinkage has also opened doors for positive outcomes, such as increasing living space and decreasing rental costs, traffic, and air pollution (Janicki 2017), but such outcomes depend on the ability of private and public decision makers to affect the actual and subjective outcomes of shrinkage.

Local actors' attitudes toward shrinkage

In Lithuania, regional policy and development decisions are still strongly centralized, and municipalities have very limited financial resources for investment in activities that are outside the field of their direct responsibility.

The regional policy of Lithuania has been assigned very limited resources since its origin in the early 2000s. Most of its available funds are related to EU structural funding and follow its requirements. Unlike many other cities in post-communist countries, such as Łódź or Wałbrzych in Poland (Stryjakiewicz *et al.* 2012, Stryjakiewicz and Jaroszewska 2016), which were mostly renovated and renewed using EU funding, Šiauliai, as formally a "large" city, missed out on EU funding support schemes in Lithuania as they were targeted at problem areas (rural municipalities mostly) or secondary growth poles (medium cities). Although Šiauliai has not gone completely without EU funding (for example, Šiauliai is currently renovating its squares, parks, and engineering networks, thanks in part to EU structural funds), it has received less attention and funding compared to rural areas.

The present government has prioritized regional development and included it government programs (Seimas of the Republic of Lithuania 2016), but given no special attention to cities like Šiauliai. Even the current government program, despite providing support for young families to purchase a home outside metropolitan areas, such as suburban areas of the sprawling city of Kaunas, ignores Šiauliai.

Municipalities in Lithuania cannot freely dispose of their own land and property and furthermore need to reconcile new projects with the national government and adjust them to national or strategic plans. Therefore, the abilities of municipalities to attract investors are limited. On the other hand, the municipality still holds budgetary powers and can regulate its spending to some extent. It can also play an active role in city promotion and image creation, implement local scale strategies, set property tax rates, and influence investment priorities of EU funding for the city (within certain limits).

The field study and interviews of this study focused on revealing how the local government, entrepreneurs, and active citizens (all hereafter referred to as local actors) understand the problems in the city, what future trends were expected, and what role local actors play in coping with shrinkage. Research results identify three main approaches the city has adopted to manage urban shrinkage: (1) keeping existing or attracting new young residents; (2) economic development; and (3) infrastructure renewal. These approaches are presented in more detail below.

First approach: attracting new young residents to the city

Given that the city is aging fast, attention has focused on attracting and retaining young people and young families. In 2019, the municipality arranged a program that covers students' rental and study fees (if the study program is paid) on the condition that after their studies, the student continues to live and work in Šiauliai for at least three years after graduation. Additionally, the municipality offers young families financial support toward the purchase of a first home. The support is in the form of covering the costs of notaries and Center of Registers services, which may total about 1,500 euros. As well, the municipality provides a starter pack for every newborn child. Recently, the municipality decided to offer rent support for young specialists who come to the city to work in institutions that lack staff, such as health care services. In addition, the municipality has been running a program where some schools offer special programs for children of returning emigrants. Furthermore, very recently, the municipality launched the platform "Global Šiauliai", which provides a range of necessary information for returning emigrants and newcomers. While these actions are very recent and, in some cases, still in development, they demonstrate an aspiration to deal with depopulation.

In general, efforts to attract new residents are noticeable in many shrinking cities; however, the ways these efforts are implemented differ. For instance, the main strategy to attract people to the formerly shrinking city of Leipzig was set to make the city more socially diverse by tackling social issues related to immigrant integration, low-income households, or unemployment (Cortese *et al.* 2013). Various projects were implemented toward these aims, such as the "urban development plan (STEP)", "Rebuilding the City – East" ("Stadtumbau Ost" Wiechmann and Pallagst 2012), or "Social city" (Cortese *et al.* 2013). These measures led to the renovation of abandoned industrial houses and entire neighborhoods, upgrading of social infrastructure, and ultimately regrowth, as Leipzig became attractive not only to immigrants but also to young German families and students.

Meanwhile, shrinking cities in Poland set the urgent task to halt the outflow of young people by creating attractive residential places for families, offering better-paying jobs, and improving social facilities such as kindergartens (Stryjakiewicz and Jaroszewska 2016). In order to improve the labor market situation in Poland's shrinking cities, various agencies were established, such as the Lower Silesian Agency for Regional Development or the Lower Silesian Science and Technology Park in Wałbrzych (Stryjakiewicz *et al.* 2012). Conversely, local governments in France decided to take measures to attract middle-class residents from suburban zones back to the core of the cities, including renovating old city centers, rebuilding cultural heritage objects, renewing public and green areas, and constructing new homes. Although these initiatives aimed to change the image of cities, a focus on attracting middle-class households to the city could contribute to gentrification (Fol 2012).

Compared to aforementioned, the main difference of the Lithuanian strategy, at least in the case of Šiauliai, is that the reaction of the government toward urban shrinkage and its efforts to attract new residents are more declarative than effective. As the programs are more recent and still with problems to be worked out, the desired inflows of young people have not yet materialized.

Second approach: economic development

Another major effort to "wake-up" Šiauliai is based on strengthening its economy. Eight years ago, Šiauliai established a Free Economic Zone

(FEZ), a well-established tactic to attract businesses by offering reduced state and municipal fees. The practice of establishing Special Economic Zones (SEZ) is common in postsocialist shrinking cities in Poland, where the aim has been to accelerate economic development, attract Polish and foreign investors, create new jobs, redevelop post-industrial infrastructure, and foster modern technologies and innovations (Stryjakiewicz *et al.* 2012). These zones, which were established in the 1990s and eventually became significant areas of investment, could serve as successful examples for managing shrinkage by economic means.

Lithuania is currently home to six operating FEZs (Lithuanian Association of Free Economic Zones 2020). However, under EU legislation, actual tax reductions offered to companies are very limited and thus do not give much impetus for companies to relocate. For a long period, the Šiauliai FEZ was not viable and only existed on paper, while other cities fared much better. In 2015, a newly elected municipal board began working more actively on economic regeneration projects, among them the establishment of a FEZ. Therefore, in 2019, four companies (producing medical equipment, plastic windows, and advertising signs) received support from the municipality to relocate to the FEZ and brought 9 million euros in investment with them. This created 200 new workplaces and added value to the city. However, although the local government proudly considers the Šiauliai FEZ to be the most rapidly growing FEZ in Lithuania, the zone is not working at full capacity: 107 of 133 hectares of land remain to be developed. In addition, in 2005, the municipality, accepting the government's call, started preparing infrastructure for another national "grand project" known as the "Industrial Park". Since FEZs are, in fact, not free of taxes, actual differences between "Parks" and FEZs are minimal and tend to be more related to the timing of the projects. The actual differences in terms of infrastructure and location play a more decisive role in attracting investments. The "Park" in Šiauliai is the biggest of five "Parks" located in Lithuania and, similar to FEZs, offers reduced fees for relocating manufacturing companies. Eleven companies together with the municipality have invested 45 million euros and created some 6,000 new jobs in this Šiauliai "Park".

In 2016, the municipality also launched a program for entrepreneurs, in which local authorities support the business plans of small companies and start-ups. This program was originally dedicated to young entrepreneurs (up to 29 years old), but at present, two additional groups (those older than 50 and middle-aged residents) are eligible for funding. This program seeks to motivate city residents to stay, bring new ideas, and create for the city. Currently, the local authority is developing a "Strategy for Economic Breakthrough". The main idea of this strategy is to highlight the path of economic development which the city should follow. The program involves different sectors, however, the main emphasis is placed on logistics, the expansion of the existing FEZ, "Industrial Park" zones, and exploiting the potential of the airport. An interview with an entrepreneur involved in the creation of the strategy indicated that the strategy remained unclear and still at its formative stage. The respondent was moreover dissatisfied with this new strategy and expected it would be another expensive document destined to be forgotten like its predecessors.

Third approach: infrastructure renewal

From the middle of the 20th century until recent times, Šiauliai has retained the same Soviet appearance. A few years ago, however, the local government decided to renew the main pedestrian zone in the city center, modernize the main city square, replace old pipes, upgrade the streets, renew parks, and clean and adapt the lake for recreation. Meantime, the city is encouraging the use of alternative transport (such as bikes and electric scooters) and improvement of urban amenities.

To reduce the number of brownfield sites, the municipality increased property taxes for unused buildings. This led to the restoration and reuse of an old abandoned building near the lake as an elderly care home. High occupancy and long queues indicate the further potential to develop the city's "silver economy". However, there remain a large number of dilapidated Soviet-era industrial buildings that should be demolished or reused.

How is Šiauliai willing to present itself?

The interviewed entrepreneurs, NGOs, and local activists emphasized that there is no clear vision or robust strategy guiding the city. Rather, it seems the local government wants to do everything at once: strengthen the economy, attract foreign investment, and invite young people to the city. However, these ideas are predominantly top-down and have not yielded the desired outcomes. A bottom-up initiative of local leaders to develop the city around the idea of "Šiauliai as a land port" was rejected ten years ago. This is regrettable because examples from other countries, such as Flint in the United States (Pallagst *et al.* 2017), show that plans developed via bottom-up collaborative processes involving citizens' suggestions can have a positive effect on city development. The city's current slogan, "Šiauliai strong economic center" (Šiauliai City Municipality 2016), seemed like a joke to some respondents, who suggested it is too late to regenerate the city's economy and that there is a need for alternative concepts for the city's development.

Local authorities underlined their own visions and ideas for the city. They, and especially the mayor, primarily see the city as a "family-friendly place" with well-developed infrastructure. Several times it was pointed out that the term "family" actually means "young family", which suggests that Šiauliai's development is still linked toward a younger generation even though because of aging, the city has become a residential location for the older generation. However, while authorities are promoting the idea of attracting young people to the city, the instruments to do so remain unclear and under discussion. The only clear vision is to maintain the status of a "university city". Šiauliai is one of four cities in Lithuania with a university, and respondents believe this institution still attracts young people from around the country. Pallagst *et al.* (2017), analyzing Kaiserslautern in Germany as well as Flint and Youngstown in the US, underlined the capacity of universities to attract young people to cities as well as to serve as a hub for hi-tech industries or start-up companies (Stryjakiewicz *et al.* 2012).

As the majority of Šiauliai's former factories have closed, the city has become less polluted. Therefore, another idea of the local government is to present the city as a "green city" and to take advantage of its green areas, parks, and lakes. While it remains unclear how this will help to change demographic and economic trends, the idea of "greening" the city (for example, by restoring or creating parks) has been raised in many shrinking cities as a strategy to attract or retain families (Fol 2012, Stryjakiewicz *et al.* 2012, Pallagst *et al.* 2017).

Šiauliai gained a strong military function due to the construction of a major military airport during the Soviet era, which has been used by NATO for more than a decade. The local government has proudly presented the city as a home for employees of the military and made efforts to attract investment toward strengthening this potential. For example, it has widely promoted the national government's 2019 decision to build a "military village" in Šiauliai.

Survey respondents identified a mismatch between authorities and residents about priorities for city development and raised questions about a reliance on foreign immigration for growth. The analysis of interviews furthermore shows a gap between authorities and residents, especially among entrepreneurs, about what steps would best manage urban shrinkage. Local authorities blame businesses for not joining the "fight" against economic emigration, whereas the entrepreneurs point out that the municipality does not allow them to join city development discussions. According to entrepreneurs, the local authorities take on too many initiatives but lack the financial and human resources to realize their aims. While the mayor declared the situation in the city is still good enough, other respondents underlined depopulation, economic backwardness, residents' dissatisfaction, and increasing social segregation as processes as matters of concern for the city. However, almost all respondents agreed that shrinkage is a natural process and not exclusive to Šiauliai.

A fundamental question was also raised about the expectation that foreign immigrants will facilitate regeneration. Even though some local municipal leaders enthusiastically declared that the population of Šiauliai has recently increased by 8,000 inhabitants, most of these individuals are newcomers from Ukraine and Belarus. Some respondents pointed out that foreign immigration is not a solution as it increases the number of employees but not necessarily the number of skilled workers. Indeed, the majority of immigrants are employed as truck drivers and construction workers. The respondents were more in favor of restrictions on immigration and willing to elect local authorities who would focus on retaining existing residents rather than attracting foreigners or returning emigrants. In general, the respondents argued that Šiauliai should follow the path of becoming a smaller, resident-friendly city.

Discussion and conclusion

The postsocialist transformation of Lithuania has proved challenging. Residents and authorities have had to adapt to changes in the political system and the transition from a socialist to a market economy (Burneika 2012). Moreover, throughout this period, the country has been confronted with demographic changes including rapid depopulation and urban-rural shrinkage. While Lithuania is not an exception in a European context of rapid population decline (Haase *et al.* 2013, Wiechmann and Bontje 2013, Wolff and Wiechmann 2018), major European countries, such as Germany, are in a stronger position to deal with urban shrinkage as they are generally favored by immigrants. Trends of shrinkage in Lithuania better resemble those observed elsewhere in CEE, especially the other Baltic States of Latvia and Estonia, which along with Lithuania were among the world's fastest shrinking countries during the last decades (since restoring the Independence in 1990) (United Nations 2015). Not only have peripheral rural areas been shrinking, but also 95 percent of cities and even capitals have been losing their populations; for instance, Riga lost 29.6 percent of its population since 1989 (Ubarevičienė 2018). Meanwhile, suburban areas surrounding capitals (in addition to two other major cities in Lithuania and Estonia) have recorded population increases. Such results suggest a strong trend of metropolization in the Baltic States. This unequal development has caused polarization among the Baltic States to become more evident.

Shrinkage might be explained by general macro-level political, economic, and socio-demographic changes (Raagmaa 1996, Krišjāne 2001, Berzins and Zvidrins 2011, Daugirdas and Pociūtė-Sereikienė 2018). Many theories of divergent regional development also predict similar trends of concentration and polarization of the economy under the conditions of the free market (Dawkins 2003). The geographical factor plays a major role as well. In general, the Baltic States are "on the edge" of the EU. In the case of Šiauliai, its location at the "periphery of the periphery" contributes to its stigmatization. Indeed, even though Šiauliai is Lithuania's fourth-largest city, its peripheral location is one reason why it has not managed to become an interregional center that attracts jobs and people from areas beyond Lithuania's northern region. And while Šiauliai struggles with low investment rates, major cities including Vilnius, Kaunas, Klaipėda, and Riga have become increasingly attractive for international companies.

The drivers and consequences of shrinkage in the case of Šiauliai are similar to those identified in shrinking cities of other European countries, such as Ostrava, Bytom, Timişoara, and Wałbrzych, or even Liverpool, Leipzig, and Saint-Étienne (Rink *et al.* 2010, Fol 2012, Stryjakiewicz *et al.* 2012). In all countries, the greatest role was played by economic decline and deindustrialization, which in turn was echoed in unemployment, emigration, negative natural population change, aging, and abandonment of residential and industrial buildings.

The management of shrinkage in Lithuania has been rather weak. In general, CEE countries lack comprehensive city regeneration strategies (especially for shrinking cities) and usually deal with shrinkage by separate programs or plans (Stryjakiewicz and Jaroszewska 2016). In Lithuania, as in other CEE countries, shrinkage is often neglected (Batunova and Gunko 2018), or the dominant approach is based on overcoming its negative effects rather than fostering the development in the conditions of shrinkage (Stryjakiewicz *et al.* 2012). Rather, planners ought to understand shrinkage as a normal phenomenon; one which requires reconsideration of the city as a holistic space for reconstruction and multi-scaler policy responses (Wolff and Wiechmann 2018).

Lithuania has not launched special programs or policies to cope with shrinkage. While the federal government has recognized depopulation as a problem in several national documents (most recently in the "Lithuanian Regional Policy White Paper" (Ministry of Interior 2017) and "Lietuva 2030" strategy (Ministry of the Environment 2019)), its management remains more conceptual than practical, and largely targeted at rural areas. The regional policy strategies presented in both the "White Paper" and "Lietuva 2030" aim to develop Lithuania more equally by strengthening peripheral regional centers. Accordingly, Šiauliai qualified as one of ten developing regional centers, and it was proposed that the city becomes a regional "intermediate center" serving northern Lithuania. However, the policies lack information about instruments to improve the city's socio-economic and demographic situation or guidelines on how to attract investment and improve residents' quality of life.

Currently, the local government in Šiauliai operates according to a strategic development plan (Šiauliai City Municipality 2016) that includes few practical measures to deal with urban shrinkage. The plan emphasizes, however, the importance of increasing the population of young people, strengthening economic potential through investment, as well as improving social, leisure, and physical infrastructure. While these aims represent small "steps" to improving life in Šiauliai, they are insufficient "instruments" for solving the challenges of shrinkage.

Although Šiauliai lost a great share of its inhabitants and nearly all of its former industries, the city nevertheless has the potential to maintain its regional center status, as all the necessary infrastructure is there. Currently, the city is working to reduce its "periphery label" and looking for ways to attract companies and investment, for example, by opening coordination centers for start-ups and young entrepreneurs. Interviews with local authorities revealed the city is willing to change its "industrial face" and become a green, compact and comfortable city attractive for families as well as tourists. Alas, following the populist idea to "make the city great again" would be one of the worst scenarios for the national and local government, though it could be an attractive idea among aging voters. Instead, the city should focus on becoming friendlier and more convenient for existing residents. Promising strategies to develop the city according to the idea of "shrinking smart" (Rink *et al.* 2010, range from making investments in human capital to strengthening educational opportunities to taking advantage of the city's geographical position (for example, as a possible logistic center), to converting abandoned public housing into comfortable and homely nursing homes for the elderly. With the right attitude, cooperation between national and local actors, and proper use of EU funds (Stryjakiewicz *et al.* 2012), Šiauliai might transition to a brighter future and serve as another example of the positive side of shrinkage.

Acknowledgment

This project has received funding from European Social Fund (Project No. DOTSUT-149 (09.3.3-LMT-K-712-02-0062)] under a grant agreement with the Research Council of Lithuania (LMTLT).

References

Audirac, I. 2014. Shrinking cities in the fourth urban revolution? In Pallagst, K. Wiechmann T. & Martinez-Fernandez, C. (eds.) *Shrinking Cities International Perspectives and Policy Implications* (pp. 42–58). New York: Routledge.

Baliutavičienė, S. & Baliutavičius, V. 1999. *Šiaulių istorija iki 2000 metų* [*Šiauliai History till 2000*]. Vilnius: Naujoji Rosma.

Batunova, E. & Gunko, M. 2018. Urban Shrinkage: An Unspoken Challenge of Spatial Planning in Russian small and Medium-sized Cities. *European Planning Studies*, DOI: 10.1080/09654313.2018.1484891

Berzins, A. & Zvidrins, P. 2011. Depopulation in the Baltic States. *Lithuanian Journal of Statistics*, 50(1), 39–48.

Burneika, D. 2012. Transformations in Lithuania—factors of change and regional patterns. In Gorzelak G., et al. (eds.) *Adaptability and Change: The Regional Dimensions in Central and Eastern Europe* (pp. 267–283). Drelow: Poligraf.

Burneika, D. 2019. Ekonominės gerovės regioniniai pokyčiai ir jų įtaka miestų sistemos transformacijai Lietuvoje [The Regional Changes of Economic Well-Being and the Impact of Them for Urban Transformations in Lithuania]. In Daugirdas, V. & Burneika, D. (eds.) *Netolygaus regioninio vystymosi problema Lietuvoje: socio-ekonominiai gerovės aspektai* (pp. 39–52). Vilnius: Lithuanian Social Research Centre.

Cortese, C., et al. 2013. Governing Social Cohesion in Shrinking Cities: The Cases of Ostrava, Genoa and Leipzig, *European Planning Studies*, DOI: 10.1080/09654313. 2013.817540

Daugirdas, V. & Pociūtė-Sereikienė, G. 2018. The Tendencies of Depopulation and Territorial Development in Lithuania. *Regional Statistics*, 8(2), 1–23.

Dawkins, C. J. 2003. Regional Development Theory: Conceptual Foundations, Classic Works, and Recent Developments. *Journal of Planning Literature*, 18(2), 131–172.

Dietzsch, I. 2009. Perceptions of Decline: Crisis, Shrinking and Disappearance as Narrative Schemas to Describe Social and Cultural Change. *Anuarul Institutului de Istorie "George Baritiu din Cluj-Napoca, Series Humanistica*, 7, 7–22.

Fol, S. 2012. Urban Shrinkage and Socio-Spatial Disparities: Are the Remedies Worse Than the Disease? *Built Environment*, 38(2), 259–275.

Grossmann, K., et al. 2013. Shrinking Cities: Notes for the Further Research Agenda. *Cities*, 35, 221–225.

Haase, A., et al. 2013. Varieties of Shrinkage in European Cities. *European Urban and Regional Studies*, 23(1), 86–102.

Haase, A., et al. 2014. Conceptualizing Urban Shrinkage. *Environment and Planning*, 45, 1519–1534.

Hall, T. 1998. *Urban Geography*. London: Routledge.

Hospers, G.J. 2012. Coping with Shrinkage in Europe's Cities and Towns. *Urban Design International*, 18, 78–89.

Janicki, W. 2017. Depopulation as Opportunity for, Not a Threat to Cities and Regions: A Paradigm Change. *Europa XXI*, 32, 89–96.

Jasilionis, D., et al. 2015. *Lietuvos demografinių procesų diferenciacija [Differentiation of Demographic Processes in Lithuania]*. Vilnius: Lithuanian Social Research Centre.

Krišjāne, Z. 2001. New Trends in the Development of Small Towns in Latvia. *Geografiski Raksti*, 9, 33–47.

Lang, T., et al. (eds.) 2015. *Understanding Geographies of Polarization and Peripheralization: Perspectives from Central and Eastern Europe and Beyond*. New York: Palgrave Macmillan.

Leetmaa, et al. 2015. Strategies to Cope with Shrinkage in the Lower End of the Urban Hierarchy in Estonia and Central Germany. *European Planning Studies*, 23(1), 147–165.

Li, H. & Mykhnenko, V. 2018. Urban Shrinkage with Chinese Characteristics. *The Geographical Journal*, 184, 398–412. DOI: 10.1111/geoj.12266

Lithuanian Association of Free Economic Zones [online]. 2020. Available from: https://lafez.lt/fez-overview/?lang=en [Accessed 23 April 2020].

Ministry of Interior of the Republic of Lithuania. 2017. *Lithuanian Regional Policy White Paper* [online]. Available from: https://vrm.lrv.lt/uploads/vrm/documents/files/ENG_versija/Lithuanian%20Regional%20Policy%20(White%20Paper).pdf [Accessed 15 December 2019].

Ministry of the Environment of the Republic of Lithuania. 2019. *Lietuva 2030. Comprehensive plan of the territory of the Republic of Lithuania* [online]. Available from: http://www.bendrasisplanas.lt/ [Accessed 15 December 2019].

Mykhnenko, V. & Turok, I. 2008. East European Cities—Patterns of growth and Decline, 1960–2005. *International Planning Studies*, 13, 311–342.

Nagy, G. & Turnock, D. 1998. The Future of Eastern Europe's Small Towns. *Regions, The Newsletter of the Regional Studies Association*, 213, 18–22.

Oswalt, P. & Rieniets, T. (eds.) 2006. *Atlas of Shrinking Cities*. Ostfildern: Hatje Cantz Publishers.

Pallagst, K., Fleschurz, R. & Said, S. 2017. What Drives Planning in a Shrinking City? Tales from Two German and Two American Cases. *Town Planning Review*, 88(1), 15–28.

Pallagst, K., Martinez-Fernandez, C. & Wiechmann, T. 2014. Introduction. In Pallagst, K., Wiechmann T. & Martinez-Fernandez, C. (eds.) *Shrinking Cities International Perspectives and Policy Implications*. New York: Routledge, 3–13.

Pociūtė-Sereikienė, G. 2019. Peripheral Regions in Lithuania: The Results of Uneven Development. *Regional Studies, Regional Science*, 6(1), 70–77.

Raagmaa, G. 1996. Shifts in Regional Development of Estonia during the Transition. *European Planning Studies*, 4(6), 683–703.

Reckien, D. & Martinez-Fernandez, C. 2011. Why Do Cities Shrink? *European Planning Studies*, 19(8), 1375–1397.

Rink, D., et al. 2010. *Addressing Urban Shrinkage Across Europe—Challenges and Prospects Shrink Smart Research Brief No. 1*, November 2010 on behalf of the Shrink Smart consortium [online]. Helmholtz Centre for Environmental Research – UFZ, Leipzig. Available from: http://www.ufz.de/export/data/400/39030_D9_ Research_Brief_FINAL.pdf [Accessed 17 September 2019].

Rink, D., Haase, A. & Bernt, M. 2009. *Specification of Working Model Shrink Smart Workpackage 1*, September 2009 on behalf of the Shrink Smart consortium [online]. Helmholtz Centre for Environmental Research–UFZ, Leipzig. Available from: https://www.ufz.de/export/data/400/39013_WP1_Paper_D1_D3_FINAL300909. pdf [Accessed 17 September 2019].

Seimas of the Republic of Lithuania. 2016. *Nutarimas dėl Lietuvos Respublikos Vyriausybės programos* [The decision on the programme of the Government of the Lithuanian Republic]. No. XIII–82. Vilnius: LR Seimas.

Šešelgis, K. 1996. Teritorijų planavimo raida Lietuvoje [The Development of Spatial Planning in Lithuania]. *Urbanistika ir architektūra*, 1(21), 4–19.

Šiauliai City Municipality. 2016. *Šiaulių miesto strateginis planas 2015–2024* [*Šiauliai City Strategic Development Plan 2015–2024*]. No. T-325. Šiauliai: Šiauliai City Municipality.

Sireika, J. (ed.) 2007. *Šiaulių miesto istorija 1940–1995 m.* [*The History of Šiauliai City 1940–1995*]. Šiauliai: Saulės delta.

Smętkowski, M. 2017. The Role of Exogenous and Endogenous Factors in the Growth of Regions in Central and Eastern Europe: The Metropolitan/non-metropolitan Divide in the Pre- and Post-crisis Era. *European Planning Studies*, 26(2), 256–278.

Soja, E. W. 2000. *Postmetropolis: Critical Studies of Cities and Regions*. Malden, MA: Blackwell.

Statistics Lithuania. 2019. *Official Statistics Portal. Database of indicators* [online]. Available from: https://osp.stat.gov.lt/pradinis [Accessed 23 September 2019].

Stryjakiewicz, T., Ciesiólka, P. & Jaroszewska, E. 2012. Urban Shrinkage and the Post-socialist Transformation: The case of Poland. *Built Environment*, 38(2), 197–213.

Stryjakiewicz, T. & Jaroszewska, E. 2016. The Process of Shrinkage as a Challenge to Urban Governance. *Quaestiones Geographicae*, 35(2), 27–38.

Sýkora, L. & Bouzarovski, S. 2012. Multiple Transformations: Conseptualising the Post-communist Urban Transition. *Urban Studies*, 49(1), 43–60.

Ubarevičienė, R. 2018. City Systems in the Baltic States: The Soviet Legacy and Current Paths of Change. *Europa Regional*, 25(2), 15–29.

United Nations. 2015. *World Population Prospects: The 2015 Revision*. New York: United Nations, Department of Economic and Social Affairs, Population Division.

Vanagas, J., et al. 2002. Planning Urban Systems in Soviet Times and in the Era of Transition: The Case of Estonia, Latvia and Lithuania. *Geographia Polonica*, 75(2), 75–100.

Wiechmann, T. & Bontje, M. 2015. Responding to Tough Times: Policy and Planning Strategies in Shrinking Cities. *European Planning Studies*, 23(1), 1–11. DOI: 10.108 0/09654313.2013.820077

Wiechmann, T. & Pallagst, K. 2012. Urban Shrinkage in Germany and USA: A Comparison of Transformation Patterns and Local Strategies. *International Journal of Urban and Regional Research*, 36(2), 261–280.

Wolff, M. & Wiechmann, T. 2018. Urban Growth and Decline: Europe's Shrinking Cities in a Comparative Perspective 1990–2010. *European Urban and Regional Studies*, 25(2), 1–18.

Part V

Conclusions, policy implications, and research

21 Postsocialist shrinking cities

Policy themes and future research

Chung-Tong Wu, Tadeusz Stryjakiewicz,
Maria Gunko, and Kai Zhou

The chapters in this book canvas topics on shrinking cities in postsocialist countries and their policy implications in the specific cultural context and institutional settings of the region under discussion. This chapter brings together a cross-national perspective by identifying the key policy issues and the major processes relating them to the relevant literature and highlighting the comparative aspects, as well as potential lessons. Emerging topics raised are identified and, where appropriate, placed in the context beyond postsocialist countries to encourage further comparisons and research. This discussion is organized in four sections: (1) national and global changes that foster urban shrinkage, (2) national policies with direct impacts on urban shrinkage, (3) local level initiatives to manage shrinkage, and (4) topics for further research. Each section will identify the policy implications and approaches taken. Given the kaleidoscope of the rich diversity of the postsocialist experience covered by the chapters in this book, this discussion is, by necessity, a skeletal discussion by which we emphasize the need to examine policies and issues through a shrinking city lens.

Policy attention to shrinking cities is uneven among the countries canvassed in this book, ranging from none to a multitude of approaches adopted at various levels of governance. The differences are partly driven by reactions to the national and global processes impacting on the whole society; the attention to urban shrinkage (or lack thereof) on the national and regional level; the awareness and attitude of local policymakers toward urban shrinkage; as well as policymaking capacity and autonomy of shrinking cities.

Table 21.1 summarizes the issues and the policy responses discussed in each section of this chapter.

National and global processes impacting urban shrinkage

To understand the issues related to urban shrinkage in postsocialist countries, it is necessary to identify the processes that were set in motion as these countries transformed their economic, social, and political environments. The profound social, economic and political implications resulting from the

DOI: 10.4324/9780367815011-26

Table 21.1 Summary of policy issues and responses

Causes	Examples	Outcomes	Policy approaches	Section in chapter
National & global processes [direct relation to urban shrinkage] National transformations Multinational & global transformations	Postsocialist transformations: --economic --social --political Globalization "third wave" urbanization	Industrial decline Closures of enterprises Unemployment Outmigration (domestic and international) New environmental legislation Fertility decline, rising mortality Centralization/ Decentralization of authorities & responsibilities Polarization Formation of mega-regions	Regrowth Immigration policies Pro-natal policies Embracing shrinkage	Resource depleted cities Regrowth Embracing shrinkage Governance Polarization
National policies [direct relation to urban shrinkage]	Population resettlement Industrial upgrading	Population decline (change of labor force) Industrial relocation	Resettlement Industrial restructuring	Deliberate shrinkage
National policies indirect impacts [often unintended]	Health care	Services centralization	Service & infrastructure rationalization	Services

transformation from a planned economy to a market-oriented economy are multi-faceted with national and local specificities. Engaging with the global economy, and for some, opening their borders brought significant changes at both the local and national levels that are exemplified by the plight of many resource dependent communities and the more recent phenomenon of polarization.

Resource dependent communities

Adopting market-oriented economic policies and mechanisms at the national level meant that many local industries that were highly subsidized during the socialist era were confronted with substantial reforms. This led to closures and large unemployment in many of the communities that were reliant on such industries. Eight chapters (Chapters 4, 5, 10, 12, 14, 17, 18, and 20) in this book reported on what took place in resource dependent and industrial company towns in China, Russia, and parts of postsocialist Europe. The legacies and path-dependency are particularly acute for resource dependent communities many of which face additional pressures stemming not only from the depletion of the resource but also from the increasing environmental concerns impacting on their operations, economic viability, and often, the need to reform state-owned enterprises. China identified 262 resource depleted cities in 2013 and subsequently implemented programs to assist 69 of them (26% of the total) (Chapter 5). While other postsocialist countries included in this book faced the same problem with the majority of their resource dependent cities, China has had the highest number and one of the largest national programs aimed at dealing with their industrial decline. Similarly, Russia adopted a state priority program, titled *Integrated development of monofunctional settlements for the period 2016–2025,* along with a list of 119 possible measures to support slightly over 300 company towns—*monogorods* (Ministry of Economic Development of Russia, 2016). The support measures were intended to help *monogorods* to balance their budgets and tackle other "burning issues", but they failed to address their fundamental problems, namely existing structural dependency on a single, aging industry and the threat that such dependency represents for residents (Crowley, 2016, Gunko et al., 2021). In 2019, because of widespread criticism, the program was abolished before the funding period ended. In Poland, the national programs dealing with industrial decline and/or industrial restructuring have been sectorally or regionally oriented, but they have also had impacts on shrinking cities located in the areas of industrial decline. This decline was caused not so much by the depletion of local resources but rather by the economic inefficiency of their extraction following the implementation of the market system and full external openness of the national economy (Chapter 17).

While the decline of resource dependent, old industrial and company towns in postsocialist countries is due to economic reforms within the transition from state socialism to variations of capitalism, urban shrinkage is not necessarily acknowledged as a direct consequence of this process by the nation states, leading to a lack of specific legislative and financial support (Chapter 14). This is certainly the case in China and Russia, where urban shrinkage is not part of the policy discourse. The programs being promoted at the national level are about economic revival and diversification without acknowledgment of urban shrinkage and its related issues. In this respect, they follow the regrowth strategy, which will be discussed in a subsequent section. This approach is also present among the postsocialist European countries included in this book. However, it is slowly changing in those countries that became European Union (EU) members through attempts to follow earlier West European experiences and best practices with EU funding.

Polarization and urban shrinkage

Uneven spatial development and polarization have been identified in several chapters as a cause of urban shrinkage. On the global, national, and regional scales, the concentration of resources and power in selected places runs parallel to the departure of capital and the state in others, leading to spatial destruction which "operates in disjointed and uneven ways, destroying some places and regions more so than it does others and creating sacrifice zones" (Gordillo 2014, p. 80, Gordillo, 2014, Harvey, 2006, Smith, 2008). In the case of Russia, the reference is largely about the expansion of Moscow and St. Petersburg metropolitan regions that are growing much faster than the rest of the country. Depending on the specific topic, it could also refer to polarization within the region in which the specific city or community is located. For example, in the discussion of medical care for expectant mothers in Russia (Chapter 13), it is the concentration of the best facilities and cares in regional capitals that matter.

In some postsocialist European countries, polarization of the national economy is an important factor for internal migration, leading to urban shrinkage in the regions where residents are moving to the more prosperous cities (Stryjakiewicz, 2009, Lang et al., 2015). Equally, or even more important though, it is the outmigration internationally due to polarization within the EU that would be the decisive factor for urban shrinkage for specific communities and may be for the whole country, for example, Lithuania, Bulgaria, or Romania. Polarization between the postsocialist EU and non-EU members is best exemplified by the nearby countries of Slovenia, and Bosnia- Herzegovina, discussed in Chapter 16.

In the chapters on China, polarization refers to the concentration of growth in specific regions, chiefly those on the east coast of the country and the metropolitan areas within a province. For specific cities or communities, dependent on their proximity to the rapidly growing regions and their own resource endowments, their experience with urban shrinkage would be dependent on a different combination of regional factors.

As yet, the discussions of polarization and urban shrinkage have not engaged with the discourses on the "third wave" of urbanization, articulated by Scott (Scott, 2011), and the emerging megaregions, for example, in China (Harrison and Gu, 2021) partly as a result of government policy and/or the discourse about the negative impacts of these emerging trends (Meagher, 2013). While discussions of the other side of the "third wave" focus chiefly on the negative social impacts, no one has examined urban shrinkage and its associated implications. Similarly, studies of the emerging megaregions of China have not considered the potential urban shrinkage implications of this phenomenon (Yeh and Chen, 2020). The emergence of "perforated" metroregions (Chapter 7) point to the potential of this taking place in ever larger megaregions, which in China's case at least, according to Harrison and Gu (2021), is based chiefly on territorial politics rather than on economics. It is through examining the emergence of urban shrinkage at these multiple scales that the studies of shrinking cities can contribute to the broader literature on urban geography, regional development, and planning policies.

While polarization is an important factor related to urban shrinkage in all the postsocialist countries, it is important to identify the scale differences. Potential policies to deal with the impacts of polarization must specify the scale issue and frame the strategies accordingly; otherwise, it would either become futile and ineffective or even delusional. For example, no small community can be expected to develop effective programs to counter polarization pull at the EU scale without concomitant policies at the national and EU levels. Alternatively, programs at the EU scale, to be effective, need to reach the local community level and adapt to local specificities to have any hope of being successful.

Differences in the spatial outcome of polarization in different regions are worthy of note though. While Russia and most of postsocialist Europe are experiencing polarization chiefly around pre-existing large cities, basically a reinforcement of pre-existing spatial inequalities, the same is not necessarily the case in China. The Chinese development policies of the past several decades have engendered regions of rapid growth that were either small towns or urban areas unconnected to former industrial cores. In addition to some pre-existing core centers, new patterns of uneven spatial development and of population shifts leading to shrinking cities have emerged. Shenzhen and the southern China industrial zone, close to Hong Kong, is an example of a completely new growth pole becoming the focus of internal migration flows, which have led to urban shrinkage in many source regions, but part of this new conurbation is now experiencing urban shrinkage. This phenomenon exemplifies the need to consider and compare polarization patterns with care.

Urban shrinkage due to policies

Policymakers do not tend to acknowledge urban shrinkage for a variety of reasons including the perception that it is not politically popular to admit that one's own community is in decline (Bernt et al., 2014); moreover, shrinkage may be viewed as a threat to the political order in general especially in "politically sensitive" areas like borderlands (Balzer and Repnikova, 2010, Dzenovska, 2020, Jia, 2012). At the same time, policies are often made without regard to their significant direct or indirect impacts on urban shrinkage. Two examples are included in this book. Chapters 6, 11, and 14 identify the implications of two such policies in China and Russia.

Resettlement policy

While Russia pursues a policy of promoting population growth, it has, since the beginning of the 2000s, also adopted a policy to resettle population from the Far North suffering from harsh climatic conditions (Chapter 11). During state socialism, the large-scale development of the Soviet Far North was due to the need for natural resources extraction and establishment of control over a vast sparsely populated area with new cities and towns being established where no permanent settlements have ever been before (Josephson 2014). With the collapse of state socialism, a reduction of state support for industries, investments in science, and military activities caused a structural crisis and outmigration from northern cities and towns in Russia. The state resettlement policy was aimed at institutionalizing this outmigration. Between 2003 and 2017, the state resettlement policy was responsible for 6% of the total outmigration from the Russian Far North. While there are good reasons for this policy such as helping people resettle into a more comfortable environment, this is an example of a policy that was implemented without regard to the consequences of depopulation for local communities which must manage daily life among ruins, abandonment, declining services, and infrastructures (see Chapter 14). The rhetorical question is, could this be a case of deliberate disregard?

Industrial restructuring

Unintended consequences of well-intentioned policies could result in urban shrinkage. Chapter 6 presents the case study of Dongguan in southern China, an example of a deliberate policy of industrial upgrading that led to urban shrinkage, which is known as "Phoenix Nirvana" ("凤凰涅槃") or the "adjustment-type" of shrinkage (Zhang et al., 2017). Policymakers did not intend it to result in medium- or long-term urban shrinkage because they made the decision in the context of rapid economic growth in China. It was a national policy to encourage movement up the industrial ladder, thus making way for more complex and high value-added activities in the part of China which pioneered economic reforms and rapid economic growth due chiefly to export-oriented industrialization. The industrial upgrading strategy is partly based on the laudable idea of sharing growth with the less developed regions. National policymakers expected temporary economic slowdown due to the time lag for the old to exit and the new to be established, but the onset of global economic slowdown, bi-lateral political processes, and international competition resulting in a loss of competitive edge led to urban shrinkage as the expected industrial upgrading was slow to materialize and employment plummeted. These events were further exacerbated by the onset of the global pandemic. The local policymakers, for a variety of reasons, were not enthusiastic about the upgrading policies either (Fang and Hung, 2019). It is an example of public policy without any plans to deal with contingencies or unintended consequences. It is also a cautionary tale about making policies based on assumptions beyond the policymakers' control, underestimating the unexpected and imposing national policies without local buy-in. As China comes to terms with the "new normal", it may find more communities could be faced with urban shrinkage due to policies imposing industrial upgrading.

The deliberate industrial restructuring policies that China has recently adopted should not be confused with the national policies of industrial restructuring in the postsocialist countries of East and Central Europe that were implemented soon after transformation to more market-oriented economic policies. The policies of industrial upgrading were aimed at overcoming or slowing down the negative impacts of urban shrinkage through new industrial zones such as the Walbrzych Special Economic Zone discussed in Chapter 17.

Over the past several decades, numerous new policies were introduced, often in the guise of pilot schemes, with clear intentions of adaptations as more experience emerged, which was the case with the Special Economic Zones (SEZs). At the same time, new institutions and governance arrangements were introduced to oversee the changes, which partly reflects the change of national leadership but also the need to adapt to new conditions. Institutional evolution is part of the endogenous transformation process. The economic success story of China is in stark contrast to other former socialist states. Deliberate policies of welcoming foreign investment in designated locations and regions stimulated vast internal migration from the less developed regions to the rapidly growing regions, exacerbating regional differences—a result of the "let some become prosperous first" strategy. The party-state-led economic reforms touch on all aspects of life but most negatively on those cities and towns reliant on highly subsidized industries in regions such as the northeast of China.

Services and infrastructure restructuring

As the population shrinks, many public services such as schools, healthcare, and public transportation are forced to reassess their functioning due to dwindling usage and resources. The tendency to rationalize services based on population size and centralization in larger cities becomes the common strategy. Although there are studies on managing the impacts of population shrinkage on infrastructure and services (Maes et al., 2012, Moss, 2008, Bierbaum, 2020), this topic has not attracted as much attention as it deserves. In countries where population decline becomes more prominent and imminent, managing public services and infrastructure demands innovative solutions.

Chapter 13 highlights the issues by drawing attention to healthcare provision through a case study of contradictory policymaking in Russia—the collision of pro-natalist policies and "optimization" of healthcare (with an example of maternity care). The study was conducted in a shrinking region impacted by centralized decisions without regard to their spatial implications. Despite the supposedly patient-focused policies, the result is an added burden on those who need the care as well as negative impacts on the professionals responsible for providing the care. These policies encourage those patients, as well as the professionals, who have the choice to move away, leading to further centralization and this perpetuates a vicious circle of service deterioration and shrinkage in the smaller towns and cities.

In the context of smart cities, since smart shrinkage is becoming a topic that attracts increasing attention, the issues of how smart infrastructure impacts on access, usage in everyday life, and access to learning new technology are also issues germane to shrinking cities (Tuitjer and Muller, 2021). In the emerging e-commerce based villages in China, research on the rise and fall of Taobao villages cautions that the lack of suitable infrastructure is a key issue for why many fail (Fu et al., 2019). The ways urban shrinkage impact on the population's right to healthcare is especially noteworthy because it deals with life and death issues, but other public services such as education, social services, public transport, and housing are equally important to the quality of life in a shrinking city. The case study in Chapter 14 also cautions about too much reliance on smart infrastructure and standard-driven policies because it is the everyday life, the mundane that residents experience most directly. In the context of governance becoming rescaled upward, implications of national policies and standards need to be scrutinized through the lens of social-spatial equity.

Managing urban shrinkage

With few postsocialist countries accepting shrinking cities as a reality that requires adaptation rather than a phenomenon to counteract, planning and policymaking become based on variations of the growth paradigm dominant in the 20th century. Since shrinking cities have connotations that few politicians in any political setting can embrace, spawning innovative programs such as the "legacy cities" (Lincoln Land Institute) are an attempt to try to entice communities and politicians to get on board. Even in the supposedly "new normal" embraced by Chinese policymakers, few, if any policies state that some communities need to embrace urban shrinkage as a permanent feature rather than focus exclusively on turning it around. Among the Central and Eastern European (CEE) countries, "regrowth" is a popular approach adopted by many communities (Chapter 19) though their attitudes and actions are gradually changing as more EU funding is dedicated to alternative programs, such as the green deal, silver economy or urban (qualitative) renewal. Similarly, "soft" strategies, aiming at image improvement of shrinking cities, are gaining importance (Chapter 17). Active policies or programs to manage urban shrinkage are still rare in Russia (Chapters 11 and 14); and the establishment of development zones in parts of Russia, including the Russian Far East shows that the growth paradigm runs deep.

Regrowth

In the cases cited in Chapters 5, 18, 19, and 20, the relevant policymakers adopted the "regrowth" approach seemingly without question. Taking these examples together and with the knowledge of cases in other postsocialist

European countries, the impression is that many policymakers do not undertake deep introspection about the relevance of a regrowth strategy to their specific situation. Furthermore, there does not seem to be much comparative review of how their approach is differentiated from other cities to gain any competitive advantage. The survey results reported in Chapter 20 on Šiauliai, Lithuania provide a glimpse of the disconnect between the perspectives of officials and those of the local businesses and community. The general approach of establishing industrial zones, providing tax incentives or other subsidies to attract investors (both domestic and international) is replicated by various cities trying to compete. One potential outcome is a "race to the bottom" and failure for many, the more so when cities are strongly dependent on decisions at the national level. It is evident that not all communities will grow again, at least not in the same way some policymakers believe. What can be done to encourage policymakers to deal with reality and consider policies that may be more appropriate to their specific circumstances? An example could be devising ways to accept a smaller city to enhance the quality of life of their inhabitants, which may have the positive effects of attracting new residents. In the post-Covid world, new opportunities such as attracting workers who can work remotely but who may seek a better living environment for themselves and their families may present opportunities for some. As Chapters 18 and 20 have concluded, a shift toward making the communities more attractive places to live and work may be more successful than the usual tax cuts and subsidies to attract new industries. Alternative strategies that follow such an approach are being discussed and tried in countries in the EU and elsewhere, focusing on green growth (Ortiz-Moya, 2020), attracting cultural and creative sectors (Musterd et al., 2010, Stryjakiewicz et al., 2014), developing social entrepreneurship and co-working spaces, and other directions that should be of great interest to policymakers everywhere, especially for postsocialist countries.

Embracing shrinkage

The literature on policymakers' attitudes toward urban shrinkage generally lists "disregard" or "trivializing", "acknowledge and ignore", and "acceptance" and "utilizing" among the type of reactions adopted (Hospers, 2014, Pallagst et al., 2017) but none has identified active embrace of urban shrinkage as a policy response. However, embracing shrinkage is what some communities in China have adopted in the first instance to cope with poverty (Zhou, Yan & Zhao et al. 2019), and secondly to encourage the return of the outmigrants to reinvent the local economy (Zhang et al. 2017). These strategies were discussed in Chapter 6, which reported on the case of Taoyuan County in China when it was faced with vast numbers of the working age underemployed population emigrating to the rapidly growing regions of China to seek employment. Instead of regarding this outmigration as a

problem, they identified opportunities. Policymakers established and promoted training programs to prepare their own residents for job opportunities in other regions. Employment agencies were established both by the
government and by the private sector to place the trained worker in appropriate employment. Through these strategies, Taoyuan assisted residents to
find relevant employment in labor shortage regions. The short to medium
term expectation is that the outmigrants will send remittances home to support the local economy and alleviate poverty. The longer-term expectation
is that some, if not many, will return with their skills and knowledge to help
their hometown restructure and grow in what is known as the "making a
nest for the Phoenix" ("筑巢引凤"). This approach is fundamentally different from what most communities confronted with urban shrinkage do when
faced with outmigration of the working age population. The social, cultural,
and economic context that made this approach possible may not be available elsewhere, and to date, there are few other studies of local governments
in China adopting such strategies. Comparative studies within China and
internationally will be valuable contributions to the literature on shrinking
cities policy approaches.

Whether this approach is germane in other contexts requires investigation. It would be easy to dismiss this as only applicable to the Chinese context where economic growth was and still is relatively high compared to
most other countries. The key to this approach is about taking a long-term
view of the issues rather than just reacting to the immediate. Whether local
officials have the agency to make long-term commitments will be discussed
under the topic of governance.

Local culture and managing shrinkage

A reference to local culture is an aspect of the case studies from Slovakia
(Chapter 18) worthy of wider attention. The policymakers of Banska Štiavnica,
Slovakia, bowed to the local culture that supports a mining community
and staged the closure of silver and gold mining to moderate the impacts.
In this case, it is about the deep roots of mining in the local culture and the
respect for what the industry has done for the local community through the
unions that made the difference. It is not a topic that has attracted much
attention in the literature but is worthy of attention given the likely impacts
on coal mining communities of recent policies toward zero carbon emissions by 2050 adopted by an increasing number of countries. It is about how
to tap into local culture to manage urban shrinkage when policies change,
and the impacts are going to be profound. It is remarkable that postsocialist
ructions did not erase the local culture of respect for the mining community
and its positive influence on moderating the community's adjustment to a
new reality.

In their review of Šiauliai's (Lithuania) programs of managing urban
shrinkage (Chapter 20), the authors noted the non-government interviewees'

views of the issues and the policies requiring attention were quite different from those of the officials. The differing views point to the need for a deeper understanding of the issues based on local knowledge before seeking solutions. This perspective is also the starting point of a research project on Portuguese shrinking cities (Guimaraes et al., 2016). While seeking local input may be appropriate for some settings, there are questions regarding whether sub-national scale research results always have any competence to make meaningful inputs (Engels, 2021, Towers, 2000). This is especially true when we are confronted with regimes that are increasingly centralized and authoritarian, as the next discussion explores.

Governance

The literature on strategies to manage urban shrinkage includes discussions about how policymakers ignore, accept, and/or actively manage urban shrinkage with the unstated assumption that policymakers at the local level have the capacity and the autonomy to make policies appropriate to the circumstances (Bernt et al., 2014). Widespread centralization of political systems over the last few decades means that policymakers at the local or even regional level lack agency to deal with the issues they confront as a result of very restricted decision-making spaces or the upscaling of such decisions. Such impacts on local governance and the ability to manage urban shrinkage are cited in chapters in the Russian and postsocialist European sections, while in the case of China, all chapters noted the pivotal role of the central government.

The most obvious example is the disposal of land or even control over the use of land. In the Russian case, the chaotic privatization of land and real estate after the collapse of state socialism makes it increasingly difficult to manage abandonment and vacancies produced by shrinkage. Local governments who oversee urban planning policy and manage the built environment have no right to redevelop brownfield sites or carry out any kind of activity there if they are not in municipal ownership. Obtaining property rights can be a matter of longstanding negotiations and court cases (especially when the owner is unknown, which is often the case), as well as requiring significant funding. Moreover, in Russia as well as some CEE countries such as Lithuania, local government cannot dispose of or develop government owned land without permission from higher-level authorities. Local governments in these settings have limited ability to deal with urban shrinkage with respect to usage of land and renewal. These examples are in sharp contrast to Chinese city governments, which have the authority to "sell" the development rights of land since urban land is state owned and collectively owned in rural areas. In fact, the designation of land for urban use and the subsequent selling of land is one of the main sources of income for local governments in China. Chinese city governments have comparatively a great deal more freedom to manage and use land, including for the purposes of managing urban shrinkage. But this level of autonomy and the desire to generate revenues from land can lead to what are called "overdraft" developments creating new urban shrinkage in the form of ghost towns with empty commercial and residential developments—projects that were largely finance rather than demand driven (Zhang et al., 2017). The above examples highlight the need to go beyond the obvious issue of central vs local decision-making to identify solutions.

In postsocialist Europe, although the issue of governing shrinking cities strongly depends on governance systems in particular countries, a few common features can be identified: a) the competencies of different levels of governance (local, regional, national) are blurred, including the division of power between central governments and self-government administrations b) multilevel coordination of actions dealing with shrinkage is weak, and sometimes the goals of these actions are contradictory (for example, the decision-makers at the regional level would like to limit suburbanization and the outflow of population from a core city to its suburbs, whereas the decision-makers in the surrounding communes do their best to encourage people to move in) c) the regulations concerning spatial planning as well as demographic and economic policies are not stable, which makes the implementation of long-term initiatives difficult (for example, in Poland, the existing local spatial developments plans were abolished almost overnight in 2003, opening the way to neo-liberal land-use management with an excessive role of private developers).

The case studies included in this book are grounded in the context of postsocialist transformation initiated at least three decades ago, but the transformation may also take different paths, especially with respect to land ownership even in other more recent postsocialist settings, such as Cambodia (Collins, 2016, Flower, 2019) and Vietnam (Le and Le, 2018). With this diversity in mind, we argue that policy studies need to consider a wide spectrum of governance context by including different settings in their discussion or at least clearly state their assumptions about the governance context to make clear where their theoretical or policy discussion would apply.

The review of policies at the national level highlighted three aspects. First, it is essential for national policies to consider impacts on localities and what is required to assist them to cope with the consequences of the policies. Second, if the basic assumption of policies is that potential unintended consequences can occur then what remedies are required. Third, national policies have spatial impacts, potentially negative for specific localities, so it is imperative such plans should consider potential impacts and mitigation strategies. All the policies reviewed in this volume are reactive in that they try to deal with urban shrinkage after the fact. As the chapters show, not all policymakers are enthusiastic about acknowledging urban shrinkage and provide the leadership and funding that are required to manage the phenomenon, even though many are trying in their own ways and in their specific context.

One of the themes of this book is that legacies and local context matter. While the range of policies canvassed above emphasizes that there is no policy that would be "one size fits all", it is not an argument against considering general policy approaches applicable across a whole range of situations. What we are advocating are policy approaches that are sufficiently robust and suitably flexible to be adapted to the local context, a challenge for all who are interested in grounded and effective shrinking cities policies.

Topics for further research

Soon after we commenced the preparation of this book, the COVID-19 pandemic struck, and it is still raging around the globe as we prepare the final version of the manuscript. The pandemic has brought upheavals to the economic, social, health care, and political systems in all the countries it touches. All communities, urban or rural, have been affected and facing diverse problems, both old and new. The evolving situation is too raw for firm conclusions to be drawn, though there are developing patterns of the impacts on cities, including postsocialist shrinking cities, and societies in general. Many research questions are emerging and will continue to emerge from the impacts of the pandemic. However, regrettably, it is beyond the remit of this book to explore in detail. Where appropriate, we raise comments and questions associated with the ongoing pandemic impacts that are relevant to the several topics we have identified for further research. We acknowledge that these are preliminary and likely to be modified as more data become available.

Informal economy

During the rapid economic restructuring, many postsocialist cities, irrespective of whether they are shrinking or not, found their informal economy blossomed as individuals sought ways to cope as shown in Chapter 12 on the Russian local labor market. It is a phenomenon replicated in all the economies that went through major transformations whether abrupt, reluctant, or gradual, but few studies have documented this in the context of shrinking cities. Chapter 8 briefly referenced the informal economy associated with the boom and bust of trade hubs in the border regions of China and Russia. There is a substantial literature on the informal economy in diverse disciplines, ranging from the urban geography classic on street peddlers in Southeast Asian countries (McGee, 1975); the informal sector in postsocialist countries (Polese and Rodgers, 2011); and studies of informal cross-border trade (Ryzhova, 2008, Stryjakiewicz, 1998, Stryjakiewicz and Kaczmarek, 2000), attesting to the significance of the informal economy in countries spanning economic, social and political systems and continents. While the informal sector conjures up street peddlers and stallholders, it encompasses a great deal more, ranging from piece workers working from home, workers in the cash economy, and more broadly, many in the gig economy of the 21st century (Ng'weno and Porteus, 2018). In the case of shrinking cities, the likelihood is that individuals had no choice but to find employment in the informal sector to cope with poverty—what is known as "exclusion" from the formal sector rather than voluntary "exits" to pursue one's own opportunities (Bromley and Wilson, 2018). Examples of such exclusion from the formal sector were also observed in some shrinking cities of postsocialist Europe at the beginning of economic transformation, for example, the so-called poor shafts in Walbrzych, where the former miners who lost their jobs were extracting coal illegally to maintain their families (Chapter 17). The informal sector still plays a relatively big role in some postsocialist Balkan countries, such as Bosnia and Herzegovina or Serbia (Antonić and Djukić, 2018). Audric et al. (2012) have also noted the role of the informal sector in absorbing the poor as industries shifted from Latin American industrial zones and old industrial centers in the United Kingdom. The pandemic has made economies with large informal sectors especially vulnerable (Elgin et al., 2021, page 38) but the pandemic has also quickly intensified connections between the informal sector and the gig economy. An example is the rapid rise in demand for home deliveries in urban areas, giving the gig economy an unexpected boost and providing employment opportunities for many who have lost their jobs due to the pandemic.

Whether the rise of the informal sector and now the gig economy becomes a permanent feature of the new reality and how this development may have evolved is seldom studied in the literature on shrinking cities. While some policymakers may regard the informal economy as a negative feature, the potential of the informal sector in helping to underpin both the economic and social stability of shrinking cities, especially smaller cities, should not be ignored. Indeed, there are those who regard the informal sector as the future (Ng'weno and Porteus, 2018), not just for the subalterns (Ohnsorge and Yu 2021). If social enterprises are included in the informal economy, then the potential of social enterprises operating at a small scale to provide services for a smaller population should be considered. The informal sector has also evolved, and in the postpandemic society and economy, the potentials of the informal sector and the gig economy should be given more attention (Georgieva, 2019). Return migrants to their home villages and towns in China using e-commerce platforms to establish enterprises to export products from grass-root enterprises is one example of the potential of the informal sector yet to be explored in other countries and shrinking regions (Wang et al., 2021). The so-called "Taobao villages" using an e-commerce platform have blossomed since the mid-2010s to about 5,500 engaged in enterprises ranging from producing local specialties, processed agriculture goods, labor-intensive products to those with more technical content (Ali Research, 2020). Their rapid spread in many regions of China has sparked expectations of reviving declining rural regions (Luo et al., 2019). Cautions have been raised by research on the social implications, especially the potential exploitation of labor (Fan, 2019) and the uneven distribution of adequate infrastructure to enable their development (Fu et al., 2019). The application of their experience to other parts of the world is yet to be researched, but they underscore the multiple aspects of the informal economy and how it has rapidly evolved. However, more attention needs to be paid on how it manifests itself in shrinking cities. The pandemic has stimulated e-commerce in ways not seen before. Whether this trend is benefitting the Taobao villages or merely reinforcing the established manufacturers and retailers is amongst a set of emerging research questions. Regional polarisation may grow because not all peripheral locations have reliable access to up-to-date digital infrastructure. Consequently, small peripheral shrinking cities or towns can be particularly impacted by the pandemic.

Gender and urban shrinkage

The case study of healthcare services in a shrinking region of Russia (Chapter 13) points to the feminization of medical services in parts of Russia. While this issue is beyond the scope of the case study, it highlights a topic seldom explored in the literature on shrinking cities. Much has been written about the loss of the working age and educated population from shrinking cities, but few of these studies differentiate the gender aspects of the outmigration or the gender issues associated with those who are left behind, for example, the medical professionals in the Russian case study. A study of the central region of Russia, which includes the region studied in Chapter 13, confirms the significance of young working age migration but offers no gender breakdown (Kashnitsky, 2020). In most cases in other settings, it is the aged and the young that are left behind by family members who move to find employment, as well as those who are in insecure employment or working for themselves. Are there differences between countries, whether postsocialist or otherwise? What are the implications? The Russian case study hinted at the implications for the services provided to patients, but there are few studies of implications, if any, for other service areas. It is a topic that requires investigations from the perspective of providers and recipients of services.

Disproportionate impacts on female workers as a result of industrial restructuring is another aspect to be considered. At the beginning of postsocialist transformation, the highest growth in the unemployment rate was among females in the shrinking cities dependent on industries such as garment and textile production, thus raising issues about poverty and those left behind by structural and economic transformation.. These issues are exacerbated by the pandemic across the world and shrinking cities are not exempt. There is increasing recognition of high representation of women in the health sector (World Bank 2020), disproportionate employment loss due to the pandemic (United Nations 2020; McKinsey 2020) and a need to tackle challenges faced by women (OECD 2021).

Indeed, the informal economy is often female dominated partly because it may be home-based work, include enterprises that involve irregular but flexible hours and/or be a very small enterprise (Fan, 2019). This is the case both in more industrialized or less industrialized economies (Carre, 2017, UN Women, n.d.). The significance of this sector to women is demonstrated by the UN estimate that in South Asia, over 80% of women in non-agricultural jobs are in informal employment; 78% applies in East and South East Asia (excluding China) and 21% in Central and Eastern Europe and Central Asia (UN Women, 2015). These statistics are not specifically about the situation in shrinking cities where the informal sector may be more important than what is observed in the overall economy, but they should alert us to the importance of the topic and the urgency for better research of this topic in the context of shrinking cities.

Rural-urban relations and shrinkage

Small towns, less urbanized and rural areas in China that are the main origins of outmigration to the more prosperous regions of the country have experienced varying degrees of shrinkage, leading to the phenomenon of "hollow villages" with abandoned housing, fallow fields, and small towns left with largely the young and the aged. Even smaller cities in the more prosperous regions have experienced growth and shrinkage (Wu et al., 2014). A few provinces, such as Heilongjiang Province located in the northeast, have also experienced both low birth rates and large outmigration, raising questions, for some, of the province's long-term sustainability (Heilongjiang, 2015). In a time dominated by the prosperous urban economy, in large parts of China, some rural areas and their settlements are dilapidated. The collective ownership of rural land has limited the scope of traditional agriculture to scale up. How to respond to the shrinking of both population and economy in rural regions of China remains neglected. It is a topic that has not yet attracted sufficient research interest. It was only in mid-2019 that the Chinese government started to acknowledge shrinking cities in a State Council document. Formally connecting the rural shrinkage problem with wider regional growth and the shrinkage pattern may take some time.

While European countries have been studying and making policies and programs to deal with shrinking regions (Simon and Mikesova, 2014), the discussion has not been clearly linked to policies about shrinking cities. In Japan, for example, housing abandonment (Suzuki and Asami, 2019) as well as shrinking rural regions (Matanle and Sato, 2010) are beginning to attract some attention. Nevertheless, it seems that concerns about urban shrinkage are yet to be fully linked to rural shrinkage. The relationship between the two phenomena and what policies approaches will be required need consideration. Cities do not end at their administrative borders, the centripetal and centrifugal effects of massive urban growth and their impacts on rural and small-town shrinkage need to be incorporated into urban shrinkage research. The pandemic has spurred many who are able to work from home to move to outer suburbs or distant small towns. Many are seeking more affordable and spacious accommodation to incorporate working from home, to isolate and access opportunities for outdoor activities far from crowds. For others, it is an opportunity to be able to live in a lower-cost location and remain employed in their professions, based in high-salary countries. The Lithuanian expressions is "lives in Lithuania, works abroad" (Marmaitė. 2020). In turn, the housing demands of these individuals may bring higher rental or pricing pressures on the existing residents. The examples above may alter the importance of some determinants of urban shrinkage in post-socialist countries. One may expect a growing role of suburbanization or urban sprawl at the periphery and a diminishing role of international and interregional migrations (Mierzejewska & Wdowicka, 2021). Irrespective of whether these are permanent or temporary changes, they highlight the important linkages between cities and their regions and beyond, as well as the potential impacts on shrinking regions.

Spatio-temporal aspects of growth and shrinkage

In their often-cited study of long-term urban changes in Europe, Turok and Mykhnenko (2007) pointed out that in the long history of European urbanization, many cities grew and declined, and many grew again. It is an excellent caution, therefore, not to ignore the temporal aspects of urban growth and decline and the need to know about the historical changes of any city to understand the process it is undergoing. Doringer et al. (2019), in their criticism of temporality being absent in many empirical studies, made the plea to take a measured view of the temporal changes of shrinking cities. The shifting urban shrinkage in some postsocialist European cities due to increasing suburbanization just outside of administrative boundaries and the emergence of perforated metropolitan areas in China and elsewhere lend additional credence to their call. While it is important to understand the temporal changes of specific shrinking cities and make comparisons, expanding this understanding to the regional scale is equally essential. Chapter 2 points out that China embarked on its transformation a decade or more ahead of other postsocialist countries with specific spatial strategies, which created new regions of rapid development and, conversely, produced shrinking cities and regions. New spatial patterns of growth and decline emerged.

In the space of three decades, many Chinese towns and cities grew from a small community to a booming manufacturing or trade hub, for example, Dongguan (Chapter 6), and the cases of consumer goods cities (Wang, 2006a, Wang 2006b). They then lost most of their industrial activities and experienced severe depopulation either due to further economic changes or direct policy intervention. Starting in the first decade of this century, similar processes of boomtown industrialization are taking place in other

Asian postsocialist countries such as Cambodia and Vietnam (Yang, 2016). In postsocialist Europe, for example, Lithuania also experienced regional shifts of growth and decline that point to the significance of understanding temporal changes beyond specific shrinking cities. Each nation's interconnected and contingent processes of migration, industrialization, globalization, economic restructuring, and public policies will likely result in distinctive patterns of urban shrinkage. Comparative studies of these cities and regions will be invaluable to our understanding of the formation and change of shrinking cities. Cases from other countries would be equally important, but the Chinese cases took place within a compressed time frame, potentially allowing better access to relevant data and information as well as individuals who lived through the process.

Migration (im)balance and urban shrinkage

One of the key indicators of a shrinking city is absolute population decline due to changes in fertility and mortality rates and, importantly, due either to outmigration to other regions in the same country or outmigration internationally. While this may still be the case for some postsocialist countries, using absolute population loss as a measure of decline is being questioned by studies in the UK where the concept of relative decline is regarded as more appropriate (Pike et al., 2016). The causes of outmigration are many, including pull factors such as better economic and education opportunities, lifestyle, and a more clement climate. Push factors may include poor economic prospects, social and political conflicts, environmental emergencies, and climate change. The impacts of outmigration of the young and educated are well documented; so are the impacts of international outmigration, which, in some cases, can lead to the shrinkage of an entire nation's population. The significance of any of the above types of migration varies between the postsocialist countries, but it is known that the impact of international migration on urban shrinkage in postsocialist Europe is much more significant than in Russia or China.

The global pandemic crisis of 2020 has upended many trends, including, it seems, the migration of population away from shrinking and declining regions. Emerging reports of large numbers of populations moving from some large urban centers to small towns and communities and the return flow of international migrants back to their countries of origin seem to herald important changes are underway. The significance of what is taking place is yet to be fathomed, but it seems to portend global shifts. Anecdotal evidence points to at least two types of return migration or outmigration from the established centers. The first includes those international migrants who have lost employment due to the shutdowns necessitated by the pandemic returning home seeking a social safety net. A second group is those individuals who are employed but found it possible to work from home, so they have moved to locations with lower rents or housing costs for more space, possibly a better environment and other factors such as family support. The flow of the second type of individuals and their families has caused glowing optimism in many towns and communities as stories of increasing population, rising demand for housing and services buoy the hopes of many.

Data on the massive return of migrants in Europe to their home countries include "an estimated 1.3 million Romanians went back to Romania… Perhaps 500,000 Bulgarians returned to Bulgaria—a huge number for a country of 7 million" (World Bank, 2020). Lithuania has seen more citizens arriving than leaving for the first time in years—the monthly average number of returning Lithuanians increased from 1,700 in 2019 to 2,000 in 2021 (Statistics Lithuania, 2022). Politicians in eastern Europe had long complained of a 'brain drain' as their brightest left in search of higher wages in the west. Now the pandemic, a shifting economy and changing work patterns are bringing many of them back. A 'brain gain' has begun" (Economist, 2021, p. 23). This turning point in the European international population movement is a potent reminder to include migration as a crucial research problem affecting shrinking cities in postsocialist Europe. However, the unexpected benefits, if any, of the Covid crisis are global (Le Coz and Newland, 2021). Positive news stories in Australia (Dusevic, 2021), Japan (Tetsushi, 2020, Su, 2020), the United States (Lambert, 2020, Austin, 2021), and other countries (Bacchi, 2020) raise expectations that many of these migrants or return migrants will stay permanently, bringing new energy, positive economic impacts, and population to some moribund communities. More cautious analysis argues the phenomena may be temporary and that the trend is merely an extension of what has been taking place for some localities (Kotkin and Cox, 2020, Davies, 2021).

The conflicting data and observations point to at least two possibilities. In the Global East, those who lost their employment due to shutdowns in many countries returned home to wait out the crisis. In the short term, the shrinking communities from which they originated are now faced with population increase and a drought of remittances. The World Bank estimated a global decline of remittances flow of 20% in 2020 (World Bank, 2020). Some may decide to remain and start something new based on what they have learned abroad much like the Chinese return migrants in Chapter 6. In the Global East and Global North, many professionals and those who had a choice relocated temporarily or permanently because working from home is now possible and acceptable to their employers, so they are choosing locations in regional centers and communities which may or may not have been shrinking. In any event, they could potentially represent a new population and new economic activities to some hitherto moribund communities.

In addition to the issue of reliable data, many potential research questions for shrinking cities in relevant postsocialist countries arise. For example, what is the size of return migration in postsocialist countries and how much of the return migration is permanent? What does the slowing down of remittances from migrants who had relocated internationally mean for shrinking cities? Do the return migrants represent a "brain gain"? Which shrinking cities will benefit from the "brain gain"? What policies could enhance the potential for "brain gain"? Does the stream of return migrants reinforce policies aiming to make shrinking cities more livable? If the outmigration/ return migration is permanent, what are the policy lessons from both the Global East and the Global North?

Climate policies and urban shrinkage

Among postsocialist shrinking cities, resource dependent ones have attracted substantial research and policy attention because mining cities are among the first that experienced the brunt of economic transformations. Shrinkage of resource dependent cities is a theme of more than one chapter in each section of this book (Chapters 4, 5, 10, 17, 18, and 19). Indeed, socialist era policies of single-minded development and subsidies of resource development, especially energy bases such as coal mining, have created a class of cities, notably the Soviet era mono-industry towns, seemingly destined for failure once heavy subsidies are withdrawn and more efficient industrial practices come into effect. Additional impacts of more stringent environmental regulations also played a part. Coal-mining cities becoming shrinking cities is not restricted to postsocialist countries. Japan, South Korea, Canada, Australia, Germany, France, and the United States, just to name a few, all have similar examples. In the postsocialist setting, a great deal more policy attention was paid to these issues, partly because the issue affects so many communities but also because mining culture (and the social fabric that developed around it) has much stronger roots within some communities. Sometimes these local communities, supported by trade unions in mining and power generation industries, are the strongest opponents against the programs of climate protection and a zero emissions economy.

In 2020, several countries, including China and Japan, pledged to achieve a net zero emissions economy by 2050–2060. The UN has called for a phasing out of coal by 2040 (United Nations, 2021). Under the Paris Agreement on Climate Change, thermal coal production is supposed to be halved by 2030. Sixty-five countries have committed to coal phase out at the Glasgow COP 26 (UN Climate Change Conference UK, 2021. page 9). Coal mining communities are likely to be among those significantly impacted by these decisions seen as critical to combat climate change. Experience shows that many will become shrinking cities and regions. Both policymakers and researchers must identify where and how the impacts will fall and what policies and programs will need to be implemented to prepare the communities and alleviate the most significant impacts. In some EU members, this has begun. The case studies in this collection and in the larger literature on shrinking cities are starting points.

Comparisons of how postsocialist countries deal with shrinking mining cities could have lessons well beyond the postsocialist realm. The Chinese government has instituted a series of programs to deal with resource depleted cities (Chapters 4 and 5), chiefly aimed at "regrowth" but these programs also include only a small group among the resource depleted cities (69 cities out of 262 resource depleted cities). More recently, a few cities such as Fuxin City (Chapter 4) began its transition from reliance on coal mining to explore new energy industries such as wind power. Multilateral organizations such as the Asian Development Bank (ADB) have been working with China on strategies aimed at restructuring and diversifying coal economies in Gansu and Heilongjiang provinces (Asian Development Bank, 2006, Asian Development Bank, 2017).These examples tend to be broadly based with aims directed toward the green economy or at least more sustainable economies for these resource depleted communities. As the relevant chapters in this book indicate, there are many more resource depleted cities in China that have not received special central government funding to reinvent their economies. These and other communities will also search for models for change. The varied success of programs aimed at reviving resource-based cities amongst the case studies (for example, Chapters 18 and 19) in this book point to the critical importance of designing policies and programs that consider the social and cultural context in addition to the economic and political.

With the world turning its attention to climate change, hopefully in a much more proactive way, and if countries do follow through with their commitment to "zero emissions" by 2050, it will mean coal mining communities across the world will be confronted with significant impacts. The International Energy Agency reports that the share of coal in global power generation is estimated to fall to 28% by 2030, down from 37% in 2019, if energy demand returns to pre-pandemic levels and stated government policies are implemented (International Energy Agency, 2020). European Union countries have already started examining the impacts of what are called "coal phase out regions" (ESPON, 2020). The call, since the 1990s, for "just transition" for mining communities becomes even more urgent, but the experience of postsocialist countries dealing with these issues could provide valuable lessons (Jakob et al., 2020, United Nations Climate Change Conference (COP24), 2018). When the Berlin Wall fell, few were prepared for the urban shrinkage that would follow, especially in resource dependent cities and towns. What transpired was reactive policies with varying degrees of success. The question that confronts us now is how best to prepare a city for urban shrinkage and a resilient future. The new challenge is to pivot from reactive to proactive management of urban shrinkage.

References

Ali Research. 2020. *China Taobao Village Research Report* 阿里研究院 2020 *中国淘宝村研究报告*. Alibaba Research Institute.
Antonić, B. & Djukić, A. 2018. The Phenomenon of Shrinking Illegal Suburbs in Serbia: Can the Concept of Shrinking Cities be Useful for Their Upgrading?. *Habitat International*, 75, 161–170.

Asian Development Bank. 2006. People's Republic of China: Preparing the Gansu Baiyin urban development project. *Technical Assistance Report*. Manila: Asian Development Bank.

Asian Development Bank. 2017. *People's Republic of China: Heilongjiang Green Urban and Economic Revitalization Project Investment Project Components (Section A)*. Manila: Philippines.

Audirac, I., Cunningham-Sabot, E., Fol, S. & Moraes, S. T. 2012. Declining Suburbs in Europe and Latin America. *International Journal of Urban and Regional Research*, 36(3), 226–244.

Austin, J. C. 2021. With techies fleeing the coasts, America's heartland has a shot at economic revival–if we save its higher education institutions. *Brookings*. Available from: https://www.brookings.edu/blog/the-avenue/2021/02/11/with-techies-fleeing-the-coasts-americas-heartland-has-a-shot-at-economic-revival-if-we-s [Accessed February 11, 2021].

Bacchi, U. 2020. Escape from the city? *Londoners Lead Europe in COVID-Inspired Dreams of Flight*. Available from: https://news.trust.org/item/20201118230536-uv6o2 [Accessed January 20, 2021].

Balzer, H. & Repnikova, M. 2010. Migration between China and Russia. *Post-Soviet Affairs*, 26, 1–37.

Bernt, M., Haase, A., Grobmann, K., Cocks, M., Couch, C., Cortese, C. & Krzysztofik, R. 2014. How Does(n't) Urban Shrinkage Get onto the Agenda? Experiences from Leipzig, Liverpool, Genoa and Bytom. *International Journal of Urban and Regional Research*, 38(5), 1749–1766.

Bierbaum, A. H. 2020. Managing Shrinkage by "Right-sizing" Schools: The Case of School Closures in Philadelphia. *Journal of Urban Affairs*, 42(3), 450–473.

Bromley, R. & Wilson, T. D. 2018. The Urban Informal Economy Revisited. *Latin American Perspectives*, 45(1), 4–23.

Carre, F. 2017. Applying the concept of the informal economy to labor market changes in developed countries: What can be learned. WIEGO Working Paper No. 36. Cambridge: WIEGO.

Collins, E. 2016. Postsocialist Informality: The Making of Owners, Squatters and State Rule in Phnom Penh, Cambodia (1989–1993). *Environment and Planning A*, 48(2), 2367–2382.

Crowley, S. 2016. Monotowns and the Political Economy of Industrial Restructuring in Russia. *Post-Soviet Affairs*, 32(5), 397–422.

Davies, A. 2021. Has COVID really caused an exodus from our cities?: In fact, moving to the regions is nothing new. The Conversation.

Doringer, S., Uchiyama, Y., Penker, M. & Koshaka, R. 2019. A Meta-analysis of Shrinking Cities in Europe and Japan: Towards an Integrative Research Agenda. *European Planning Studies*, 28(9), 1693–1712.

Dusevic, T. 20–21 February 2021. Moment is here for our regional renaissance. *Weekend Australian*.

Dzenovska, D. 2020. Emptiness: Capitalism without people in the Latvian countryside. *American Ethnologist*, 47(1), 10–26.

Economist. 2021. Eastern Europe's brain gain: How the pandemic reversed old migration patterns. London: Economist.

Elgin, C., Kose, M. A., Ohnsorge, F. & Yu, S. 2021. Understanding the informal economy: Concepts and trends. In Ohnsorge, F. & Yu, S. (eds.) *The Long Shadow of Informality: Challenges and Policies*. Advance Edition. License: Creative Commons Attribution CC BY 3.0 IGO. Washington, D.C.: World Bank.

Engels, B. 2021. All Good Things Come From Below?: Scalar Constructions of the "Local" in Conflicts Over Mining. *Political Geography*, 84, 102295

ESPON. 2020. Structural change in coal phase-out regions. Available from: https://www.espon.eu/structural-change-coal-phase-out-regions [Accessed January 17, 2021].

Fan, L. 2019. *Taobao Villages: The Emergence of a New Pattern of Rural Ecommerce in China and its Social Implication*. Jakarta: Freidrich Ebert Stiftung.

Fang, Z. & Hung, H.-F. 2019. Historicizing Embedded Autonomy: The Rise and Fall of a Local Developmental State in Dongguan, China, 1978–2015. *Sociology of Development*, 5(2), 147–173.

Flower, B. C. R. 2019. Legal Geographies of Neoliberalism: Market-oriented Tenure Reforms and the Construction of an "Informal" Urban Class in Post-socialist Phnom Penh. *Urban Studies*, 56(12), 2408–2425.

Fu, Z., Luo, Z. & Qiao, Y. 2019. Disappearing under the Growth: Spatial Distribution Patterns and Evolution Mechanisms of Taobao Villages. 傅哲宁 罗震东 乔艺波 (2019) 增长下的消失: 淘宝村空间分布格局与演进机制研究, 美丽乡村. *Beautiful Villages*, 24–30.

Georgieva, K. 2019. The informal economy and inclusive growth. *Measuring the Informal Economy*. Washington, D.C.: IMF.

Gordillo, G. 2014. *Rubble: The Afterlife of Destruction*, Durham, NC: Duke University Press.

Guimaraes, M. H., Nunes, L. C., Barreira, A. P. & Panagopoulos, T. 2016. Residents' Preferred Policy Actions for Shrinking Cities. *Policy Studies*, 37(3), 254–273.

Gunko, M., Kinossian, N., Pivovar, G., Averkieva, K. & Batunova, E. 2021. Exploring Agency of Change in Small Industrial Towns Through Urban Renewal Initiatives. *Geografiska Annaler, Series B: Human Geography*, 103(3), 218–234.

Harrison, J. & Gu, H. 2021. Planning Megaregional Futures: Spatial Imaginaries and Megaregion Formation in China. *Regional Studies*, 55(1), 77–89.

Harvey, D. 2006. *Spaces of Global Capitalism: Towards a Theory of Uneven Geographical Development*. London and New York: Verso.

Heilongjiang. 2015. *Heilongjiang facing depopulation turning point*. 黑龙江省将现人口负增长拐点 [online]. Available from: http://www.renkou.org.cn/china/heilongjiang/2015/2253.html [Accessed May 13, 2017].

Hospers, G.-J. 2014. Policy Responses to Urban Shrinkage: From Growth Thinking to Civic Engagement. *European Planning Studies*, 22(7), 1507–1523.

International Energy Agency. 2020. *World Energy Outlook 2020 Executive Summary*. International Energy Agency.

Jakob, M., Steckel, J. M., Jotzo, F., Sovacool, B. K., Cornelsen, L., Chandra, R., Edenhofer, O., Holden, C., Loschel, A., Nace, T., Robins, N., Suedekum, J. & Urpelainen, J. 2020. The Future of Coal in a Carbon-constrained Climate. *Nature Climate Change*, 10, 702–707.

Jia, Y.-M. 2012. A Study of Population Security and Economic Social Development in Border Areas. 贾玉梅 (2012) 边境地区人口安全与经济社会发展研究以黑龙江省边境地区为, 人口学刊. *Population Journal (Renkou Xuekan)*, 5, 22–29.

Josephson, P. 2014. *The Conquest of the Russian Artic*. Cambridge, MA: Harvard University Press.

Kashnitsky, I. 2020. Russian Periphery is Dying in Movement: A Cohort Assessment of Internal Youth Migration in Central Russia. *GeoJournal*, 85(1), 173–185.

Kotkin, J. & Cox, W. 2020. America after Covid: What Demographics Tell Us. Available from: https://chiefexecutive.net/america-after-covid-what-demographics-say/.

Lambert, L. 2020. Are cities really on the edge of mass exodus? New York: *Fortune.*

Lang, T., Henn, S., Sgibnev, W. & Ehrlich, K. 2015. *Understanding Geographies of Polarization and Peripheralization: Perspectives from Central and Eastern Europe and Beyond.* London: Palgrave Macmillan.

Le, T. T. H. & Le, T. T. H. 2018. Privatization of Neighborhood Governance in Transition Economy: A Case Study of a Gated Community in Phu My Hung New Town, Ho Chi Minh City, Vietnam. *GeoJournal*, 83, 783–801.

Le Coz, C. & Newland, K. 2021. *Rewiring Migrant Returns and Reintegration after the COVID-19 Shock.* Washington, D.C.: Migration Policy Institute.

Luo, Z., Chen, F. & Shan, J. 2019. Towards the Version 3.0 of Taobao Village: A Feasible Road to Rural Revitalization. 罗震东, 陈芳芳, 单建树 (2019) 迈向淘宝村 3.0 乡村振兴的一条可行道路, 小城镇建設. *Development of Small Cities & Towns*, 37(2), 43–49.

Madgavkar, A., White, O., Krishnan, M. & Mahajan, D. 2020. *COVID-19 and gender equality: Countering the regressive effects.* Boston: McKinsey Global Institute.

Maes, M., Loopmans, M. & Kresteloot, C. 2012. Urban Shrinkage and Everyday Life in Post-socialist Cities: Living with Diversity in Hrusov, Ostrava, Czech Republic. *Built Environment*, 38, 229–243.

Marmaitė, R. 2020. *A new trend that allows you to earn more: lives in Lithuania-works abroad* [online]. Available from: https://www.tv3.lt/naujiena/lietuva/nauja-tendencija-kuri-leidzia-uzdirbti-daugiau-gyvena-lietuvoje-dirba-uzsienyje-n1060898 [Accessed 10 January, 2022].

Matanle, P. & Sato, Y. 2010. Coming Soon to a City near You!: Learning to Live "Beyond Growth" in Japan's Shrinking Regions. *Social Science Japan Journal*, 13(2), 187–210.

McGee, T. 1975. *Hawkers In Selected Asian Cities; The Comparative Research Study Outline, Findings and Policy Recommendations.* Ottawa: IDRC.

Meagher, S. M. 2013. The Darker Underside of Scott's Third Wave. *City*, 17(3), 395–398.

Mierzejewska, L. & Wdowicka M. (eds.) 2021. *Miasta i regiony w obliczu pandemii COVID-19 i innych wyzwań współczesnego świata (Cities and Regions Facing the Pandemic COVID-19 and Other Challenges of Contemporary World).* Poznań: Bogucki Wydawnictwo Naukowe.

Ministry of Economic Development of Russia. 2016. *Unified List of Support Measures for Single-Industry Municipalities of the Russian Federation Moscow.*

Moss, T. 2008. "Cold Spots" of Urban Infrastructure: "Shrinking" Processes in Eastern Germany and the Modern Infrastructural Ideal. *International Journal of Urban and Regional Research*, 32(2), 436–451.

Musterd, S., Brown, J., Lutz, J., Gibney, J. & Murie, A. 2010. *Making Creative-Knowledge Cities. A Guide for Policy Makers.* Amsterdam: University of Amsterdam.

Ng'weno, A. & Porteus, D. 2018. Let's be Real: The Informal Sector and the Gig Economy are the Future, and the Present, of Work in Africa [online]. *CGD Notes.* [Accessed 23 January, 2021].

OECD. 2021. *Towards Gender-inclusive Recovery.* Paris: OECD.

Ohnsorge, F. & Yu, S. (eds.) 2021. *The Long Shadow of Informality: Challenges and Policies.* Advance Edition. License: Creative Commons Attribution CC BY 3.0 IGO. Washington, D.C.: World Bank.

Ortiz-Moya, F. 2020. Green Growth Strategies in a Shrinking City: Tackling Urban Revitalization Through Environmental Justice in Kitakyushu City, Japan. *Journal of Urban Affairs*, 42(3), 312–332.

Pallagst, K., Fleschurz, R. & Said, S. 2017. What Drives Planning in a Shrinking City?: Tales from two German and two American cases. *Town Planning Review*, 88(1), 15–28.

Pike, A., MacKinnon, D., Coombes, M., Champion, T., Bradley, D., Cumbers, A., Robson, L. & Wymer, C. 2016. *Uneven Growth: Tackling City Decline*. York, United Kingdom: Joseph Rowntree Foundation.

Polese, A. & Rodgers, P. 2011. Surviving Post-socialism: The Role of Informal Economic Practices. *International Journal of Sociology and Social Policy*, 31(11/12), 612–618.

Ryzhova, N. 2008. Informal Economy of Translocations: The Case of the Twin City of Blagoveshensk-Heihe. *Inner Asia*, 10(2), 323–351.

Scott, A. J. 2011. Emerging Cities of the Third Wave. *City*, 15(3–4), 289–321.

Simon, M. & Mikesova, R. (eds.) 2014. *Population Development and Policy in Shrinking Regions: The Case of Central Europe*. Prague: Institute of Sociology, Academy of Sciences of the Czech Republic.

Smith, N. 2008. *Uneven Development: Nature, Capital, and the Production of Space*. Athens: University of Georgia Press.

Statistics Lithuania, 2022. *Official Statistics Portal. Database of indicators* [online]. Available from: https://osp.stat.gov.lt/statistiniu-rodikliu-analize#/ [Accessed 10 January, 2022].

Stryjakiewicz, T. 1998. The Changing Role of Border Zones in the Transforming Economies of East-Central Europe. *GeoJournal*, 44(3), 203–213.

Stryjakiewicz, T. 2009. The Old and the New in Geographical Pattern of the Polish Transition. *Acta Universitatis Palackianae Olomucensis. Facultas Rerum Naturalium, Geographica*, 40(1), 5–24.

Stryjakiewicz, T. & Kaczmarek, T. 2000. Transborder co-operation and development in the conditions of great socio-economic disparities. In Parysek, J. J. & Stryjakiewicz, T. (eds.) *Polish Economy in Transition: Spatial Perspectives*. Poznan: Bogucki Wydawnictwo Naukowe.

Stryjakiewicz, T., Meczynski, M. & Stachowiak, K. 2014. Role of Creative Industries in the Post-socialist Urban Transformation. *Quaestiones Geographicae*, 33(2), 19–37.

Su, X. 27 July 2020. Tokyoites leaving bustle behind as pandemic proves catalyst for change. *Japan Today*.

Suzuki, M. & Asami, Y. 2019. Shrinking Metropolitan Area: Costly Homeownership and Slow Spatial Shrinkage. *Urban Studies*, 56(6), 1113–1128.

Tetsushi, K. 6 November 2020. As Japan moves to revive its countryside, pandemic chases many from cities. *Japan Today*.

Towers, G. 2000. Applying the Political Geography of Scale: Grassroots Strategies and Environmental Justice. *The Professional Geographer*, 52(1), 23–36.

Tuitjer, L. & Muller, A.-L. 2021. Re-thinking urban infrastructures as spaces of learning. *Geography Compass*.

Turok, I. & Mykhnenko, V. 2007. The Trajectories of European Cities, 1960–2005. *Cities*, 24(3), 165–182.

UN Climate Change Conference, UK. 2021. *COP 26 The Glasgow Climate Compact*. Glasgow, UK.

UN Women. 2015. *Progress of the World's Women 2015–2016*.

UN Women. n.d. *Women in Informal Economy* [online]. Available from: https://www.unwomen.org/en/news/in-focus/csw61/women-in-informal-economy [Accessed 17 January, 2021].

United Nations. 2020. *Policy Brief: The Impact of COVID-19 on Women*. New York: United Nations.

United Nations. 2021. *2021 a "Crucial Year" for Climate Change*. UN Chief tells Member States. New York: UN.

United Nations Climate Change Conference (COP24). 2018. Solidarity and Just Transition Silesia Declaration. December ed. New York: UN.

Wang, J.-C. 2006a. China's Consumer-goods Manufacturing Clusters, with Reference to Wenzhou Footwear Cluster. *Innovation: Management, Policy, & Practice*, 8, 160–170.

Wang, J.-C. 2006b. Global impact and local spirit in "supply-chain cities" of China. *IILS Research Conference–Decent Work*. Geneva, Switzerland: Social Policy and Development.

Wang, C. C., Miao, J. T., Phelps, N. A. & Zhang, J. 2021. E-commerce and the Transformation of the Rural: The Taobao Village Phenomenon in Zhejiang Province, China. *Journal of Rural Studies*, 81(1), 159–169.

World Bank. 2020. *COVID-19 Crisis through a Migration Lens*. Washington DC: The World Bank Group.

Wu, C.-T., Zhang, X.-l., Cui, G.-h. & Cui, S.-p. 2014. Shrinkage and expansion in peri-urban China: Exploratory case study from Jiangsu Province. In Pallagst, K., Wiechmann, T. & Martinez-Fernandez, C. (eds.) *Shrinking Cities: International Perspectives and Policy Implications*. New York: Routledge.

Yang, C. 2016. Relocating Labour-intensive Manufacturing firms from China to Southeast Asia: A Preliminary Investigation. *Bandung Journal of Global South*, 3(3).

Yeh, A. G. O. & Chen, Z. 2020. From Cities to Super Mega City Regions in China in a New Wave of Urbanisation and Economic Transition: Issues and Challenges. *Urban Studies*, 57(3), 636–654.

Zhang, J., Feng, C. & Chen, H. 2017. International Research and China's Exploration of Urban Shrinkage. 张京祥 冯灿芳 陈浩 2017 城市收缩的国际研究与中国本土化探索, 国际城市规划 *Urban Planning International*, 32(5), 1–9.

Zhou K., Yan Y. & Zhao Q. H. 2019. Planning Policy Responses in Population Contraction Scenarios: A Case Study of Hunan Province. *Beijing Planning and Construction*, (3), 12–19. //周恺, 严妍, 赵群荟, 2019. 人口收缩情景下的规划政策应对:基于湖南案例的探讨. 北京规划建设, (3), 12–19.

Index

Locators in **bold** and *italics* denote tables and figures; n after locators denote notes.

Printed in the United States
by Baker & Taylor Publisher Services

Printed in the United States
by Baker & Taylor Publisher Services